IONIC POLYMERIZATION

UNSOLVED PROBLEMS

IONIC
POLYMERIZATION

UNSOLVED PROBLEMS

FIRST JAPAN—U.S. SEMINAR ON POLYMER SYNTHESIS

J. Furukawa
Department of Synthetic Chemistry
Kyoto University
Kyoto, Japan

O. Vogl
Polymer Science and Engineering
University of Massachusetts
Amherst, Massachusetts

MARCEL DEKKER, INC.   New York and Basel

MARCEL DEKKER, INC.

270 Madison Avenue, New York, New York 10016

LIBRARY OF CONGRESS CATALOG CARD NUMBER: 76-11944

ISBN: 0-8247-6476-5

Current Printing (last digit)
10 9 8 7 6 5 4 3 2 1

PRINTED IN THE UNITED STATES OF AMERICA

CONTENTS

iii

Preface

A Japan-U.S. Seminar entitled "Unsolved Problems in Ionic Polymerization" was held at the Fuji-View Hotel near Lake Kawakuchi on October 15-19, 1974, under the cosponsorship of the U.S. National Science Foundation and the Japan Society for the Promotion of Science. Judging from the comments of all the participants, it was a highly successful meeting.

This was the first Japan-U.S. Seminar on Polymer Synthesis.

The organization of this meeting was inspired by a meeting in Kyoto in 1967 after the IUPAC Symposium for Macromolecules in Tokyo/Kyoto and by an International Colloquium on Polymer Synthesis held in Amherst in 1971 after the IUPAC Symposium for Macromolecules in Boston.

Other Japan-U.S. Seminars concerned with Polymer Science and Engineering have been held in the past in Polymer Physics, Rheology, and Morphology. I would like to acknowledge the advice that we received from Professor R. S. Stein and Professor S. Onogi from their experience in arranging a successful U.S.-Japan Seminar.

It was, however, the determination and dedication of Professor J. Furukawa, Professor Saegusa, and Professor Hayashi with the able assistance of Dr. Kiji and Dr. Ito which made the Seminar possible. In less than 2 months time the location was selected; requests for preprints, biographical sketches, and full papers were distributed; the preprints and biographical sketches were printed, bound, and ready for distribution at the meeting.

In the meantime, all participants were alerted and timetables for the United States participants in Japan were aranged and firmed up. Special thanks are due to Professor J. P. Kennedy who cooperated with me in the organization of the meeting but, because of unforeseen circumstances, was unable to attend the Seminar. His spirit, however, was very much in evidence during the meeting.

We also acknowledge the help of Dr. B. Bartocha, Dr. E. O'Connell, Dr. M. Cziesla, and Mrs. L. Trent of the National Science Foundation.

v

The Seminar was held in a remote location which provided very close interaction between the individual participants. The time was equally divided between presentation and discussion; this was only possible because of the limited number of participants. Emphasis was placed on the "Unsolved Problems" in the individual presentation. New acquaintances and friendships were developed and established relations were reinforced.

We all left the Seminar with a sense of accomplishment and deep satisfaction.

O. Vogl
Co-Chairman (U.S.)
Japan-U.S. Seminar

Preface

The Japan-U.S. Seminar on "Unsolved Problems in Ionic Polymerization" was held on October 15-19, 1974, in Fuji Hakone, Japan; it was sponsored by the Japan Society for the Promotion of Science and the U.S. National Science Foundation. This was the first seminar on Polymer Synthesis and 13 United States and 20 Japanese polymer chemists attended.

It has been recognized that ionic polymerization is more complicated than radical polymerization, but it is also more versatile for the preparation of new kinds of polymers. One of the reasons is that ionic polymerization is influenced by stereo and sequence regulation; isomerization and grafting may also occur during polymerization.

Vogl reviewed general aspects of ionic polymerization and the polymerization of chloral and related compounds. In cationic polymerization, P. Dreyfuss read the paper of Kennedy on cationic bigrafting of styrene, α-methylstyrene, or isobutene polymers on bromide and chloride terminated hydrocarbons. Kunitake proposed a mechanism of syndiotactic polyaddition of α-methylstyrene which takes into account the influence of radius of counteranion on the cationic polymerization.

In ring-opening polymerization, Saegusa described a new alternating copolymerization of oxazoline and acrylic monomer through their addition product which exists as a zwitterion. Litt reported on the polymerization of N-substituted oxazoline and pointed out the sharp increase in the molecular weight of this polymer due to chain branching. M. P. Dreyfus explained the polymerization and oligomerization of cyclic ethers by oxonium intermediates. The ring-opening polymerization of spiro acetals prepared from alkylene oxide and lactone was reported by Bailey, in which volume expansion accompanies the polymerization. Harwood found by careful NMR analysis that the polymerization of N-carboxylic acid anhydride proceeds through a carbamate anion.

In anionic polymerization, Ise described the importance of triple
anions as active species for the polymerization of styrene in addition
to free anions and ion pairs in mixed solvents such as dimethyoxyethane
and benzene. Stereoregulation of α-phenyl as well as α-methyl acrylate
polymerizations were reported by Yuki. Modification of alkylmetal
catalyst by ethers and amines was reported by Tsuruta and potassium-
substituted compounds of pentadiene derivatives were discussed by
Tani. Anionic polymerization of aromatic and heteroaromatic vinyl
monomers was reported by Pearson and Iwakura (read by Toda), respectively. Ionic polymerization by radiation was investigated by
Hayashi. Lenz reported the NMR study of poly-α-chloroacrylic ester
prepared with Grignard reagent as an initiator and proposed a template polymerization.

The mechanism of alternating copolymerization of olefins or diole-
fins with acrylic monomer in the presence of Lewis acid was discussed
by Hirai and Furukawa, emphasizing the importance of donor-acceptor
complex intermediates. Stille reported the role of donor-acceptor
complexes in the polymerization of vinyl ether or dioxene as donors
and quinones or vinylidene cyanide as acceptors.

Polymerizations with transition metal catalysts, particularly alkyl
nickel dipyridyl, was investigated by Yamamoto, and ESR studies of a
titanium alkoxide-metal alkyl system was discussed by Takeda. Otsu
described the isomerization polymerization of β-olefins which yields
polymers of α-olefins. Calderon reviewed several hypotheses for the
metathesis polymerization. Finally, an ionic condensation of aromatic
amines with dicarboxylic acid by means of alkyl phosphite was reported
by Yamazaki.

All problems discussed are of current interest. Some of them seem
to have been solved, but new problems are developing which may be
important contributions for the creation of new macromolecules. This
seminar contributed significantly to the identification of "Unsolved
Problems in Ionic Polymerization," and the discussions shed light
for their possible solution.

Junji Furukawa
Co-Chairman (Japan)
Japan-U.S. Seminar

PARTICIPANTS

M. Asada (Sumimoto Chemical Industries Company)
William J. Bailey (University of Maryland)
Nissim Calderon (Goodyear Tire and Rubber Company)
M. G. Cziesla (NSF Representative in Tokyo)
M. Peter Dreyfuss (B. F. Goodrich Company)
Patricia Dreyfuss (University of Akron)
Junji Furukawa (Kyoto University)
H. James Harwood (University of Akron)
Koichiro Hayashi (Osaka University)
Hidefumi Hirai (Tokyo University)
Jihei Inomata (Mitsubishi Chemical Industries Company)
Norio Ise (Kyoto University)
Yoshio Iwakura (Seikei University)
Toyoki Kunitake (Kyushu University)
Robert W. Lenz (University of Massachusetts)
Morton Litt (Case Western Reserve University)
Hiroshi Maki (Mitsui Petrochemical Industries Company)
Hiroyuki Morikawa (Mitsubishi Petrochemical Company)
Takayuki Otsu (Osaka City University)
Rudolf Pariser (E. I. du Pont de Nemours and Company)
James M. Pearson (Xerox Company)
Takeo Saegusa (Kyoto University)
Ryozo Sakata (Bridgestone Tire Company)
John K. Stille (University of Iowa)
Taro Suminoe (Japan Synthetic Rubber Company)
Masatami Takeda (Tokyo Science University)
Hisaya Tani (Osaka University)
David P. Tate (Firestone Tire and Rubber Company)
Fujio Toda (Tokyo University)
Teiji Tsuruta (Tokyo University)
Otto Vogl (University of Massachusetts)
Akio Yamamoto (Tokyo Institute of Technology)
Naboru Yamazaki (Tokyo Institute of Technology)
Heimei Yuki (Osaka University)

PARTICIPANTS

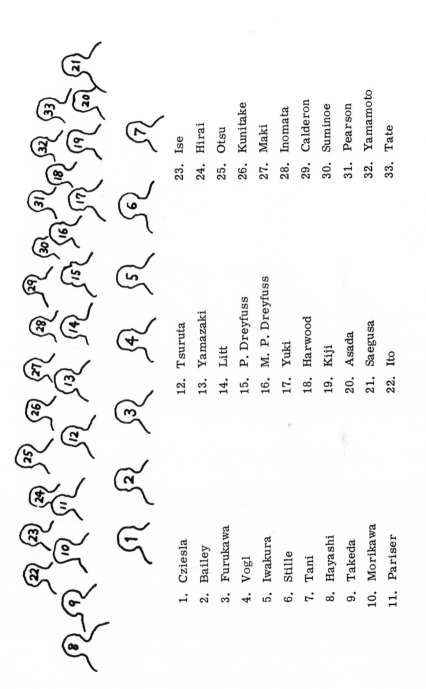

1. Cziesla
2. Bailey
3. Furukawa
4. Vogl
5. Iwakura
6. Stille
7. Tani
8. Hayashi
9. Takeda
10. Morikawa
11. Pariser

12. Tsuruta
13. Yamazaki
14. Litt
15. P. Dreyfuss
16. M. P. Dreyfuss
17. Yuki
18. Harwood
19. Kiji
20. Asada
21. Saegusa
22. Ito

23. Ise
24. Hirai
25. Otsu
26. Kunitake
27. Maki
28. Inomata
29. Calderon
30. Suminoe
31. Pearson
32. Yamamoto
33. Tate

Review and Future of Ionic Polymerization
with Special Emphasis on Carbonyl Polymerization

O. VOGL

Polymer Science and Engineering
University of Massachusetts
Amherst, Massachusetts 01002

REVIEW

Ionic polymerization received prominence about 35 years ago
when isobutylene was commercially polymerized by two processes
which, with some modifications, are still used today [1]. One
process uses aluminum chloride as the initiator and the other
uses boron trifluoride; both cationic polymerization processes
are carried out at low temperatures. A number of additional
commercial processes based on cationic and anionic polymerization have since been developed. Cyclic ethers, most prominently
tetrahydrofuran, are polymerized cationically to relatively low
molecular weight hydroxyl terminated polyethers which have
found important uses in polyurethanes. Trioxane is copolymerized
with a small amount of ethylene oxide to form a useful copolymer
of polyoxymethylene. Other products which are of interest are the
polymers of caprolactone and epichlorohydrin and polymers of
various epoxides, mainly those of glycidyl ethers which are most
commonly known as epoxy resins. Anionic polymerization on a
commercial scale has developed along the lines of styrene and
isoprene polymers. Stereorubber, stereoregular 1,4-cis isoprenes,
are based on lithium initiators and were introduced in the middle
1950s. Triblock polymers based on A-B-A block polymers of

1

isobutylene with styrene as endblocks and prepared from living polymers have been known since the early 1960s.

Some of these developments in ionic polymerization were the results of fundamental scientific investigations, but others, particularly the earlier work, were based on intuitive work of ingenious inventors.

It is clear that the general scope of ionic polymerizations involves a number of different fields of interest. It is the purpose of this symposium to bring together investigators involved in investigations of various disciplines in order to disseminate the information and cross-fertilize the ideas developed in their own area of interest.

Ionic polymerization can be looked upon from the traditional mechanistic or historical as well as from the practical point of view, and it has been treated in numerous ways. In cationic polymerization, isobutylene polymerization has often been compared with styrene polymerization; carbenium ion (carbocation) polymerization has competed or coexisted with oxonium polymerization; and coordinative ionic polymerization has profited from classical ionic polymerization.

Cationic and anionic polymerization are chain growth polymerizations which frequently involve bond-opening reactions of carbon-carbon double bonds with carbenium (carbocations) or carbanions as polymeric propagating species. Heteroatom-containing monomers, such as aldehydes, polymerize by bond-opening polymerization of carbonyl groups, and cyclic ethers by ring-opening polymerizations with oxonium ions or alkyoxide ions at the end of the growing polymer chain. Lactam and lactone polymerizations involve the ring opening of the carbonyl group, but the methylene group attached to the ether oxygen may also be the point of attack [2].

In addition to the above-mentioned chain growth polymerization to form high molecular weight polymers by ionic processes, polymer formation by a step growth polymer process which involves ionic intermediates has frequently been investigated. It involves electrophilic and nucleophilic substitution reactions on aliphatic or aromatic compounds to form high molecular weight polymers. Discussions of these polymerizations are usually not included when ionic polymerization is discussed.

Cationic polymerization of olefins is carried out with olefins which are substituted with electron-donating groups. The best studied examples are isobutylene, styrene, butadiene, and isoprene. Methyl substitution, especially disubstitution on one carbon atom, as for example in isobutylene, causes the carbon-carbon double bond to be properly polarized for the addition of the electrophile.

Cationic carbenium ion polymerization is most easily discussed with isobutylene as the example [3]. The electrophile, for example a proton, acts as the initiator, and termination of the growing polymeric cation is primarily by proton transfer to another iso-butylene molecule to form an olefin-terminated polyisobutylene and a new tertiary butyl carbenium ion which is capable of further propagation.

Initiation: $R^{\oplus}$ + $CH_2=C(CH_3)_2$ $\longrightarrow$ $R-CH_2-\overset{\oplus}{C}(CH_3)_2$

Transfer: $P-CH_2-\overset{\oplus}{C}(CH_2)_3$ $\longrightarrow$ $P-CH=C(CH_3)$ + $H^{\oplus}$

or $P-CH_2-\underset{\underset{CH_3}{|}}{C}=CH_2$ + $H^{\oplus}$

+ $CH_2=C(CH_3)$

$P-CH=C(CH_3)_2$ + $(CH_3)_3C^{\oplus}$

Fundamental knowledge has been acquired about cationic carbenium ion polymerization over the years, particularly the influence of coinitiators [4], the influence of temperatures on molecular weight, and, most recently, the importance of the equilibrium between carbenium ions and nonionic species as in the case of the polymerization of styrene with perchloric acid [5]. A number of other examples of equilibrium between ionic and nonionic intermediates has recently become known in the ring-opening polymerization of heterocyclic monomers [6, 7].

A parallel to the proton initiation and proton transfer in cationic olefin polymerization is the fluoride ion initiation and fluoride ion transfer in anionic fluorolefin polymerization as, for example, in the case of the oligomerization of hexafluoropropylene. The initiating fluoride ion adds to the less hindered CF_2 group of the olefin to form the heptafluoroisopropyl anion which is then capable of adding a new molecule of hexafluoropropylene. Because of the relative ease of fluoride loss under these conditions, polymerization to high molecular weight does not occur and the formation of highly branched dimeric and trimeric fluoro olefins is the result [8].

$$F^{\ominus} + CF_2{=}CF{-}CH_3 \longrightarrow CF_3\ \overset{\ominus}{C}F\ CF_3$$

$$(CF_3)_2CF^{\ominus} + CF_2{=}CF{-}CF_3 \longrightarrow (CF_3)_2CF{-}CF_2\overset{\ominus}{C}F{-}CF_3$$

$$\downarrow -F^{\ominus}$$

$$(CF_3)_2CF{-}CF{=}CF{-}CF_3$$

This reaction is very similar to the oligomerization of propylene in the presence of protic acid.

Ring-opening polymerizations proceed only by ionic mechanisms, the polymerization of cyclic ethers mainly by cationic mechanisms, and the polymerization of lactones and lactams by either a cationic or anionic mechanism. Important initiators for cyclic ethers and lactone polymerization are those derived from aluminum alkyl and zinc alkyl/water systems. It should be pointed out that substitution near the reactive group of the monomer is essential for the individual mechanism that operates effectively in specific cases; for example, epoxides polymerize readily with cationic and anionic initiators, while fluorocarbon epoxides polymerize exclusively by anionic mechanisms [9]. Polymers of fluorocarbon epoxides can be readily obtained but fluoride transfer is sometimes excessive; hexafluoropropylene oxide polymerizes anionically with a growing alkoxide which easily loses a fluoride ion to form acid fluoride-terminated polymers with the fluoride initiating a new polymer chain. This is a severe limitation to obtaining high molecular weight materials.

$$F^{\ominus} + \underset{\underset{O}{\diagdown\diagup}}{\overset{\overset{CF_3}{|}}{CF}}{-}CF_2 \longrightarrow CF_3\ CF_2{-}CF_2{-}O^{\ominus}$$

$$R_f{-}\underset{\underset{CF_3}{|}}{CF}{-}CF_2O^{\ominus} \xrightarrow{\ -F^{\ominus}\ } R_f{-}\underset{\underset{CF_3}{|}}{CF}{-}CF{=}O\ +\ F^{\ominus}$$

Another example of the importance of substitution near the reactive group of the monomer is exemplified by the polymerization of β-propiolactone as compared to dimethyl-β-propiolactone (pivalolactone).

While β-propiolactone polymerizes with both cationic and anionic initiators, pivalolactone apparently polymerizes with typical anionic initiators but not with aluminum alkyl/water systems as initiators.

The polymerization of aldehydes can be accomplished by cationic and anionic initiators [10, 11]. Substitution in the α-position of the carbonyl group determines which polymerization mechanism is more facile or predominant. In the case of formaldehyde polymerization, weak nucleophiles are capable of polymerizing formaldehyde to polyoxymethylene but relatively strong acids are necessary for its polymerization. Aliphatic higher aldehydes are easier polymerized with cationic initiators, and relatively weak electrophiles will produce high molecular weight polymers; however, strong nucleophiles such as alkoxides must be used for the initiation of the anionic polymerization of aliphatic aldehydes.

$$\underset{\substack{| \\ H}}{\overset{\substack{R \\ |}}{C}}=O \quad \xrightarrow[\text{METHYLCYCLOHEXANE}]{(CH_3)_3COLi, \; -78°C} \quad \underset{\substack{| \\ H}}{\overset{\substack{R \\ |}}{C}} - \left[O - \underset{\substack{| \\ H}}{\overset{\substack{R \\ |}}{C}} \right]_n O - \underset{\substack{| \\ H}}{\overset{\substack{R \\ |}}{C}} - O$$

The stereochemistry of the aldehyde polymers is also influenced by the nature and type of initiators. Ionic initiators generally favor the formation of crystalline isotactic polymer while most cationic initiators produce amorphous polymers of low tacticity. The tacticity increases, especially in the case of cationic initiators, when the length of the aliphatic side chain is increased.

On the other hand, when the α-position of the aldehyde is substituted with an electron-withdrawing group, such as a trichloromethyl group as in the case of chloral, the anionic polymerization of haloaldehydes becomes more facile. Chloral polymerizes readily with weak nucleophiles, and even chloride ions are very effective initiators. Many other alkoxides, amines, and phosphines are effective initiators for the anionic polymerization of chloral [12].

$$R^{\ominus} + \underset{\substack{| \\ H}}{\overset{\substack{CCl_3 \\ |}}{C}}=O \longrightarrow R - \underset{\substack{| \\ H}}{\overset{\substack{CCl_3 \\ |}}{C}} - O^{\ominus} + n\underset{\substack{| \\ H}}{\overset{\substack{CCl_3 \\ |}}{C}}=O \longrightarrow$$

INITIATION PROPAGATION

$$R \left(\underset{\substack{| \\ H}}{\overset{\substack{CCl_3 \\ |}}{C}} - O \right)_n \underset{\substack{| \\ H}}{\overset{\substack{CCl_3 \\ |}}{C}} - O^{\ominus}$$

Strong acids, such as sulfuric acid and particularly trifluoro-methane-sulfonic acid as well as selected Lewis acids, have been found to be initiators for the polymerization of chloral [13].

In all cases, polychloral has been obtained only in its isotactic form.

Fluoral, on the other hand, can be polymerized with anionic and cationic initiators to the crystalline polymer or a mixture of crystalline and amorphous polymer.

The replacement of oxygen by sulfur in aldehydes leads to thioaldehydes which have a substantially different electron density in the thiocarbonyl bond and also a longer C—S bond length. In general, the thiocarbonyl compounds are easier to polymerize than the corresponding carbonyl compounds. For example, thioformaldehyde is so reactive that it has not been prepared in its pure form; even at -120°C thioformaldehyde polymerizes rapidly. It is apparently initiated by the trace impurities which are normally present during its preparation. On the other hand, many thiocarbonyl compounds can be polymerized whose corresponding carbonyl compounds have resisted polymerization. Examples of the polymerization of such thiocarbonyl compounds are thioacetone, thiohexafluoroacetone, thiocarbonyl fluoride, and a number of fluoroacyl fluorides.

Few cases of established radical polymerizations have been reported involving thiocarbonyl compounds, for example, thiocarbonyl fluoride and thioacetone.

NEW DEVELOPMENTS

Cryotachensic Polymerization of Chloral

Many years ago chloral was polymerized with such typical anionic initiators as pyridine or other tertiary amines to an insoluble and infusible powder which could not be formed into a useful shape. Cationic polymerization with aluminum chloride and sulfuric acid as initiators gave the same polymer.

We have now found that chloral may be initiated above the threshold temperature of polymerization which, in the case of neat chloral, is 58°C [12]. Above the threshold temperature initiated chloral can be handled and transferred and no polymerization occurs. However, as soon as the initiated mixture is

cooled below the threshold temperature, polymerization occurs
(cryotachensic polymerization):

cryos   tacheitis
cold   acceleration

(1) Initiate above T_c (ceiling temperature)

(2) Polymerize by cooling (removal of heat of polymerization)

Chloral polymerization can be carried out neat or in the
presence of inert diluents (solvents) which do not interfere with
the polymerization.

The threshold temperature depends on the concentration of chloral.
The ceiling temperature of a 1-$\underline{M}$ solution of chloral in methylcyclohexane has been determined to be 18°C.

The rate of chloral polymerization depends on the efficiency of
the cooling bath which surrounds the container or mold containing
the initiated chloral solution; it also depends on the thickness of the
piece and consequently on the heat transfer through the already
polymerized polychloral.

Polychloral is formed from the initiated monomer in the form of
a homogeneous gel. At about 2 to 5% conversion the gel is already
rigid, and during the course of the polymerization the polychloral
matrix becomes more and more rigid until polymerization is
complete. Above 50 to 60% conversion the polymerization of
chloral becomes strongly diffusion controlled, and it comes to a
complete stop at approximately 85% conversion if no diluent is
used for the polymerization. With about 20% diluent, such as
hexane, conversions of more than 95% may be obtained. The
progress of the polymerization can be very conveniently followed
by the disappearance of the chloral monomer signals by nuclear
magnetic resonance spectroscopy (Fig. 1).

Numerous nucleophiles are initiators for chloral polymerization.
Good nucleophiles such as lithium t-butoxide add quantitatively to
1 mole of chloral above the threshold temperature of polymerization
and no further addition to more chloral occurs.

$$(CH_3)_3CO^{\ominus}Li^{\oplus} \ + \ CCl_3CHO \rightleftharpoons$$

$$(CH_3)_3CO-\underset{\underset{H}{|}}{\overset{\overset{CCl_3}{|}}{C}}-O^{\ominus}Li^{\oplus}$$

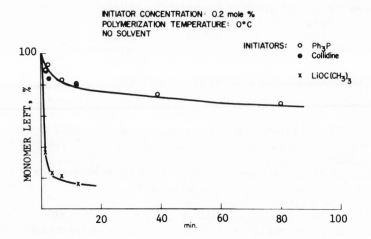

FIG. 1. Rate of chloral polymerization (determined by NMR). Initiator concentration: 0.2 mole %. Polymerization temperature: 0°C. No solvent.

Poor nucleophiles such as chloride ion or 2,4,6-collidine do not add to a significant degree to chloral monomer and no addition product could be identified by NMR in the mixture. However, when the initiated chloral mixture is cooled, polychloral is formed rapidly, indicating that although the equilibrium of the initiation reaction does not favor addition, propagation proceeds rapidly.

Triphenylphosphine is also a good initiator for chloral polymerization, but triphenylphosphine is not the actual initiator. It reacts almost instantaneously with 1 mole of chloral to form a quaternary phosphonium salt whose anion, chloride, is the actual initiator.

Polymers of chloral prepared under various conditions and with various initiators have identical IR spectra and according to x-ray studies are isotactic; no soluble atactic form has ever been isolated.

Chloral has also been polymerized cationically with strong acids, particularly trifluorosulfonic acid and sulfuric acid, as the initiator. A few selected Lewis acids such as aluminum chloride and antimony pentachloride also act as initiators for the chloral polymerization, but many other typical cationic initiators, Lewis acids, and cationic salts are ineffective (Table 1).

TABLE 1. Polymerization of Chloral with Lewis Acids[a]

Initiator	Yield (%)	Remarks
$AlCl_3$	78	Rate comparable to anionic rate
$FeCl_3$	4	—
$TlCl_4$	2	Time: 14 days
$SbCl_5$	13	Gel
$SbCl_5 + CH_3COCl$ (5 mole %)	8	Gel
$SbCl_5 + Et_2O$ (5 mole %)	—	Dark solution

[a]Initiator concentration: 2 mole %. Polymerization: Bulk. Temp: -5°C. Time: 3 days.

A proposed scheme for the initiation, propagation, and termination of cationic chloral polymerization is described in Scheme 1.

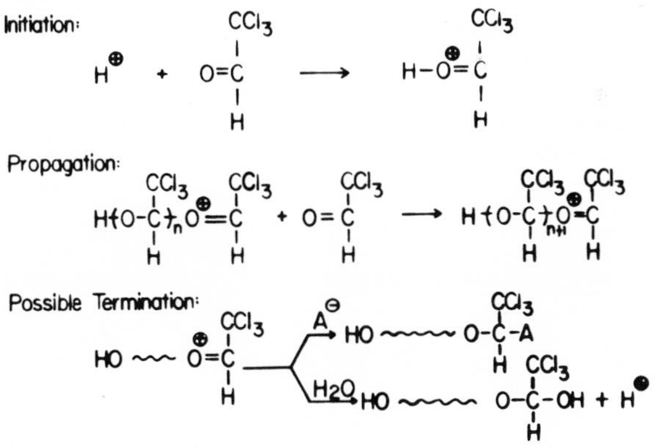

SCHEME 1. Cationic polymerization of chloral.

Very little is known about the end groups of chloral polymers prepared by cryotachensic polymerization. Some of the polymers are quite unstable and have a maximum degradation rate temperature near 150°C while the best stabilized polychloral samples show a maximum degradation rate at 385°C. On the other hand, homopolymers of chloral cannot be endcapped by the usual acetylation techniques known to be effective for the stabilization of hydroxyl-terminated polyoxymethylene. Earlier reports by other investigators seem to have demonstrated that chloral polymers also have hydroxyl end groups which can be acetylated. We have prepared many chloral polymers under different conditions but our acetylation attempts under various conditions were unsuccessful.

Since polychloral is insoluble, few techniques for the characterization of polymer samples are available. One of the most convenient techniques to compare various polymer samples is the determination of the thermal stability of individual chloral polymers by differential thermogravimetric analysis (DTG). Polychloral samples prepared with different initiators and under different reaction conditions gave a distinct and characteristic spectrum of polymer stability. In this way general types of end groups could be identified on the basis of the thermal degradation pattern of the polychloral samples. Generally, anionically prepared polychloral samples were less stable to thermal degradation, and it is believed that these samples may have occluded living ends which have not terminated but are not accessible for the capping reagents and consequently cannot be acetylated.

Since the acetylation was not possible, we attempted to increase the thermal stability of the polychloral sample by trying to carry out chain transfer reactions during the polymerization. To our surprise we found that anionic chloral polymerization can be carried out in the presence of up to 2 mole % of acyl chlorides, for example, benzoyl chloride and acetyl chloride, when the polymerization is initiated with triphenylphosphine.

Lithium tertiary butoxide could not be used as initiator for these systems because it reacted immediately with the acid chloride and acetylation of the tertiary butoxide anion occurred.

It was initially believed that polymerization of chloral in the presence of benzoyl chloride, acetyl chloride, or acetic anhydride was governed by a chain transfer reaction whereby the growing alkoxide reacted with the carbonyl carbon of the acetyl chloride or the acetic anhydride to form an acetyl-terminated polychloral with the loss of chloride or acetate ion which then initiated a new polymer chain. More extensive studies showed that termination and virtually no chain transfer actually occurs toward the end of the polymerization [14]. In a mixture containing two types of carbonyl compounds, the monomer chloral and chain transfer agent acetyl chloride, there is competition throughout the reaction for the growing alkoxide to add to the carbonyl carbon of either of the two carbonyl compounds. Apparently the chloral very effectively competes with the acetylating agent for the addition because analysis of the polymers at different conversions showed that although the conversion of monomer to polymer is high even after a few minutes, the concentration of carbonyl end groups in the polymer increases very slowly and the maximum acylation is not obtained for several hours. An estimation of the number-average molecular weight showed that a DP of about 300 of the stable fraction was obtained.

Chloral can be successfully copolymerized with isocyanates, particularly with aromatic isocyanates, and with ketenes [15].

$$
\begin{array}{c}
\underset{\substack{| \\ H}}{\overset{\substack{CCl_3 \\ |}}{R^{\ominus} + C=O}} \longrightarrow \underset{\substack{| \\ H}}{\overset{\substack{CCl_3 \\ |}}{R-C-O^{\ominus}}} + \underset{\substack{\| \\ O}}{\overset{\substack{R' \\ |}}{C=N}} \longrightarrow
\end{array}
$$

$$
\underset{\substack{| \quad \| \\ H \quad O}}{\overset{\substack{CCl_3 \quad R' \\ | \qquad |}}{R-C-O-C-N^{\ominus}}} + \underset{\substack{| \\ H}}{\overset{\substack{CCl_3 \\ |}}{C=O}} \longrightarrow
$$

$$
\underset{\substack{| \quad \| \quad | \\ H \quad O \quad H}}{\overset{\substack{CCl_3 \quad R' \ CCl_3 \\ | \qquad | \quad |}}{R-C-O-C-N-C-O^{\ominus}}}
$$

This polymerization is carried out with anionic initiators to form polymers containing urethane or ester linkages. As expected, copolymers of chloral with isocyanates and ketenes have a substantially improved thermal stability. Recent studies indicate that even the more reactive isocyanates, aromatic isocyanates, are not incorporated into the polymer from the very beginning of the copolymerization but are incorporated in the later stage.

Homopolymers of chloral prepared under various conditions and with various initiators have identical IR spectra and consequently have the same chemical structure and stereochemistry. According to x-ray studies, polychloral is isotactic and no soluble atactic form of chloral has yet been observed.

Sequential Polymerization and the Formation of Interpenetrating Networks

Chloral polymers cannot be mixed with other polymers by melt or solution blending because they are insoluble and infusible. We were, however, able to produce polymer blends and interpenetrating networks [16] of polychloral on the basis of a sequential polymerization. Chloral is first ionically polymerized or copolymerized with an isocyanate, for example phenyl isocyanate, which is then followed by a second polymerization of an addition monomer with a radical initiator.

Composites of Chloral Polymers and Copolymers with Addition Polymers

A. Dissolve addition polymer in chloral and polymerize chloral.
B. Prepare chloral polymer, soak with addition monomer (containing initiator), and polymerize monomer.
C. Mix addition monomer (and its initiator) with chloral.
    1. Carry out chloral polymerization.
    2. Carry out polymerization of addition monomer.

In principle, polymer blends can also be prepared by dissolving a chloral soluble polymer in chloral, initiating the chloral polymerization above the threshold temperature of polymerization and then cooling the mixture to perform the actual chloral polymerization. Polymers of up to 10% of their weight can be incorporated into chloral polymers; they include ethylene/propylene rubber, various acrylates, and soluble polyolefins. One of the limitations of this technique is the solubility of the high molecular weight addition polymer in chloral monomer.

Another method of preparing polymer blends with polychloral consists of the preparation of appropriate polychloral samples, imbibing the polymer with the addition monomer, and then polymerizing the addition monomer within the polychloral matrix. This

method also has its limitations because the amount of monomer which
can be imbibed into the polychloral matrix does not exceed 30%.
Addition monomers which can be used for this technique are of a
much wider scope because it does not make any difference if these
monomers interfere with the chloral polymerization. Acrylates,
styrene, and dienes have been used with radical initiators, and
cyclic ethers with cationic initiators.

The most flexible and chemically unique technique is the sequential polymerization of chloral followed by the polymerization of the
addition monomer. As mentioned earlier, chloral can be polymerized in the presence of inert solvents. Potential monomers which
do not interfere with the polymerization of chloral can be used for
the sequential polymerization.

$$
\begin{array}{c}
CCl_3 \\
|\\
C{=}O \\
|\\
H
\end{array}
\quad\xrightarrow{\ R^{\ominus}\ }\quad
\begin{array}{c}
CCl_3 \\
|\\
\sim CH{-}O\sim
\end{array}
$$

$$
\begin{array}{c}
R \\
|\\
CH_2{=}C \\
|\\
R'
\end{array}
\quad\xrightarrow{\ R^{\ominus}\ }\quad
\begin{array}{c}
R \\
|\\
\sim CH_2{-}C\sim \\
|\\
R'
\end{array}
\qquad
\begin{array}{l}
R \;\; = H,\ CH_3 \\[4pt]
R' = COOR \\
\qquad\ \ | \\
\qquad\ \ Ph
\end{array}
$$

It has been found that homopolymerization of chloral or copolymerization of chloral with isocyanates can be carried out in the usual
manner in the presence of monomers which can then be polymerized
in a second step, preferrably by a radical mechanism.

The actual preparation of the polymer blends is carried out as
follows: Chloral, a monomer which can be polymerized by a radical
mechanism, and a corresponding radical initiator, for example,
AIBN, are mixed and heated above the threshold temperature of
chloral polymerization. The chloral initiator is then added and
chloral polymerization, which is carried out by cooling to 0°C, proceeds rapidly to a homogeneous gel. It is believed that the chloral
polymerization goes to more than 95% conversion because it has
been determined in different experiments that chloral polymerization
in the presence of 20% of heptane gives a 96% conversion to polychloral. After the chloral polymerization is complete, the sample

is heated to the temperature at which the radical initiator is activated and the second polymerization proceeds by a radical mechanism.

Polymer blends of chloral with a number of addition polymers were prepared by the sequential polymerization technique. They include styrene, methyl methacrylate, methacrylate, and α-methylstyrene. No grafting occurred during this polymerization and an intimate mixture of polychloral with the addition polymer was obtained. The addition polymer could be extracted from the polychloral matrix to more than 95%. The initiated monomer mixture may be cast directly in films or sheet, and the composite compositions may be translucent to opaque depending on the exact reaction conditions and also depending on the individual addition monomers used to prepare these blends.

Some time ago, Bamford [17] described a grafting of addition monomers by radical mechanism onto polychloral. In this work, manganese carbonyls were used as the grafting initiators and their reaction with the trichloromethyl group of the polychloral initiated the radical graft polymerization. We have used AIBN and similar radical initiators and have found no evidence of grafting in our system.

Sequential polymerization, the anionic polymerization of chloral followed by the radical polymerization of addition monomers, also permitted the preparation of interpenetrating polymer networks [18]. Instead of preparing a linear polymer from the radically initiated monomer, it was only necessary to use cross-linking agents to cross-link the second polymer network within the polychloral matrix. We have been able to prepare interpenetrating networks of polychloral as one phase and polystyrene cross-linked with divinylbenzene or poly(methyl methacrylate) cross-linked with ethylene dimethacrylate as the other phase. It should be mentioned, however, that our attempts to prepare interpenetrating networks required the use of 8 to 10 mole % of cross-linking agents in order to form a properly cross-linked second phase; with less than 5% cross-linking agent, the polymer can still be extracted. This amount of cross-linking agent is substantially greater than normally needed to form a lightly cross-linked polymer if this cross-linking reaction is carried out in solution.

We believe that it is remarkable to perform a sequence of two polymerizations, one ionic polymerization or copolymerization of chloral followed by a radical polymerization of a suitable addition monomer, without interference of either polymerization with each other and without observing radical grafting on polychloral.

Preparation and Polymerization of Other Haloaldehydes

Aliphatic aldehydes polymerize to isotactic polymers primarily with alkali metal alkoxide and to some extent with specific aluminum organic compounds and to atactic polymer. Chloral polymerizes to one type of polymer only—isotactic polychloral. The polymerizations may be initiated by anionic or cationic initiators. Simple initiators such as tetranalkyl ammonium chloride or pyridine as well as metal alkoxides are efficient initiators. Since there is no coordination necessary to form isotactic polymer, it is believed that the size of the trichloromethyl side group is alone responsible for the formation of the isotactic polymer. Bromal (tribromoacetaldehyde) has not yet been polymerized, and fluoral (trifluoroacetaldehyde) has been polymerized to crystalline, presumably isotactic polymer, and under different conditions to an amorphous soluble polyacetal.

It was of interest to change the size and shape of the side groups of halogenated acetaldehydes in order to determine the size and shape of the trihalomethyl side groups which can be tolerated to form isotactic polymers. We were interested in determining basically two limits: how many chlorine atoms in trichloroacetaldehyde can be replaced by bromine before the side group becomes too sizeable and the corresponding haloaldehyde does not polymerize any more, and how small can the trihalomethyl side group be made without the formation of atactic polyhaloaldehydes. This can be readily done by systematically substituting chlorine atoms in chloral by fluorine [19].

Difluorochloroacetaldehyde and fluorodichloroacetaldehyde were prepared by lithium aluminum hydride reduction of the corresponding difluorochloroacetate and fluorodichloroacetate. This reduction must be carried out at -78°C in order to avoid further reduction.

$$CCl_3COOCH_3 \xrightarrow[\substack{\text{Reflux, 15 hrs.} \\ (\sim 120\ °C)}]{SbF_3 \ - \ Br_2} CCl_2FCOOCH_3$$

53%

$$CCl_2FCOOCH_3 \xrightarrow[-78°,\ Ether]{LiALH_4} \xrightarrow{H_2O} CCl_2FCH(OH)_2$$

45%

$$CCl_2FCH(OH)_2 \xrightarrow{conc.\ H_2SO_4} CCl_2FCHO$$

73%

The aldehydes were initially isolated as the aldehyde hydrates, which were dehydrated with sulfuric acid followed by several careful distillations from phosphorous pentoxide. Both monomers polymerized with cationic and anionic initiators but, as expected, more readily with anionic initiators as indicated in Tables 2 and 3.

Ceiling temperatures of chlorofluoroacetaldehydes and their boiling points are shown in Fig. 2.

TABLE 2. Polymerization of $CClF_2CHO$ with Ph_3P

Initiator amount (mole %)	Polymerization		Polymer yield (%), acetone	
	Temp (°C)	Time (hr)	Soluble	Insoluble
0.8	25	1	0	54
0.5	−78	1	19	21
0.02	−78	20	73	0

TABLE 3. Polymerization of $CClF_2CHO$

Initiator	Amount (mole %)	Polymerization		Polymer yield (%), acetone	
		Temp (°C)	Time (hr)	Soluble	Insoluble
$Al(C_2H_5)_3$	0.8	−78	1	0	91
$LiOC(CH_3)_3$	0.2	−78	1	11	81
$SbCl_5$	0.3	−78	1	56	0
$TiCl_4$	0.3	−78	1	0	5
H_2SO_4	0.4	25	480	3	0

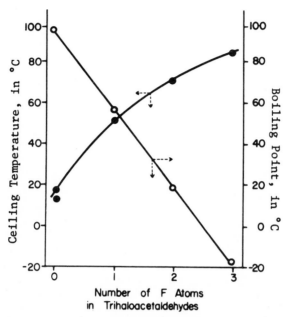

FIG. 2. Ceiling temperatures and boiling points of chlorofluoroacetaldehydes.

Dichlorobromoacetaldehyde and chlorodibromoacetaldehyde
were prepared by bromination of the corresponding chlorinated
diethylacetal. The preparation of dibromochloroacetaldehyde via
the acetal route is more convenient and gives a much better yield
of the pure aldehyde than the dichlorobromoacetaldehyde.

$$Cl_2CH-(OEt)_2 \quad + \quad Br_2 \quad \xrightarrow[\sim 5\ hrs]{80°}$$

$$\underset{\underset{OH}{|}}{Cl_2CBr-CH-OEt} \quad \xrightarrow{H_2SO_4} \quad Cl_2CBr-CHO$$

$$\sim 30\%$$

For the preparation of pure polymer grade dichlorobromoacetaldehyde, a more laborious route was found more desirable.

$$Ph_3\overset{\oplus}{P}-O-CH=CCl_2 \quad + \quad Br_2 \quad \xrightarrow[CH_2Cl_2]{O^\circ}$$
$$Cl^{\ominus}$$

$$Ph_3\overset{\oplus}{P}-O-CHBr-CCl_2Br \quad \xrightarrow{H_2O}$$
$$Cl^{\ominus}$$

$$Cl_2CBr-CHO \quad + \quad Ph_3PO \quad + \quad HBr \quad + \quad HCl$$

50−60%

Triphenylphosphine reacts instantaneously with 1 mole of chloral to form triphenyldichlorovinyloxy phosphonium chloride. This compound could be brominated to the dibromo compound. Upon hydrolysis a good yield of pure dichlorobromoacetaldehyde was obtained in addition to triphenylphosphine oxide. Purification by distillation from phosphorous pentoxide gave a polymer grade material. Dichlorobromoacetaldehyde polymerized with cationic and anionic initiators to an insoluble polymer unlike the gel-like polymer which was obtained during the chloral polymerization and no soluble polymer was found.

$$R^{\ominus} \quad + \quad \underset{H}{\overset{Cl_2CBr}{\underset{|}{\overset{|}{C}}=O}} \quad \xrightarrow[\sim -30^\circ C \atop \sim 3 \text{ hrs.}]{\text{bulk}} \quad \left(\!\!\left(\underset{H}{\overset{Cl_2CBr}{\underset{|}{\overset{|}{C}}-O}}\right)\!\!\right)_n$$

$$R = (CH_3)_3CO$$

also used as initiators

All attempts to polymerize dibromochloroacetaldehyde have failed.

Polymerization of n-Aliphatic Aldehydes with C_3 to C_8 Side Chains

Anionic polymerization of aliphatic aldehydes with C_3 to C_7 side groups have been reinvestigated with particular emphasis on the polymerization of n-heptaldehyde. Reaction conditions have been very carefully defined, and it was found that the polymerization of n-heptaldehyde is most conveniently carried out with lithium t-butoxide as the initiator in methylcyclohexane as the solvent at a temperature of about -60°C.

$$
\underset{\substack{\\ \mathrm{H}}}{\overset{\substack{CH_3 \\ | \\ (CH_2)_5 \\ | \\ C=O \\ |}}{}} \quad \xrightarrow[\substack{toluene \\ -78°}]{LiOC(CH_3)_3} \quad \underset{\substack{| \\ H}}{\overset{\substack{CH_3 \\ | \\ (CH_2)_5 \\ |}}{\sim\!\!+\!C\!-\!O\!+\!\!\sim}}
$$

The reaction is essentially complete in half an hour; workup and endcapping can subsequently be performed to stabilize the polymer. According to our 300 MHz NMR investigation, the polymer is isotactic.

We had earlier reported in a qualitative way that polyaldehydes with linear aliphatic side chains showed a peculiar melting behavior [20]. This characteristic has now been confirmed and quantitatively evaluated. A careful DSC study and a study of the thermal mechanical behavior of C_4 to C_8 isotactic polyaldehydes showed that, beginning with poly(n-valeraldehyde), two major transition regions may be defined [21]: a lower temperature transition which is apparently related to the melting of the aliphatic side chains and a higher temperature transition around 140 to 150°C which is related to the melting of the main polymer chain (Fig. 3). Similar characteristics have been observed by other workers in polyacrylates and polyolefins, but the behavior of our polyaldehydes is the first well-studied case of side chain crystallization of stereoregular polymers where both the side chain and the main chain crystallized separately.

Solvent Mobility in Rigid Polymeric Matrices

The problem of the solvation of a growing polymer chain, especially in ionic polymerization, and the competition of solvent and monomer during the polymerization to become involved in the structure of the transition state has been the concern of almost everyone involved in ionic polymerization. We have recently studied the line width of the proton signals of a number of solvents

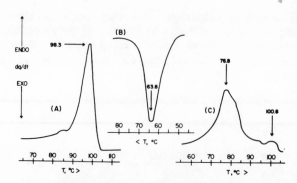

FIG. 3. Poly(n-heptaldehyde) DSC studies. First melting endotherms (A and C) and crystallization exotherm (B). Polymer insoluble in CH_2Cl_2. $\overline{M}_n$ = 6600. Heating rate: 10°C/min under nitrogen.

during the polymerization of chloral and found that the line width of various solvents is not constant throughout the polymerization. As the polymer matrix becomes more immobilized, solvent molecules become immobilized in various ways.

The example in Fig. 4 shows that the aldehydic proton of chloral retains the mobility throughout the polymerization even at high conversion. Toluene, on the other hand, shows a distinctly different behavior. The phenyl protons become very restricted after approximately 40% conversion while the methyl group is still relatively mobile.

This finding may give some important insight into the interaction between solvent molecules and particular parts of specific polymer chains.

FUTURE PROBLEMS

Any attempts to predict future developments are based on the subjective curiosity of the interested person and his own knowledge of needs in a certain area. A few thoughts should point toward unsolved problems and possible future developments in ionic polymerization.

A very recent development [22] describes the preparation of comb-like polymers with the possibility of preparing a variety of microphase

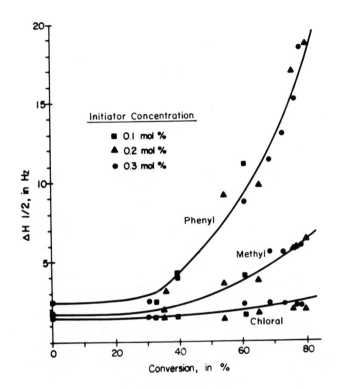

FIG. 4. Polychloral. PMR line broadening of toluene and chloral signals.

separated systems with the potential of new and unusual properties based on new products using established chemistry. Living polystyrene (and other living polymer systems such as polyisoprene) were terminated with reactive halo compounds such as allyl chloride, acrylyl chloride, vinyl chloroethyl ether, epichlorohydrin, and chlorovinyl acetate to form polymeric monomers: olefins, acrylates, vinyl ethers, epoxides, and vinyl acetates. Copolymerization with the appropriate comonomers gave a large variety of new polymers with various properties.

Complete and broad understanding of the influence of substituents of the ceiling temperature in ionic polymerization.

Solvation and definition of solvent spheres around the growing ions and their influence on rate and stereodensity of polymerization.

Control of chain transfer reactions in ionic polymerization.

Mobility and spatial arrangements in the crystal lattice in ionic solid state polymerization.

Diffusion of monomers in polymerization systems where the polymer crystallized during polymerization; effect of polymerization rate and polymerization mechanism on the morphology of the polymer (occlusion of reactive sites).

New anionic and cationic graft and block copolymers of conventional monomers using new techniques to achieve polymers with new properties.

Pure head-to-head polymers and their direct preparation.

Chlorocarbon polymers by ionic polymerization mechanism.

Preparation of stereoregular fluorocarbon polymers. (Previous reports of the successful preparation of stereoregular hexafluoropropylene polymers are doubtful, for they were probably hexafluoropropylene/isobutylene alternating copolymers, the isobutylene coming from relatively large amounts of allyl aluminum used as part of the initiator system.)

Polymerization of acid chlorides and acid fluorides or a reasonable explanation as to why this is unsuccessful.

Much of the immediate work on ionic polymerizations will be related to the preparation of polymers with new and unusual combinations of properties, and truly fundamental work for the search for knowledge without an ultimate goal will be less emphasized than previously.

ACKNOWLEDGMENTS

This work was carried out in cooperation with I. Negulescu, P. Kubisa, B. Yamada, K. Hatada, and D. Lipp, and was supported by the National Science Foundation.

REFERENCES

[1] H. Gueterbook, Polyisobutylene, Springer, Berlin, 1959.
[2] P. H. Plesch, Chemistry of Cationic Polymerization, Pergamon, Oxford, 1963.
[3] J. P. Kennedy, in Polymer Chemistry of Synthetic Elastomers (J. P. Kennedy and E. G. M. Tornqvist, eds.), Wiley-Interscience, New York, 1969, p. 291.
[4] J. P. Kennedy, J. Polym. Sci., A-1, 6, 3139 (1968).

[5] A. Tsukamoto and O. Vogl, Progr. Polym. Sci., 3, 199 (1971).
[6] T. Saegusa, H. Ikeda, and H. Fujii, Macromolecules, 6, 315 (1973).
[7] S. Penczek, IUPAC Symposium on Cationic Polymerization, Rouen, 1973.
[8] W. J. Brehm, K. G. Bremer, H. S. Eleuterio, and R. W. Meschke, U.S. Patent 2,918,501 (1961).
[9] H. S. Eleuterio, J. Macromol. Science—Chem., A6, 1027 (1972).
[10] J. Furukawa and T. Saegusa, Polymerization of Aldehydes and Oxides, Wiley-Interscience, New York, 1963.
[11] O. Vogl, Polyaldehydes, Dekker, New York, 1967.
[12] O. Vogl, U.S. Patent 3,454,527 (1969).
[13] P. Kubisa and O. Vogl, Vysokomol. Soedin., A17, 929 (1975).
[14] P. Kubisa and O. Vogl, Polym. J. (Japan), 7, 186 (1975).
[15] O. Vogl, U.S. Patent 3,668,184 (1972).
[16] J. R. Millar, J. Chem. Soc., 1960, 1311.
[17] Bamford et al., Trans. Faraday Soc., 60, 751 (1964).
[18] O. Vogl, U.S. Patent 3,707,524 (1972).
[19] C. Woolf, U.S. Patent 2,870,213 (1959).
[20] O. Vogl, J. Polym. Sci., 46, 241 (1960).
[21] I. Negulescu and O. Vogl, J. Polym. Sci., Polym. Lett. Ed., 13, 17 (1975).
[22] R. Milkovich and M. T. Chiang, U.S. Patent 3,786,116 (1974).

No Catalyst Copolymerization by Spontaneous Initiation. A New Method of Preparation of Alternating Copolymers

TAKEO SAEGUSA, SHIRO KOBAYASHI, YOSHIHARU KIMURA, and HIROHARU IKEDA

Department of Synthetic Chemistry
Faculty of Engineering
Kyoto University
Kyoto, Japan 606

ABSTRACT

A new type of copolymerization is presented which proceeds without any added catalyst. A monomer of nucleophilic reactivity (M_N) is mixed with the second monomer having electrophilic reactivity (M_E) to produce a zwitterion $^+M_N-M_E^-$ ($\underline{1}$), which is responsible for initiation as well as propagation. By the following scheme of reactions, alternating copolymer is formed.

$$\underline{1} + \underline{1} \longrightarrow {}^+M_N M_E - M_N M_E^-$$
$$\underline{2}$$

$$\underline{2} + n\underline{1} \longrightarrow {}^+M_N \left(M_E M_N \right)_{n+1} M_E^-$$
$$\underline{3}$$

Sometimes the reactions between zwitterion species $\underline{1}$, $\underline{2}$, and $\underline{3}$ with free monomer occur, which give rise to the formation of copolymer having a biased composition. Cyclic imino ethers, exoimino cyclic ether, and azetidine have been explored as the M_N comonomers. β-Propiolactone, cyclic anhydride, sultone (sulfolactone), acrylic acid, acrylamide, and β-hydroxyethyl acrylate have successfully been adopted as the M_E comonomers. Copolymerization occurred without any catalyst with all combinations of M_N and M_E monomers, and various alternating copolymers were produced. Several important results which explicate the new concept are presented and discussed.

INTRODUCTION

Usually the initiation of polymerization requires a catalyst. This paper describes a new concept of copolymerization which proceeds without any added catalyst. An ionic initiating species is generated by the interaction of two comonomers. For this type of copolymerization, the combination of two comonomers is essential. One monomer (M_N) must have nucleophilic reactivity and the other (M_E) must possess electrophilic reactivity. From M_N and M_E, a zwitterion $^+M_N-M_E^-$ is generated, which functions as the initiator:

$$M_N + M_E \longrightarrow {}^+M_N-M_E^-$$

$$\underline{1}$$

(1)

The genetic zwitterion $\underline{1}$ is responsible not only for initiation but also for propagation. The following series of reactions lead to the formation of alternating copolymer:

$$^+M_N-M_E^- + {}^+M_N-M_E^- \longrightarrow {}^+M_N-M_E M_N-M_E^-$$

$$\underline{1} \qquad\qquad \underline{1} \qquad\qquad\qquad \underline{2}$$

(2)

$$\underline{2} + \underline{1} \longrightarrow {}^+M_N - M_E M_N M_E M_N - M_E^- \qquad (3)$$

$$\underline{3}$$

Generally,

$$^+M_N \!\!\left(\, M_E M_N \,\right)_{\!n}\! M_E^- \; + \; \underline{1} \longrightarrow {}^+M_N \!\!\left(\, M_E M_N \,\right)_{\!n+1}\! M_E^- \qquad (4)$$

$$\underline{4} \qquad\qquad\qquad\qquad \underline{5}$$

First, 2 moles of the genetic zwitterion 1 react with each other to produce an oligomeric zwitterion 2 (Eq. 2). Then the oligomeric zwitterion 2 grows to be polymeric (4 and 5) by its reaction with the genetic zwitterion 1 (Eqs. 3 and 4). Hereafter, the oligomeric zwitterions (2 and 3) and polymeric ones (4 and 5) are called "macrozwitterion" which are differentiated from the genetic zwitterion 1. As to the sites of the reaction between the macro-zwitterion and the genetic zwitterion, there are two possibilities, i.e., the reaction between the cationic site of the macrozwitterion and the anionic site of the genetic one, and vice versa.

In addition to the propagation involving the genetic zwitterion, the intermolecular and intramolecular reactions of macrozwitterion may be considered to occur. The intermolecular reaction between macrozwitterions (Eq. 5) increases the molecular weight of alternating copolymer without consuming the monomer.

$$^+M_N \!\!\left(\, M_E M_N \,\right)_{\!m}\! M_E^- \; + \; {}^+M_N \!\!\left(\, M_E M_N \,\right)_{\!n}\! M_E^-$$

$$\longrightarrow {}^+M_N \!\!\left(\, M_E M_N \,\right)_{\!m+n+1}\! M_E^- \qquad (5)$$

The possibility of an intramolecular reaction between the cationic and anionic sites in a single macrozwitterion may not be excluded, which leads to the formation of a macrocyclic molecule:

$$^+M_N \!\!\left(\, M_E M_N \,\right)_{\!n}\! M_E^- \longrightarrow \boxed{\; (M_N M_E)_{n+1} \;} \qquad (6)$$

The cationic and anionic sites of zwitterions may also react with free M_N and M_E monomers, respectively (Eqs. 8 and 9). These homopropagations (Eqs. 8 and 9) are competitive with the alternating propagation with zwitterions (Eq. 7).

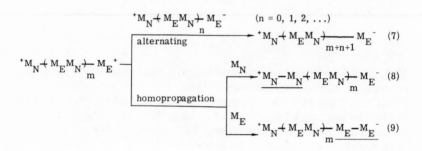

By homopropagations, the homo sequences of the diads $-M_N M_N-$ and $-M_E M_E-$ are formed, which disturb the alternating sequences of two kinds of monomeric units. The relative contributions of the alternating propagation and the two homopropagations are determined by the concentrations and reactivities of the respective species. The no-catalyst copolymerization by spontaneous initiation does not always produce alternating copolymer. Alternating copolymer is formed only when the alternating propagation occurs exclusively.

RESULTS AND DISCUSSION

Variety of Monomers and Combinations of M_N and M_E

On the basis of the above principle, four M_N monomers and six M_E monomers have been explored (Table 1) in which the structures of cationic and anionic parts derived respectively from M_N and M_E are also shown.

All M_N monomers are of cyclic structure, and the site of nucleophilic reaction consists of a nitrogen atom which is converted into the onium species to generate the corresponding cyclic onium species.

Three of four M_N monomers contain an imino ether (imidate) group as the functional group responsible for the reaction with M_E monomers. The high reactivity of imino ether may be ascribed to the stabilization of the product onium due to resonance between the oxonium and ammonium structures.

$$-O-\underset{R}{C}=N-\ \xrightarrow{E^+}\ \left[-\ddot{O}-\underset{R}{C}=\overset{+}{N}-\ \longrightarrow\ -\overset{+}{O}=\underset{R}{C}-\underset{E}{\ddot{N}}-\right] \quad (10)$$

$$E = \text{electrophile}$$

The cyclic onium ring thus formed is opened by the attack of a nucleophile (:Nu) onto the carbon atom at the α-position of the oxonium oxygen, e.g.:

$$\underset{O}{\overset{N}{\diagdown}}\hspace{-0.5em}\underset{R}{\diagup}\ +\ E^+\ \longrightarrow\ \underset{O}{\overset{N-E}{\diagdown}}\hspace{-0.5em}\underset{R}{\overset{+}{\diagup}}\ \xrightarrow{\ Nu:\ }\ \underset{RC=O}{Nu-CH_2CH_2\underset{|}{N}-E} \quad (11)$$

M_E monomers have a wider variety of functional groups. Some are cyclic compounds and some are electron-deficient olefins. By the nucleophilic attack of M_N, lactone and cyclic acid anhydride are opened to form the genetic zwitterion consisting of carboxylate anion, e.g.,

$$\underset{O}{\overset{N}{\diagdown}}\ +\ \begin{array}{l} CH_2-O \\ | \quad\quad | \\ CH_2-C=O \end{array}\ \longrightarrow\ \underset{O}{\overset{NCH_2CH_2CO_2^-}{\diagdown}}_{+} \quad (12)$$

$$\underset{\underset{\sim}{6}}{}$$

$$\underset{O}{\overset{N}{\diagdown}}\ +\ \begin{array}{l} CH_2-C{\overset{\diagup O}{\diagdown}} \\ | \quad\quad\quad O \\ CH_2-C{\diagdown}_O \end{array}\ \longrightarrow\ \underset{O}{\overset{NCCH_2CH_2CO_2^-}{\diagdown}}_{+\overset{||}{\ }O} \quad (13)$$

TABLE 1. Variety of Monomers of Spontaneous Initiation Copolymerization

M_N	$^+M_N^-$	M_E	$-M_E^-$
			$-CH_2CH_2CO_2^-$
			$-CCH_2CH_2CO_2^-$ $\parallel$ O
			$-(CH_2)_3SO_3^-$

$CH_2=CHCO_2H$ $-CH_2CH_2CO_2^-$

$CH_2=CHCONH_2$ $-CH_2CH_2C{\overset{NH}{\underset{O^-}{}}}$

$CH_2=CHCO_2CH_2CH_2OH$ $-CH_2CH_2CO_2CH_2CH_2O^-$

Sulfolactone (sultone) is opened to produce the sulfonate zwitterion:

$$\text{(structures)} \qquad (14)$$

The nucleophilic reactivity of the sulfonate group, however, is not high. Alternating copolymerizations with sultone usually require higher reaction temperatures.

Acrylic acid, acrylamide, and β-hydroxyethyl acrylate behave as M_E monomer in an interesting way. M_N monomer adds to the α,β-unsaturated bond of these M_E monomers to produce the transient carbanionic zwitterion which, following proton transfer, leads to the more stable zwitterion, e.g.:

$$\text{OZO} + CH_2=CHCO_2H \longrightarrow \left[\text{NCH}_2\bar{\text{C}}\text{HCO}_2\text{H} \right] \longrightarrow \text{NCH}_2\text{CH}_2\text{CO}_2^- \qquad (15)$$

$$\underset{7}{\qquad} \qquad \underset{6}{\qquad}$$

$$\text{OZO} + CH_2=CHCNH_2 \longrightarrow \left[\text{NCH}_2\text{CHCNH}_2 \right] \longrightarrow \text{NCH}_2\text{CH}_2\text{C} \qquad (16)$$

$$\underset{8}{\qquad}$$

$$\text{OZO} + CH_2=CHCOCH_2CH_2OH$$

$$\longrightarrow \left[\text{NCH}_2\bar{\text{C}}\text{HCOCH}_2\text{CH}_2\text{OH} \right] \longrightarrow \text{NCH}_2\text{CH}_2\text{COCH}_2\text{CH}_2\text{O}^- \qquad (17)$$

$$\underset{9}{\qquad}$$

It is quite interesting to note that the key zwitterion 6 produced from 2-oxazoline (OZO) and acrylic acid is the same as that generated from 2-oxazoline and β-propiolactone (BPL). Accordingly, the structure of the alternating copolymer 10 from acrylic acid and OZO is the same as that produced from BPL and OZO:

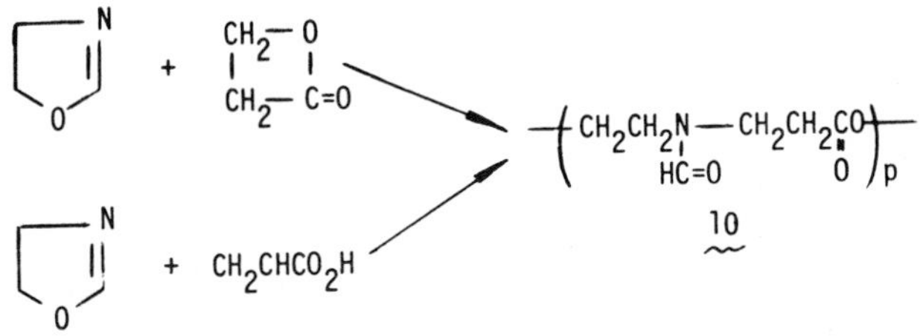

The detailed mechanism of proton transfer to the carbanion site in the transient zwitterion (7-9) has not been clarified. There are two possibilities: the direct rearrangement in the transient zwitterion and the indirect proton transfer which proceeds via the acidic hydrogen of the free monomer.

From four M_N monomers and six M_E monomers, 24 pairs of combinations can be constructed (Table 2). Fourteen of these pairs have been examined. In all combinations examined, no catalyst copolymerization occurred without an added catalyst. Alternating copolymers were not always obtained. As has been discussed before, the character of copolymerization, i.e., the arrangements of two kinds of monomeric unit in the copolymer, is dependent upon the natures of the monomers involved. In Table 2, A indicates the cases in which alternating copolymers are produced under a wide variety of reaction conditions, B means that alternating copolymers are obtained under suitable conditions of reaction, and C designates the combinations in which alternating copolymer has not been accomplished. The following part of this paper describes several experimental findings which are reasonably explained by the general principle proposed in the preceding section.

Alternating Copolymerization of 2-Oxazoline with β-Propiolactone

When OZO was mixed with an equimolar amount of BPL in an aprotic polar solvent at 40°C, the solution gradually became

TABLE 2. Copolymerizations between M_N and M_E Monomers[a]

M_N	M_E (β-lactone ring, R)	anhydride (O=C–O–C=O)	sultone (ring–SO$_2$)	CH_2=$CHCO_2H$	CH_2=$CHCNH_2$ (‖O)	CH_2=$CHCO_2$—CH_2CH_2OH
2-oxazoline (N═, O ring, R)	1 (Refs. 1–3) R = H (B) Me (C) Ph (C)	5 (Ref. 8) R = H (A)	9 (Ref. 4) R = H (C) Me (B)	13 (Ref. 9) R = H (A) Me (A)	17 (Refs. 11, 12) R = H (A) Me (A)	21 (Ref. 13) R = H (B)
dihydrooxazine (N═, O ring, R)	2 (Refs. 4, 5) R = H (B) Me (C) Ph (C)	6	10	14 (Ref. 4) R = H (A) Me (A) Ph (A)	18 (Refs. 11, 12) R = H (A)	22 (Ref. 13) R = H (A)
iminolactone (NR)	3 (Ref. 6) R = CH$_2$Ph (B)	7	11	15 (Ref. 6) R = CH$_2$Ph (A)	19	23
azetidine (N–R, R, R)	4 (Ref. 7) R = Me (C)	8	12	16 (Ref. 10) R = Me (A)	20	24

a(A): Alternating copolymers are readily produced under a wide variety of conditions.
(B): Alternating copolymers are produced under suitable conditions.
(C): Copolymerizations occur without catalyst, but alternating copolymers have not been produced.

viscous and the 1:1 alternating copolymer 10 was produced [1]:

$$\text{(structure)} + \begin{array}{l} CH_2-O \\ | \quad | \\ CH_2-C=O \end{array} \longrightarrow \quad 10 \qquad (18)$$

The structure of the alternating copolymer 10 has been established by its IR and NMR spectra as well as by alkaline (NaOH) hydrolysis experiment. The NMR spectrum of the alkaline hydrolysis product of the copolymer in D_2O was identical with that of an equimolar mixture of the authentic Na salts of formic acid and N-(β-hydroxyethyl)-β-alanine.

$$10 \xrightarrow[\text{D}_2\text{O}]{\text{NaOH}} HCO_2Na + HOCH_2CH_2NHCH_2CH_2CO_2Na$$

Figure 1 shows the composition curve of the OZO-BPL copolymerization in DMF at 40°C. The 1:1 alternating copolymer is obtained when the mole fraction of OZO in the feed monomers mixture is higher than about 0.5. At lower fractions of OZO in the monomer feed, copolymers consisting of more than 50 mole % BPL units are formed. This experimental finding is satisfactorily explained on the basis of two competitive propagations, i.e., the propagation at

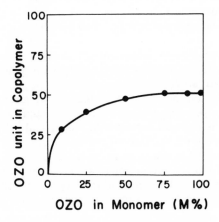

FIG. 1. OZO-BPL copolymerization in DMF at 40°C [1].

the carboxylate end of zwitterion 11 with the genetic zwitterion
(Eq. 19) and the propagation with free BPL (Eq. 20).

$$(19)$$

$$(20)$$

At lower fractions of OZO in the monomer mixture, the concentration
of the genetic zwitterion 6 is lower because its production rate is
lower and the concentration of free BPL is higher. These two
factors favor propagation with free BPL. Considering the reaction
conditions of the anionic polymerization of BPL [14], the ring-
opening process of BPL by carboxylate anion (Eq. 20) is deemed
quite possible even under these mild conditions. On the other hand,
the reaction between the oxazolinium ring and free OZO is assumed
to be negligible [15].

Table 3 shows some results which illustrate the solvent effect
upon the OZO-BPL alternating copolymerization with an equimolar
feed. Qualitatively, the rate of copolymerization is higher in polar
solvents than in nonpolar solvents. The results are compatible
with the assumption that the rate of formation of the genetic
zwitterion 6 is decisive in determining the overall rate. In
general, the formation of ionic species from two neutral molecules
is favored in polar solvents.

Figure 2, in which the average molecular weight and the quotient
of the conversion percent divided by the molecular weight are
plotted against the percent conversion, represents the growth of
copolymer molecule during the course of copolymerization with
1:1 monomer feed. The quotient is an index which is proportional
to the number of copolymer molecules. In these plots, a transient
Region A is observed. The molecular weight, as well as the number
of copolymer molecules, continues to increase up to Region A.

TABLE 3. Effect of Solvent on the Copolymerization of OZO with BPL[a] [1]

Solvent	Copolymerization yield (%)	OZO in copolymer (mole %)
DMF	70	47
CH_3CN	39	52
CH_2ClCH_2Cl	26	50
n-Bu_2O	29	45
Toluene	9	46

[a]OZO = BPL = 7.5 mmoles in 2 ml of solvent at 40°C for 3 hr.

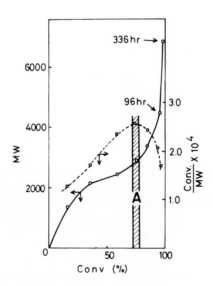

FIG. 2. OZO–BPL alternating copolymerization in acetonitrile at 25°C [11].

After Region A the increase of molecular weight becomes sharp, and
the number of copolymer molecules begins to decrease. These plots
are also compatible with the general scheme of Eqs. (1)-(5). As the
reaction proceeds, the concentration of monomers continues to
decrease and the concentration of macrozwitterions increases.
Before Region A the rate of increase of macrozwitterions by the
process of Eq. (2) prevails over the rate of decrease of macrozwitterions by the process of Eq. (5). After Region A the rate of
the process of Eq. (5) exceeds that of the process of Eq. (2), and
hence the number of copolymer molecules begins to decrease.

   In Table 4 the OZO-BPL copolymerization is compared with
the copolymerizations of BPL with 2-methyl-2-oxazoline (MeOZO)
and with 2-phenyl-2-oxazoline (PhOZO). Even with an equimolar
feed of monomers, copolymerizations with substituted 2-oxazolines
produces copolymers containing more than 50 mole % BPL units.
These results are also explained by the competitive processes of
propagation in the general scheme (Eqs. 7-9).

The ring-opening reactivities of 2-methyl- and 2-phenyl-substituted
oxazolinium rings have been found to be less than that of the unsubstituted one [15]. Thus the process of Eq. (21) predominates over
the alternating propagation of Eq. (22). A similar tendency of substituent effects has been observed in the copolymerization of BPL
with six-membered cyclic imino ethers of unsubstituted, 2-methyl-,
and 2-phenyl-5,6-dihydro-4H-1,3-oxazines [4].

   Table 5 illustrates the effect of methyl substitution at the α-carbon
of BPL on the overall rate of the OZO-BPL copolymerization. The
qualitative order of the relative rates of 1:1 alternating copolymerization of OZO with BPL, with α-methyl-β-propiolactone (MeBPL),

TABLE 4. Copolymerization of BPL with 2-Oxazolines[a] [1]

2-Oxazoline	Copolymer yield (%)	2-Oxazoline in copolymer (mole %)
OZO	70	47
MeOZO	52	11
PhOZO	23	4

[a]2-Oxazoline = BPL = 3.8 mmoles in DMF at 40°C for 3 hr.

TABLE 5. Copolymerization of OZO with β-Lactones[a] [2]

β-Lactone	Time (hr)	Yield (%)	Mol wt	β-Lactone in copolymer (mole %)
BPL	25	61	3500	50
MeBPL	27	42	4000	50
Me₂BPL	45	21	3610	56

[a]β-Lactone = OZO = 7.5 mmoles in 2.0 ml of acetonitrile at 10 to 20°C.

and with α,α-dimethyl-β-propiolactone (Me$_2$BPL) is BPL > MeBPL > Me$_2$BPL. The unfavorable effect of the methyl substituent has been reported in the anionic polymerization of substituted and unsubstituted propiolactones, in which the ring opening of lactone by the nucleophilic attack of carboxylate controls the rate [14]. The overall rate of the alternating copolymerizations of lactones with OZO is controlled by the rate of the zwitterion formation, which is also the ring-opening process of lactone by a nucleophile of OZO [2].

Alternating Copolymerizations of Acrylic Acid with M_N Monomers

Acrylic acid is copolymerized successfully with four M_N monomers to produce the respective alternating copolymers. This is an interesting example of copolymerization between heterocyclic monomers and olefinic monomers. As the key intermediates of this alternating copolymerization, zwitterions of the onium–carboxylate structures (12) are assumed, which are generated according to the following general process:

$$M_N + CH_2 = CHCO_2H \longrightarrow [M_N^+ CH_2\bar{C}HCO_2H] \longrightarrow M_N^+ CH_2CH_2CO_2^- \quad (23)$$

$$\underline{12}$$

This mechanism has been supported by several experimental findings. First, betaine (13) was isolated as a crystalline material (mp 131 to 132°C) from pyridine and acrylic acid (Eq. 24), which used to be prepared from pyridine and BPL [9]:

$$(24)$$

From a mixture of acrylic acid and 2-phenyl-5,6-dihydro-4H-1,3-oxazine (14) in acetonitrile at lower temperatures, e.g., 5°C, a zwitterion of 15 was isolated in a monohydrated form, mp 145 to 146°C. When 15 was heated at 150°C, 15 was converted quantitatively into the alternating copolymer 16, which was also produced by the direct alternating copolymerization of acrylic acid with 14 at 120°C.

Concerning the alternating copolymerization of acrylic acid with 1,3,3-trimethylazetidine (17), a cyclic ammonium salt (18) was isolated from a reaction mixture in Et_2O at a lower temperature, e.g., 10°C.

Isolation of 18 is taken as support to the intermediacy of the zwitterion 20, from which 18 is readily derived by the proton transfer between 20 and free acrylic acid. Heating of an equimolar mixture of a salt (18)

and another mole of 17 afforded alternating copolymer 19 [10].

Unlike β-propiolactone, acrylic acid is reluctant to react with carboxylate anion under moderate reaction conditions. The homo-propagation of acrylic acid (Eq. 26) scarcely occurs, and the alter-nating propagation (Eq. 25) is strongly favored in the following competitive processes:

A strong indication of support of this consideration has been obtained, e.g., in the copolymerization of 2-benzyliminotetrahydrofuran (BIT) with acrylic acid (AA) and with BPL in acetonitrile, apparent copolymerization parameters of $\gamma_{BIT} = \gamma_{AA} = 0.00$ at 100°C and $\gamma_{BIT} = 0.00$ and $\gamma_{BPL} = 0.65$ at 50°C were determined [6].

Alternating Copolymerizations of Acrylamide with Cyclic Imino Ethers

Copolymerization of acrylamide with cyclic imino ethers (21 and 22) presents another interesting example of alternating copolymerizations between heterocyclic monomers and olefinic monomers.

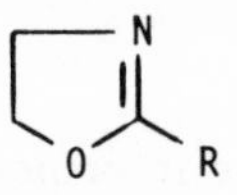

21 (R=H (OZO), R=Me) 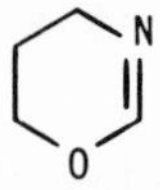   22 (OZI)

In the alternating copolymers, the acrylamide unit is incorporated as a structure of imidate group. The structures of alternating copolymers were established by IR and NMR spectra as well as the NMR identification of the alkaline hydrolysis mixture of copolymers. For example, the NMR spectrum of alkaline hydrolysis products of the alternating copolymer from acrylamide and 22 was identical with that of an equimolar mixture of the authentic salts of N-(γ-hydroxypropyl)-β-alanine (25) and formic acid. Thus the following scheme of alternating copolymerization is constructed:

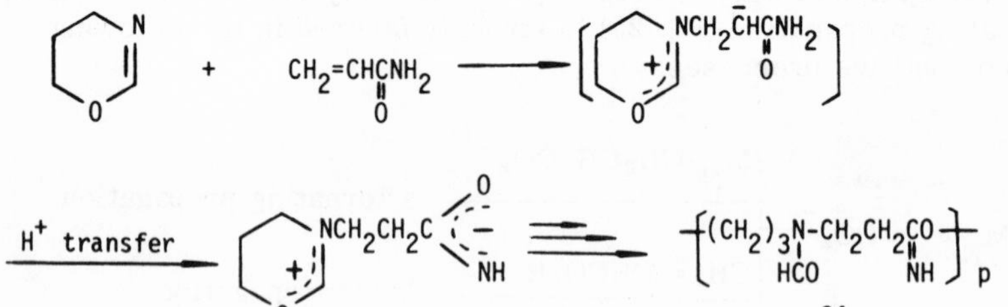

$$\underline{24} \xrightarrow[\text{H}_2\text{O}]{\text{NaOH}} \text{HO(CH}_2)_3\text{NHCH}_2\text{CH}_2\text{CO}_2\text{Na} + \text{HCO}_2\text{Na}$$

$$\underline{25}$$

The amide anion part of zwitterion is of an ambient nature, having the two canonical forms of 26a and 26b which reacts in the form of 26a in the above copolymerization.

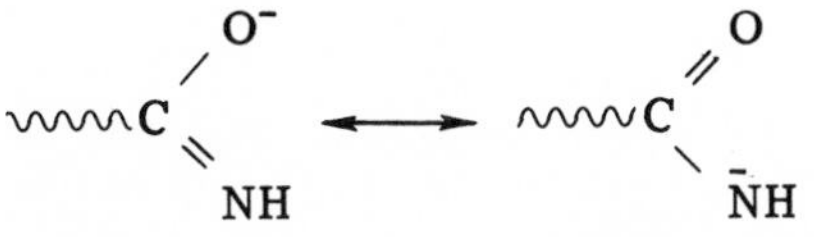

26a   26b

The above behavior of amide anion has been interestingly compared with the homopolymerization of acrylamide by strong base catalysts, in which the propagating species of amide anion reacts at the nitrogen site [12].

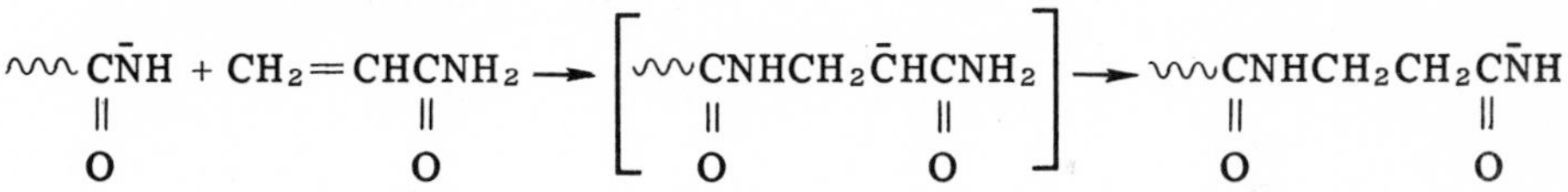

Alternating Copolymerizations of β-Hydroxyethyl Acrylate with Cyclic Imino Ethers

β-Hydroxyethyl acrylate (BHEA) is a reactive M_E monomer which is copolymerized readily with OZO and with OZI to give the corresponding alternating copolymers. The alternating copolymerization of BHEA with OZI is explained by the following reaction scheme in which the key intermediate of a zwitterion (28) is derived by the hydrogen transfer from the hydroxyl group to the carbanion center of the transient zwitterion 27. The structure of copolymer 29 corresponds to the 1:1:1 terpolymer of OZI-BPL-ethylene oxide.

In the present paper, no catalyst copolymerization by spontaneous initiation mechanism has been discussed, for the emphasis has been on the synthesis of alternating copolymers. On the basis of the general principle, there may be further possibilities of exploring other M_N and M_E monomers to develop new combinations of copolymerizations.

REFERENCES

[1] T. Saegusa, H. Ikeda, and H. Fujii, Macromolecules, 5, 354 (1972).

[2] T. Saegusa, S. Kobayashi, and Y. Kimura, Ibid., 7, 1 (1974).

[3] T. Saegusa, Y. Kimura, and S. Kobayashi, Presented at 30th Annual Meeting of the Chemical Society of Japan, Osaka, April 1974.

[4] T. Saegusa, Y. Kimura, H. Ikeda, S. Hirayanagi, and S. Kobayashi, Macromolecules, 8 (May/June 1975).

[5] T. Saegusa, Y. Kimura, and S. Kobayashi, Presented at 29th Annual Meeting of the Chemical Society of Japan, Hiroshima, October 1973.

[6] T. Saegusa, Y. Kimura, K. Sano, and S. Kobayashi, Macromolecules, 7, 546 (1974).

[7] T. Saegusa, Y. Kimura, and S. Kobayashi, Unpublished Results.

[8] T. Saegusa, Y. Kimura, and S. Kobayashi, Presented at 28th Annual Meeting of the Chemical Society of Japan, Tokyo, April 1973.

[9] T. Saegusa, S. Kobayashi, and Y. Kimura, Macromolecules, 7, 139 (1974).

[10] T. Saegusa, Y. Kimura, S. Sawada, and S. Kobayashi, Ibid., 7, 956 (1974).
[11] T. Saegusa, Y. Kimura, and S. Kobayashi, Presented at 23rd Annual Meeting of the Society of Polymer Science, Japan, June 1974.
[12] T. Saegusa, S. Kobayashi, and Y. Kimura, Macromolecules, 8 (May/June 1975).
[13] T. Saegusa, S. Kobayashi, and Y. Kimura, Unpublished Results.
[14] Y. Yamashita, Y. Ishikawa, and T. Tsuda, Kogyo Kagaku Zasshi, 67, 252 (1964).
[15] T. Saegusa, H. Ikeda, and H. Fujii, Polym. J., 4, 87 (1973).

Activated Monomers and Nucleophilic Reagents in Addition and Polymerization Reactions

T. TSURUTA

Department of Synthetic Chemistry
Faculty of Engineering
University of Tokyo
Hongo 7-3-1, Bunkyo-ku, Tokyo, Japan

ABSTRACT

A number of examples of addition and polymerization
reactions is presented with special emphasis on the
chemical behaviors of activated monomers and/or activated nucleophilic reagents. Lithium alkoxyethanolate
forms a complex with lithium alkyl. Spectroscopic
studies showed this complex to possess agent-separated
ion pairs. The nature of the complex is characterized by
the enhanced reactivity of styrene in the copolymerization
reaction with butadiene initiated by the complex. Magnesium
alkyl can be sufficiently activated by magnesium alkoxyethanolate to polymerize styrene and diene. Aluminum
alkyl and zinc alkyl are able to induce the anionic polymerization of vinyl ketones, but not of unsaturated esters
or nitriles. Aluminum or zinc alkoxyethanolates fail to
activate their corresponding metal alkyls. Bipyridyl,
sparteine, triphenylphosphine, HMPT, and related Lewis
bases, however, activate aluminum alkyl enough to react
with carbon-carbon double bonds of the unsaturated esters
and nitriles. Crotononitrile can be polymerized by the

AlR₃-HMPT system to form a colorless polymer, where
possible side reactions between CN and AlR₃ are prevented
by HMPT. Mutual activation through complex formation is
confirmed by a model system of a vinyl ketone with organozinc compounds. AlR₃-HMPT does not polymerize vinyl
ketones because of a lack of complex formation. N-Carboxy-
α-alanine anhydride (NCA) can be polymerized with zinc
alkyl as initiator. The formation of activated NCA by proton
abstraction from the NH group is shown to be the essential
stage for polymerization. Zinc alkyl is also activated by
conventional acid anhydrides. The propylene oxide ring
can be cleaved with the ZnR_2-phthalic anhydride system,
which is the initiation step in the alternate copolymerization
between propylene oxide and the acid anhydride. The propagation mechanism of the CO_2-epoxide copolymerization
is also discussed.

INTRODUCTION

Since the brilliant success of the Ziegler-Natta polymerization, a
variety of research on polymerization reaction has been carried out
with various types of organometallic compounds or nucleophilic
reagents as initiator. We compiled a large collection of experimental
results in early 1960s, but there were few studies directed to syste-
matizing the structure-reactivity relationship of vinyl monomers
and initiators in anionic polymerization. In 1961 the author proposed a correlation diagram which systematizes the reactivities of
nucleophilic initiators in terms of their ability to effectuate anionic
polymerizations of a group of vinyl monomers [1]. For instance,
magnesium alkyl was ranked as an initiator possessing a smaller
reactivity than lithium alkyl because a group of hydrocarbon
monomers such as styrene and butadiene were polymerized by
lithium alkyl but not by magnesium alkyl. The reactivity of
aluminum alkyl was still smaller because a group of acrylic
esters such as methyl methacrylate and methyl acrylate were
polymerized by magnesium alkyl but not by aluminum alkyl.
    Significant changes in the reactivity of organometallic com-
pounds have recently been observed when an appropriate complexing agent is coupled with the organometallic compounds.
Magnesium alkyl, for example, could be sufficiently activated

by magnesium alkoxyethanolate to polymerize styrene and diene. A part of the results of studies on the activated nucleophilic reagents was discussed briefly in another review article [2].

In addition to the activated nucleophilic reagents, the formation of activated monomers was confirmed in some reactions. Vinyl ketone, for instance, must be activated by the coordination of aluminum alkyl (or zinc alkyl) to accept the nucleophiles at the conjugated double bond.

Examples of activated nucleophilic reagents and for activated monomers have also been found in the polymerization reactions of ring compounds. This paper reviews the recent results of the author's studies with special reference to the chemical behaviors of activated nucleophilic reagents and activated monomers in the elementary reactions of the polymerization of vinyl and ring compounds.

ALKOXYETHANOLATE AS A COMPLEXING AGENT

The reactivity of n-BuLi—$CH_3OCH_2CH_2OLi$ in toluene is characterized by the enormous enhancement of reactivity of styrene in the copolymerization reaction with butadiene [3]. Neither CH_3OCH_2-CH_2OCH_3 nor $LiOCH_2CH_2OLi$ exhibits such an enormous enhancement of styrene reactivity as the complexing agent. An aminoethanolate showed a similar reactivity to $CH_3OCH_2CH_2OLi$ as the complexing agent [4].

Measurements [5] of the electronic spectra of 1,1-diphenyl-n-hexyllithium (DPHLi) in cyclohexane in the presence of varying quantity of $CH_3OCH_2CH_2OLi$ showed a considerable bathochromic shift (from 413 to 490 nm) in the absorption maximum as the [OLi]/[CLi] ratio increased from 0 to 2. No pronounced shift, however, was observed in the range of the [OLi]/[CLi] ratio higher than 2. The results suggest the formation of a one to two complex between DPHLi and $CH_3OCH_2CH_2OLi$.

It was also found in the electronic spectrum of one to two system of 9-fluorenyllithium—$CH_3OCH_2CH_2OLi$ that the peak assignable to the contact ion pairs has virtually disappeared, the only peak observed at 373 nm being in agreement with that expected for the solvent (or coordination agent)-separated ion pairs.

With the use of the lithium methoxyethanolate complex in a catalytic amount, diphenylmethane was easily butenylated on reacting with butadiene to form predominantly 5,5-diphenyl-cis-2-pentene (DPPE) [6]:

Metallation: $Ph_2CH_2 + n\text{-}BuLi\cdot D \longrightarrow Ph_2CHLi\cdot D + nBuH$ (1)

Butenylation: $Ph_2CHLi\cdot D + C_4H_6 \longrightarrow Ph_2CH(C_4H_6)Li\cdot D$ (2)

Transmetallation: |_____ (recycle)_____

$Ph_2CH(C_4H_6)Li\cdot D + Ph_2CH_2 \longrightarrow Ph_2CH(C_4H_6)H + Ph_2CHLi\cdot D$ (3)

where D is a complexing agent such as $CH_3OCH_2CH_2OLi$.

Table 1 shows the main feature of reactivity of the lithium methoxyethanolate complex in comparison with two other complex systems. A high yield of DPPE, despite of the low matallation yield, was characteristic with n-BuLi—$CH_3OCH_2CH_2OLi$ as the catalyst.

Magnesium alkyl can also be activated by magnesium alkoxyethanolate [7]. For instance, an equimolar mixture of n-Bu_2Mg with $CH_3OCH_2CH_2OH$ is capable of initiating styrene polymerization, in contrast with the lack of the ability of n-Bu_2Mg alone. According to a study on the influence of the added amount of $CH_3OCH_2CH_2OH$ on the extent of styrene polymerization while keeping the feed concentration of n-Bu_2Mg constant, the conversion to polystyrene was increased enormously under the condition of $[O\text{-}Mg]/[C\text{-}Mg] = 3$.

TABLE 1. Comparison of Reactivity of Three Types of Activated BuLi System[a] [6]

Complexing agent	Molar[b] ratio	Metallation yield (%)	DPPE yield (%)
$CH_3OCH_2CH_2OLi$	2	16	62
TMEDA	2	100	31
HMPT	2	37	80

[a]Reaction between diphenylmethane and butadiene.
[b]Molar ratio = [complexing agent]/[BuLi].

Colorless precipitates were formed from a mixture of n-Bu$_2$Mg and 2-methoxyethanol. The precipitates were recrystallized to colorless prisms, the analysis of which showed the composition Mg:n-Bu:CH$_3$OCH$_2$CH$_2$O:dioxane = 2.0:1.06:3.05:1.0, in agreement with the formula of n-BuMgO-CH$_2$CH$_2$OCH$_3$ · Mg(OCH$_2$CH$_2$OCH$_3$)$_2$· dioxane. Tetrafurfuryl alcoholate (THFA) was an excellent complexing agent for magnesium alkyl [8]. For instance, 100% conversion was obtained at -70°C within 1 min in the polymerization of styrene with [C$_6$H$_5$CH$_2$Mg(THFA)][(THFA)$_2$Mg] as initiator, contrary to the very poor reactivity of [C$_6$H$_5$CH$_2$MgOCH$_2$CH$_2$OCH$_3$]-[(CH$_3$OCH$_2$CH$_2$O)$_2$Mg]. The THFA complex also induced the rapid polymerization of butadiene at -70°C to form a polymer consisting of more than 90% of 1,2-enchainment. With the magnesium methoxyethanolate complex, on the other hand, practically no polymer was obtained after a 4-day reaction at temperatures as high as 50°C. An electronic spectrum of the [C$_6$H$_5$CH$_2$Mg(THFA)][(THFA)$_2$Mg] system showed λ_{max} at 458 nm, suggesting the formation of a coordination agent separated ion pairs.

LITHIUM AMIDE-SECONDARY AMINE COMPLEXES

A stereospecific addition reaction was found to take place when lithium diethylamide was reacted with butadiene in cyclohexane [9]. It was noted in this reaction that at least 2 moles of free amine should be present in the reaction system to activate 1 mole of the lithium amide to add to the butadiene double bonds. The formation of the two to one complex, 2R$_2$NH.R$_2$NLi, was indicated by the results of IR and NMR spectra. The overall reaction scheme of the addition reaction is depicted as

$$Et_2NLi + 2Et_2NH \longrightarrow Et_2NLi \cdot 2Et_2NH \qquad (4)$$

$$Et_2NLi \cdot 2Et_2NH + C_4H_6 \xrightarrow{\ k\ } Et_2NC_4H_6Li \cdot 2Et_2NH \qquad (5)$$

$$Et_2NC_4H_6Li \cdot 2Et_2NH + Et_2NH$$

$$\longrightarrow Et_2NCH_2-CH=CH-CH_3 + Et_2NLi \cdot 2Et_2NH \qquad (6)$$

(recycle)

(cis 98 to 99%)

Kinetic studies revealed that the rate (v) of the addition reaction was expressed as

$$v = k[A][\text{butadiene}] \tag{7}$$

where [A] is the concentration of the one to two complex. For diethylamine, k was 8.5×10^{-4} [liter/(mole)(sec)]; for diisobutylamine k was 2.4×10^{-3} [liter/(mole)(sec)].

Kinetics similar to Eq. (7) was obtained in the addition reaction of diethylamine to styrene [10]. In order to compare the reactivity of lithium diethylamide with that of lithium alkyl, the reactivity of styrene toward the lithium amide was examined in the presence of butadiene. It is seen in Table 2 that the presence of butadiene has no effect on the reactivity of styrene in contrast to the reaction with lithium alkyl. The mechanism of the stereospecific addition reaction of the lithium diethylamide complex to butadiene seems to be different from the usual stereospecific addition of lithium alkyl in hydrocarbon solvents.

ACTIVATION OF ALUMINUM ALKYL WITH LEWIS BASES

Contrary to lithium and magnesium alkyls, aluminum or zinc alkyl was not activated by the corresponding alkoxyethanolates. Bipyridyl (Bipy), triphenylphosphine (PPh₃), sparteine (Spar), HMPT, and some other Lewis bases activated aluminum (or zinc) alkyl enough to react with carbon-carbon double bonds of unsaturated esters and nitriles [11]. Among several $AlEt_3$-Lewis base complexes investigated, the $AlEt_3$ complex with bidentate ligand or with monodentate ligand of sufficiently high basicity showed high catalytic activity for the polymerization of AN and MMA. $AlEt_3$ and rigid bidentate Lewis bases such as Bipy and Spar form an unstable equimolar five-coordinated complex, and the complex is stabilized to assume a six-coordinated structure by the coordination of another complex or solvent to its sixth coordinate site. When monomers are present along with the aluminum complex, some of the ligands at the sixth coordination site are replaced with monomer molecules, which is the preliminary stage for the initiation reaction of polymerization [12].

The electronegativity, χ, of Al and Zn in $AlEt_3$ and $ZnEt_2$ was estimated, respectively, from the differences between the chemical

TABLE 2. Second-Order Rate Constants of Addition Reaction of $Et_2NLi.2Et_2NH$ Complex with Styrene and Butadiene[a] [10]

$\dfrac{[Et_2NH]_0}{[Et_2NLi]_0}$	Styrene, $k_s \times 10^3$ [liter/(mole)(sec)]	Butadiene, $k_b \times 10^3$ [liter/(mole)(sec)]	Styrene, $k_s \times 10^3$ [liter/(mole)(sec)]	Butadiene, $k_b \times 10^3$ [liter/(mole)(sec)]
3.0	-	0.85	-	-
10	1.6	1.4	1.5	1.6
15	2.0	-	-	-
28	2.6	-	-	-

[a] In cyclohexane at 50°C.

shifts of their methyl and methylene protons, $\Delta CH_3 - CH_2$, in NMR
spectra. The coordination of Bipy or PPh_3 to $AlEt_3$ makes the
electronegativity of Al decrease. The coordination of Bipy to $ZnEt_2$
causes the electronegativity of Zn in the complex to increase, which
may suggest a possible role of the back donation from the Zn d-orbital
to the Bipy π-orbital. Compared to the Al complexes, the zinc complexes showed a much smaller activity as initiators for MMA
polymerization.

Model reactions were studied with methyl α-isopropylacrylate to
elucidate the initiation mechanism, because the α-isopropylacrylate
has been shown to be nonpolymerizable. Two addition products, I
and II, were isolated from a reaction mixture of α-isopropylacrylate
and $(i-C_4H_9)_3Al$-Bipy. It was also confirmed that only one isobutyl

$$
\begin{array}{cc}
\text{H} & \\
| & \\
\text{i-}C_4H_9-CH_2-CCH(CH_3)_2 \qquad & CH_3-C(\text{i-}C_4H_9)CH(CH_3)_2 \\
| & | \\
CO_2CH_3 & CO_2CH_3 \\
\\
\qquad\text{I} & \qquad\text{II}
\end{array}
$$

group of $(i-C_4H_9)_3Al$ has entered the addition reaction, even in the
presence of excess (20 times) methyl α-isopropylacrylate [12].

Methyl crotonate undergoes an isomerized dimerization in the
presence of a catalytic amount of the AlR_3-Spar complex [13].
With the aid of an α-deuterated methyl crotonate, it was confirmed
that the reaction starts with proton abstraction from the β-methyl
group of the crotonate. From the facts that AlR_3 by itself did not
dimerize methyl crotonate and that the dimer of methyl crotonate
obtained by the AlR_3-Spar complex showed optical activity, it is
concluded that the coordination of Spar to AlR_3 not only activates
the Al—C bond but also affords a chiral structure to the environment around the aluminum species [14].

From the results stated above, it was concluded that AlR_3-Lewis
base complexes are capable of abstracting hydrogen from monomer
molecules. Thus, in the polymerization of AN and methyl acrylate
by an AlR_3-Lewis base complex, the initiation reaction and the
chain transfer reaction may also occur partially through the
abstraction of the somewhat acidic α-hydrogen of the monomer
[14].

Equilibrium constants for the complex formation between $AlEt_3$
and several Lewis bases were estimated from the IR data of these

systems. A parallel correlation was found between the equilibrium constants and the anionic reactivity of the binary systems [15]. HMPT exhibited a very unique behavior. Crotononitrile can be polymerized by AlEt$_3$ in HMPT to form a colorless polymer. The formation of the colorless polycrotononitrile indicates that possible side reactions between nitrile group and aluminum alkyl have been excluded because all of the coordination sites of aluminum are occupied by the strong coordination of HMPT. The behavior of the double bond of crotononitrile toward the AlR$_3$-HMPT system can reasonably be understood in terms of the concept proposed by Tsuruta [16].

Methyl isopropenyl ketone was not polymerized with the AlR$_3$-HMPT system. Since methyl isopropenyl ketone is easily polymerized with AiR$_3$ alone, HMPT seems to deactivate the aluminum alkyl in contrast to the case of crotononitrile. As stated in the next section, vinyl ketone must be activated by the coordination of AlR$_3$ (or ZnR$_2$) to accept nucleophilic reagents at the conjugated double bond.

MUTUAL ACTIVATION IN VINYL KETONE-ORGANOZINC COMPOUND SYSTEM

Anionic polymerization of phenyl vinyl ketone (PVK) was carried out in toluene at 0°C by the use of an initiator, enthylzinc 1,3-diphenyl-1-pentene-1-olate (ZC), which possesses the same structure as the growing chain end of PVK in ZnEt$_2$-initiated polymerization [17, 18]. Gas volumetric analysis confirmed that the ethyl-zinc bond in ZC, $C_6H_5(C_2H_5)CHCH=C(C_6H_5)-OZnC_2H_5$, was kept intact during the polymerization process.

A unimodal molecular weight distribution was found in the polymer obtained. The polymerization system exhibited a "living characteristic" with an initiator efficiency being 1.0. The reactions between ZC and PVK were kinetically traced under various conditions. Rate analysis showed a second-order dependence with respect to the concentration of ZC. The second-order dependence of zinc alkyl was previously observed in the reaction between vinyl ketone and dibutylzinc [19]. Tsuruta suggested a concerted mechanism for this addition reaction. The facts that ZC exhibits the highest reactivity in toluene and that its reactivity decreases in polar solvents or by the addition of Lewis bases seem to support the above mechanism.

Experimental evidences for complex formation have recently been obtained by use of a model system consisting of benzalacetophenone

(C) and ZC. Benzalacetophenone is sparingly soluble in cyclohexane, whereas a mixture of C and ZC is easily soluble in the same solvent, suggesting a significant interaction to be operative between C and ZC. The complex formation was also suggested by the UV absorbance at 302 nm of the binary system of C and ZC in cyclohexane. When the ratio $[C]/[ZC]$ was changed, a break of linearity in the UV absorbance was observed at a one to one mole ratio. In the light of the results of cryoscopic measurements in cyclohexane, it is plausible to assume that the ZC-C species were present as a dimeric form [20].

In the range of larger concentrations, ZC could undergo an addition reaction with benzalacetophenone to form ZC_2:

$$ZC + C_6H_5CH{=}CHCOC_6H_5 \longrightarrow$$

$$C_6H_5CH(Et){-}CH(COC_6H_5){-}CH(C_6H_5){-}CH{=}C(C_6H_5)OZnEt \qquad [ZC_2]$$

A series of studies on the addition reaction were carried out by changing the C/ZC ratio and other conditions. The results obtained are 1) the maximum percent conversion of reaction did not exceed 50% with respect to the minor component C or ZC, and 2) the reaction proceeded according to the second order of ZC concentration. These results can be interpreted in terms of the formation of dimeric ZC-C or ZnR_2-C species [20].

$$ZC + C \xrightarrow{\text{coordi-}\atop\text{nation}} ZC{-}C \longrightarrow \begin{pmatrix} ZC{-}C \\ C{-}ZC \end{pmatrix} \xrightarrow{\text{addi-}\atop\text{tion}} \begin{pmatrix} ZC_2 \\ C{-}ZC \end{pmatrix}$$

$$\qquad\qquad\qquad\quad \text{III}\qquad\qquad\quad \text{IV}\qquad\quad \text{V}$$

$$R_2Zn + C \xrightarrow{\text{coordi-}\atop\text{nation}} R_2Zn{-}C \longrightarrow \begin{pmatrix} R_2Zn{-}C \\ C{-}R_2Zn \end{pmatrix} \xrightarrow{\text{addi-}\atop\text{tion}} \begin{pmatrix} ZC \\ C{-}R_2Zn \end{pmatrix}$$

$$\qquad\qquad\qquad\qquad\qquad \text{VII}\qquad\qquad\quad \text{VIII}\qquad\quad \text{IX}$$

The mutual activation of vinyl ketone and the organozinc compound in IV or VIII should be the essential step for the coordination-addition type reaction.

The ^{13}C shieldings of β-carbon atom in olefinic double bonds of butadiene-type monomers were measured in 10% cyclohexane solution; δ_c values (ppm from TMS) are $CH_2=CH-CH=CH_2$ (119.16), $CH_2=CHCOCH_3$ (129.05), $CH_2=CHCOOCH_3$ (131.66), and $CH_2=CHCN$ (138.21). These NMR data are nicely correlated with the Hammett substituent constants, but they are not compatible with the reactivity of these monomers toward organozinc compounds. The reactivity of chemical species operative in the coordination-addition type reaction should not be directly correlated with spectroscopic data on the "free" monomers or "free" nucleophilic reagents.

MODE OF ACTIVATION IN SOME RING-OPENING POLYMERIZATIONS

The ring-opening polymerization of N-carboxy-α-amino acid anhydride (NCA) initiated by zinc alkyl affords us another example of the activated monomers [21]. In the reaction between n-Bu_2Zn and DL-alanine NCA, it was confirmed that the proton abstraction reaction took place rapidly and quantitatively, no sign of the occurrence of the carbonyl addition being detected:

$$2\ \begin{array}{c} CH_3-CH-NH \\ |\quad\ | \\ CO\ \ CO \\ \diagdown\ \diagup \\ O \end{array} + \text{n-Bu}_2\text{Zn} \longrightarrow \begin{array}{c} CH_3-CH-N-Zn-N-CH-CH_3 \\ |\quad\ |\qquad\quad |\quad\ | \\ CO\ \ CO\qquad CO\ \ CO \\ \diagdown\ \diagup\qquad\quad \diagdown\ \diagup \\ O\qquad\qquad O \end{array} + 2\text{n-BuH} \qquad (8)$$

$$XI$$

When n-Bu_2Zn was mixed with an equimolar mixture of DL-alanine NCA and sarcosine NCA in a 3 to 10 mole ratio, the proton abstraction reaction from the alanine NCA took place rapidly in the first stage. The IR spectra of the reaction mixture showed that the concentration of DL-alanine NCA decreased gradually with time in contrast with the unchanged concentration of sarcosine NCA. This result suggested that XI was not able to add directly to the carbonyl group of free DL-alanine NCA as well as to sarcosine NCA. An interpretation was made in terms of a reaction between the activated NCA's themselves:

$$
\begin{array}{l}
\text{CH}_3-\text{CH}-\text{C}=\text{O} \\
\quad\ \ \ |\quad\ \ | \\
\text{XZn}-\text{N}\quad\text{O} \\
\qquad\ \ \diagdown\ \diagup \\
\qquad\quad\text{C} \\
\qquad\quad\| \\
\qquad\quad\text{O}
\end{array}
\ +\
\begin{array}{l}
\qquad\text{O} \\
\qquad\diagup\diagdown \\
\quad\text{CO}\ \text{CO} \\
\quad\ |\quad\ | \\
\text{XZn}-\text{N}-\text{CH}-\text{CH}_3
\end{array}
\ \longrightarrow\
\begin{array}{l}
\qquad\qquad\qquad\text{O} \\
\qquad\qquad\qquad\diagup\diagdown \\
\qquad\qquad\text{O}\ \ \text{CO}\ \text{CO} \\
\qquad\qquad\|\quad\ |\quad\ | \\
\qquad\qquad\text{C}-\text{N}-\text{CH}-\text{CH}_3 \\
\text{CH}_3-\text{CH} \\
\qquad\ \ | \\
\qquad\ \ \text{N} \\
\qquad\diagup\ \diagdown\text{C}-\text{O}-\text{ZnX} \\
\text{XZn}\quad\ \|\\
\qquad\qquad\text{O}
\end{array}
\qquad (9)
$$

XII

n-Butylzinc-N,N-diethylcarbamide (BZC), a model compound for the growing chain end of the NCA polymerization, exhibited no reactivity toward sarcosine NCA. The first stage of the reaction of BZC with DL-alanine NCA was shown to be a proton abstraction (or metallation) reaction:

$$
\text{Et}_2\text{NCOOZnBu} +
\begin{array}{l}
\text{HN}-\text{CH}-\text{CH}_3 \\
\ \ \ |\quad\ | \\
\ \ \text{CO}\ \text{CO} \\
\quad\diagdown\diagup \\
\qquad\text{O}
\end{array}
\ \xrightarrow[(-\text{BuH})]{}\
\begin{array}{l}
\text{Et}_2\text{NCOOZn}-\text{N}-\text{CH}-\text{CH}_3 \\
\qquad\qquad\ \ \ |\quad\ | \\
\qquad\qquad\ \text{CO}\ \text{CO} \\
\qquad\qquad\quad\diagdown\diagup \\
\qquad\qquad\qquad\text{O}
\end{array}
\quad (10)
$$

XIII

XIII was activated enough to accept the nucleophilic attack at the carbonyl-carbon in the 5-position.

$$
\begin{array}{l}
\qquad\text{CH}-\text{C}=\text{O} \\
\qquad\ \ |\quad\ \ | \\
\text{Et}_2\text{NCOOZn}-\text{N}\quad\text{O} \\
\qquad\qquad\ \diagdown\diagup \\
\qquad\qquad\quad\text{C} \\
\qquad\qquad\quad\| \\
\qquad\qquad\quad\text{O}
\end{array}
\ +\
\begin{array}{l}
\qquad\text{O} \\
\qquad\diagup\diagdown \\
\quad\text{CO}\ \text{CO} \\
\quad\ |\quad\ | \\
\text{Et}_2\text{NCOOZn}-\text{N}-\text{CH}-\text{CH}_3
\end{array}
$$

$$
\xrightarrow{\qquad\qquad}\
\begin{array}{l}
\qquad\qquad\qquad\text{Et}_2\text{NCOO} \\
\qquad\qquad\qquad\quad\ | \\
\qquad\qquad\text{CH}_3-\text{CH}-\text{C}=\text{O}\qquad\text{O} \\
\qquad\qquad\qquad\ \ |\qquad\qquad\diagup\diagdown \\
\text{Et}_2\text{NCOOZn}-\text{N}\qquad\qquad\text{CO}\ \text{CO} \\
\qquad\qquad\qquad\diagdown\text{C}-\text{OZn}-\text{N}-\text{CH}-\text{CH}_3 \\
\qquad\qquad\qquad\quad\ \| \\
\qquad\qquad\qquad\quad\text{O}
\end{array}
\qquad (11)
$$

XIV

In the reaction system of polymerization initiated by R_2Zn or BZC, the activated monomer molecules should be regenerated from XII or XIV:

$$
\begin{array}{c}
\underset{\substack{| \\ }}{\text{CH}_3} \quad \underset{\substack{| \\ }}{\text{ZnX}} \\
-\text{CO}-\text{CH}-\text{N}-\text{C}-\text{OZnX} \\
\underset{\substack{\| \\ \text{O}}}{} \\
\text{XII}
\end{array}
+
\begin{array}{c}
\text{CH}_3-\text{CH}-\text{CO} \\
\underset{\substack{| \\ }}{} \qquad \text{O} \\
\text{HN}-\!\!-\text{CO}
\end{array}
$$

$$
\xrightarrow{}
\begin{array}{c}
\underset{\substack{| \\ }}{\text{CH}_3} \\
-\text{CO}-\text{CH}-\text{NH}-\text{C}-\text{OZnX} \\
\underset{\substack{\| \\ \text{O}}}{}
\end{array}
+
\begin{array}{c}
\text{CH}_3-\text{CH}-\text{CO} \\
\qquad\qquad \text{O} \quad (12) \\
\text{XZn}-\text{N}-\!\!-\text{CO}
\end{array}
$$

The acidity of the $-\text{NH}-\text{COOZnX}$ group in XII is expected to be lower than that of NH in NCA owing to the electropositive nature of the Zn atom in the carbamate group.

The concept of "an activated monomer" has been advanced for the elucidation of the mechanism of NCA polymerization initiated by strong bases, in which "an activated NCA" was assumed to add to the carbonyl group of free NCA molecules [22-24]. Seeney and Harwood [24a], however, have recently shown that both weak and strong base initiated polymerizations of NCA proceed through the attack of carbamate ion as the propagating species onto free NCA molecules. It should be noted that copolymerization between sarcosine NCA and DL-alanine NCA was caused with tertiary amines or alkali metal alkoxide, but not with dialkylzinc. The different behaviors of the strong bases and zinc alkyl in the copolymerization are also interpreted in terms of the postulated mechanism discussed above.

N-Carboxyl-α-amino acid anhydride (NCA) could be copolymerized with propylene oxide by means of dialkylzinc or trialkylaluminum as initiator [25]. The copolymers formed seem to contain many kinds of linkages such as amide, ester, ether, and urethane in the polymer chain. There were, however, significant differences between the structures of Copolymers XV and XVI formed with R_2Zn and R_3Al, respectively, as initiator. The IR spectrum of Copolymer XV indicated the presence of a considerable amount of peptide block. On the other hand, Copolymer XVI seemed to contain urethane linkages. It was also confirmed that strong bases such as sodium methoxide caused only homopolymerization of NCA even in the presence of propylene oxide.

   A previous study showed that dialkylzinc could induce alternating
copolymerization of phthalic anhydride and propylene oxide at room
temperature [26]. This result is to be noted because homopolymerization of either of these monomers is not effectuated by dialkylzinc
alone in hydrocarbon or in ether. Later studies revealed that ZnR_2
is activated by the complexation with phthalic anhydride [27]. The
activated R_2Zn cleaved the $O-CH_2$ bond of propylene oxide to form
2-pentanolate (XVII). The nature of the complex of phthalic
anhydride is presumably very similar to $[ZnR_2-C]$ as stated above.

$$
\begin{array}{c}
\\
CH_3 \\
| \\
CH_2-CH-CH_3 + Et_2Zn \xrightarrow[\text{anhydride}]{\text{phthalic}} EtCH_2-CH-OZnEt \qquad (13) \\
\diagdown\diagup \\
O
\end{array}
$$

                                    XVII

XVII reacts selectively with phthalic anhydride to form XVIII, which
reacts in turn with propylene oxide to form IX. Since carbon dioxide
is one of the acid anhydrides, it was expected that there would be a
series of reactions similar to Eqs. (14) and (15) when CO_2 was used

$$
\begin{array}{c}
CH_3 \qquad\qquad\qquad\qquad CH_3 \\
| \qquad\qquad\qquad\qquad\quad | \\
EtCH_2-CH-OZnEt + \text{(phthalic anhydride)} \longrightarrow \text{COOCHCH}_2Et \qquad (14) \\
\text{COOZnEt}
\end{array}
$$

       XVII   XVIII

$$
\begin{array}{c}
CH_3 \\
| \\
XVIII + CH_2-CH-CH_3 \longrightarrow \text{COOCHCH}_2Et \longrightarrow \text{polyester} \quad (15) \\
\diagdown\diagup \qquad\qquad\qquad \text{COOCH}_2\text{CHOZnEt} \\
O \qquad\qquad\qquad\qquad\qquad\quad | \\
\qquad\qquad\qquad\qquad\qquad\qquad\quad CH_3
\end{array}
$$

                                    IX

as the comonomer. An alternating copolymerization between CO_2
and epoxides was found possible with the use of the ZnR_2-H_2O
system as initiator [28]. Contrary to phthalic anhydride, CO_2 could

not activate ZnR_2 enough to cleave the epoxide ring. According to later studies, binary systems involving primary amine, dihydric phenol, aromatic dicarboxylic acid, or aromatic oxycarboxylic acid as the partner component with ZnR_2 formed the most effective initiators for copolymerization [29]. These initiator systems are characterized by the repetition of ZnO (or ZnN) linkages [30]. The ZnR_2-methanol system has no activity, though the methanol system was one of the most excellent initiators for homopolymerization of propylene oxide [30]. The difference in the catalytic behaviors of the various types of zinc compounds can be interpreted in terms of changes of reactivity in the Zn—OCOOR bond toward epoxide in response to variation of the structure of X [31].

$$XZnOCOR + CH_2-CH-CH_3 \longrightarrow XZnOCHCH_2OCOR \qquad (16)$$

This reaction should be the "key step" for the copolymerization with epoxide, the ZnR_2-methanol system having been confirmed as reacting readily with CO_2 in a way similar to the ZnR_2-water system. The repetition of ZnO linkages in X seems to favor the "key" reaction, though the mechanism of activation has not been detailed.

REFERENCES

[1] T. Tsuruta, Kobunshi no Gosei (Synthesis of Macromolecules), Kagaku-Dojin, Kyoto, 1961, p. 54; T. Tsuruta and K. F. O'Driscoll, Structure and Mechanism in Vinyl Polymerization, Dekker, New York, 1969, p. 31.
[2] T. Tsuruta, Progr. Polym. Sci., Japan, 3, 2 (1972).
[3] T. Narita and T. Tsuruta, Kogyo Kagaku Zasshi, 72, 994 (1969); T. Narita, A. Masaki, and T. Tsuruta, J. Macromol. Sci.—Chem., A4, 277 (1970).
[4] T. Narita, M. Kazato, and T. Tsuruta, Ibid., A4, 885 (1970).
[5] T. Narita and T. Tsuruta, J. Organometal. Chem., 30, 289 (1971).
[6] T. Yamaguchi, T. Narita, and T. Tsuruta, Polym. J., 3, 573 (1972).
[7] T. Narita, T. Yasumura, and T. Tsuruta, Ibid., 4, 421 (1973).
[8] T. Narita, Y. Kunitake, and T. Tsuruta, To Be Published.
[9] N. Imai, T. Narita, and T. Tsuruta, Tetrahedron Lett., 38, 3517 (1971); Bull. Chem. Soc. Japan, 46, 1242 (1973).

[10] T. Narita, T. Yamaguchi, and T. Tsuruta, Bull. Chem. Soc. Japan, 46, 3825 (1973).

[11] M. Ikeda, T. Hirano, and T. Tsuruta, Makromol. Chem., 150, 127 (1971).

[12] M. Ikeda, T. Hirano, S. Nakayama, and T. Tsuruta, Ibid., 175, 2775 (1974).

[13] M. Ikeda, T. Hirano, and T. Tsuruta, Tetrahedron Lett., 43, 4477 (1972).

[14] M. Ikeda, T. Hirano, and T. Tsuruta, Tetrahedron, 30, 2217 (1974).

[15] M. Ikeda, T. Hirano, and T. Tsuruta, To Be Published.

[16] T. Tsuruta and Y. Yasuda, J. Macromol. Sci.—Chem., A2, 943 (1968); T. Tsuruta, Progr. Polym. Sci. Japan, 3, 39 (1972).

[17] R. Tsushima and T. Tsuruta, Makromol. Chem., 166, 325 (1973).

[18] R. Tsushima and T. Tsuruta, J. Polym. Sci., Polym. Chem. Ed., 12, 183 (1974).

[19] Y. Kawakami, Y. Yasuda, and T. Tsuruta, J. Macromol. Sci.— Chem., A3, 205 (1969).

[20] R. Tsushima and T. Tsuruta, To Be Published.

[21] T. Makino, S. Inoue, and T. Tsuruta, Makromol. Chem., 131, 147 (1970).

[22] C. H. Bamford, H. Block, and A. C. P. Pugh, J. Chem. Soc., 1961, 2057, 4989.

[23] M. Goodman, J. Hutchison and U. Arnon, J. Amer. Chem. Soc., 86, 3384 (1964); 87, 3524 (1965); 88, 3627 (1966).

[24] M. Szwarc, Carbanions, Living Polymers, and Electron-Transfer Processes, Wiley-Interscience, New York, 1968, p. 592.

[24a] C. E. Seeney and H. J. Harwood, Japan-U.S. Seminar on Unsolved Problems in Ionic Polymerization, 1974.

[25] T. Tsuruta, S. Inoue, and K. Matsuura, J. Polym. Sci., Part C, 22, 981 (1969).

[26] T. Tsuruta, K. Matsuura, and S. Inoue, Makromol. Chem., 75, 211 (1964).

[27] S. Inoue, K. Kitamura, and T. Tsuruta, Ibid., 126, 250 (1969).

[28] S. Inoue, H. Koenuma, and T. Tsuruta, J. Polym. Sci., Part B, 7, 287 (1969); Makromol. Chem., 130, 210 (1969).

[29] M. Kobayashi, S. Inoue, and T. Tsuruta, Macromolecules, 4, 658 (1971); J. Polym. Sci., Polym. Chem. Ed., 11, 2383 (1973).

[30] M. Ishimori, O. Nakasugi, N. Takeda, and T. Tsuruta, Makromol. Chem., 115, 103 (1968); 128, 52 (1969).

[31] M. Kobayashi, Yang-Lik Tang, T. Tsuruta, and S. Inoue, Ibid., 169, 69 (1973).

Polymerization of Cyclic Imino Ethers. X
Kinetics, Chain Transfer, and Repolymerization

M. LITT,* A. LEVY,† and J. HERZ‡

Central Research Laboratories
Allied Chemical Corporation
Morristown, New Jersey 07960

ABSTRACT

The main conclusions in studies on polymerization catalysts
and the nature of chain transfer in the polymerization of cyclic
imino ethers were that there is extensive chain transfer in the
2-alkyl oxazolines to produce polymer with a reactive end.
Toward the end of polymerization, these chain-transferred
molecules repolymerize back on the active center producing
a multibranched star polymer. A theory for the data was
developed.

INTRODUCTION

The polymerization of cyclic imino ethers has been studied by
four groups [1-9] over the last few years. While the mechanism
has been recognized as a cationic polymerization from the first,
it has not been studied in detail until recently. Tomalia and Sheetz
[3] investigated the polymerization of methyl oxazoline with BF_3
etherate. In this case the counterion is destroyed. Recently,

---

*Present address: Department of Macromolecular Science, Case
Western Reserve University, Cleveland, Ohio 44106.
†Present address: Ethicon Corporation, Somerville, New Jersey
08876.
‡Present address: Central Research Laboratories, Stauffer
Chemical Corporation, Tarrytown, New York 10591.

Kagiya and Matsuda [8] studied the polymerization of phenyl oxazoline using a wide variety of catalysts. They found that perchloric acid produced the fastest rate of polymerization. There were no unusual phenomena connected with this study except that the polymerization rate constant in solvents was smaller than in bulk. Saegusa has concentrated on the polymerization of unsubstituted oxazolines [9].

Our studies were, in the main, on alkyl oxazolines and oxazines. Potential catalysts were surveyed, the better ones were looked at more closely, and finally one of the best, perchloric acid, was investigated in detail. The polymerization mechanism is much more complicated in the aliphatic oxazoline series compared to the aromatic oxazoline series. This paper summarizes our studies on catalysts, reaction rates, chain transfer, and repolymerization.

EXPERIMENTAL

Monomers

Monomers were synthesized as described previously [6, 7].

Catalysts

In the initial screening, catalysts were purchased from various supply houses and used as received. HI, HBr, and perchloric acid, obtained as aqueous solutions, and p-toluene sulfonic acid hydrate were used in the form of salts of phenyl oxazine.

Preparation of Mineral Acid Salts of 2-Phenyl Δ2-Oxazine

HClO$_4$ Salt. To a solution of 5.08 g (0.032 mole) of 2-phenyl oxazine in 30 ml of absolute ethanol was added an equivalent amount of mineral acid [perchloric acid 70% solution, 4.6 g (0.032 mole)] dissolved in 30 ml of absolute ethanol. Upon chilling to 0°C, white crystals precipitated. A further crop was obtained by adding ether. After drying under vacuum, a total weight of 6.9 g (83% yield), mp = 118.5 to 119.5°C, was obtained.

Element analysis: Calculated for C$_{10}$H$_{12}$ClNO$_5$: C, 45.90; H, 4.62; N, 5.35. Found: C, 46.21; H, 4.37; N, 5.97.

Toluene Sulfonic Acid Salt. In a 250-ml flask was placed 6.44 g (0.040 mole) of 2-phenyl oxazine and 50 ml of abso-lute ethanol. To this was added 7.48 g (0.0039 mole) of hydrated p-toluene sulfonic acid in 50 ml of absolute alcohol. Upon chilling, 9.0 g (69%) of the expected salt, a white crystalline solid, precipitated. It was filtered and dried under vacuum, mp = 157.5°C.

Analysis for $C_{17}H_{19}NO_4S$: Calculated: C, 61.24; H, 5.74; N, 4.20; S, 9.62. Found: C, 60.93; H, 6.05; N, 4.65; S, 9.7.

Heating for 2 days at 85°C under vacuum turned the compound pale pink, mp = 155.5 to 157.5°C.

HI Salt. A solution of 5.25 g (0.033 mole) in 30 ml of absolute alcohol was mixed with 8.94 g (0.033 mole) of 47% HI in 30 ml of ethanol. After chilling and adding several milliliters of ethyl ether, a yellowish crystalline solid precipitated. It was washed several times with absolute ether and gave a pale yellow product, mp = 97 to 98°C.

2-p-Chlorophenyl Δ2-Oxazolenium Perchlorate (PCOP)

To a solution of 18.16 g (0.10 mole) 2-(p-chlorophenyl) Δ2-oxazoline in 100 ml absolute ethanol was added 14.5 g of 70% $HClO_4$ solution (0.10 mole). The solution was refrigerated overnight ($\sim$4°C); the precipitate was filtered out and washed with absolute ether. Yield, 21.4 g (75%). The salt was recrystallized from absolute ethanol, mp = 193.5 to 195°C.

Polymerizations

Screening

Lewis acid initiators were screened as follows. About 5 g of pure monomer, either 2-phenyl oxazine or one of several alkyl oxazolines, was distilled into a dried weighed polymerization tube. To this was added the Lewis acid; the tube was chilled, degassed, and sealed under vacuum. When the initial screening was done on oxazolines, catalyst was added in a ratio of about 1:500 to monomer. For 2-phenyl oxazine, the Lewis acid added was weighed in and was used either at a monomer/initiator ratio of 200 or 500. The sealed tubes were heated in an aluminum block for the times indicated. Where no times are indicated, they were heated for at least 20 hr at 130°C. The results are summarized in Table 1.

Oxygen and halogen acids were tested as the free acid, ester, or as "onium" salts. The procedure described above was used. p-Toluene sulfonic acid was obtained as the hydrate. It was dried under vacuum at 150°C before use.

TABLE 1. Lewis Acids and Inorganic Salts as Polymerization Initiators

Monomer	Initiator	M/I	Time (hr)	Temp (°C)	% Polymer[a]
Phenyl oxazine	$BF_3O(Et)_2$	200	16	160	100
	I_2	200	20	130	100
	$AlCl_3$	500	24	160	15
	$FeCl_3$	500	24	160	15
	$VOCl_2$	200	48	160	0
	$LiClO_4$	500	48	160	0
	$Mg(ClO_4)_2$	200	0.5	160	100
Various oxazolines	SbF_5				+++
	$\phi N_2^+ PF_6^-$				80
	$SbCl_5$				+
	$SnCl_4$				+
	PCl_3				+

[a] +++ = good polymerization, + = poor polymerization.

Relative Polymerization Rates

The procedure was the same as in the Screening section above. All monomer to initiator molar ratios (M/I) were 200. Tubes, polymerized in a heated alunimum block, were removed every half hour and inverted. The time to complete polymerization was taken as the time when the viscosity of the polymerizate no longer increased. This is a rough estimate, but the relative order of initiators can be told by this method. The relative rates were reasonably meaningful ($\pm 25\%$).

Measurement of Polymerization Rate

An appropriate amount of the perchloric acid salt of 2-(p-chlorophenyl) oxazoline was weighed into a cleaned, flamed flask. An appropriate amount of 2-heptyl oxazoline was distilled from sodium through a spinning band column into the flask. Where solvent was used, the dried solvent was then distilled into the same flask using a second column. When the initiator had dissolved, known amounts of solution were syringed into cleaned, flamed polymerization tubes. These were then sealed under vacuum after chilling. They were polymerized by setting all the tubes in an oil bath set at a temperature of either 115, 130, or 145°C, and removing tubes at appropriate intervals.

Polymer was isolated by adding the polymerizate to dioxane, making the resulting solution acid with HCl, and then adding water. The polymer precipitated cleanly. The procedure was then repeated. It was filtered into a sintered glass funnel, washed with water, and dried under vacuum at 40°C for 60 hr. Constant weight was achieved after 30 hr drying in a separate experiment.

Runs to Complete Conversion

The perchloric acid salt of 2-(p-chlorophenyl) oxazoline was dissolved in dried, distilled acetonitrile to make up a 0.1-$\underline{M}$ solution. An appropriate amount of this solution was syringed into a cleaned, flamed polymerization tube and the acetonitrile evaporated away slowly. After drying and weighing, the tube was connected to a distilling column and the monomer, either 2-pentyl or 2-isobutyl oxazoline, was distilled from sodium, using a spinning band column, into the polymerization tube. This was chilled, sealed under vacuum, and reweighed. It was polymerized for an appropriate period of time at 160°C, usually 4 hr for those with a molar ratio of less than 10,000 and 40 hr for those with a higher ratio. Complete polymerization is about 1.5 hr at M/I = 10,000. A check was carried out by polymerizing two samples of 2-isobutyl oxazoline with M/I $\approx$ 14,500. One was heated for 7.5 hr and one for 67 hr. $\overline{X}_n$ and $[\eta]$ were unchanged within experimental error. $\overline{X}_w$ increased from 3,500 to 5,000. This is discussed later.

Characterization

Element analyses were performed by the analytical staff at Allied Chemical Corp. Reduced viscosities were normally run as a 0.52% solution of the polymer in m-cresol. When other concentrations or solvents were used, it is noted in the text.

Intrinsic viscosities were run in a dilution viscometer in various solvents. These are noted in footnotes to the tables.

Number-average molecular weights were measured in chlorobenzene using high-speed membrane osmometry.

Weight-average molecular weights were measured by light scattering in acetic acid or butanol or both. Zimm plots were made and the values given are for turbidity extrapolated to zero concentration and zero degrees.

RESULTS AND DISCUSSION

This work was done over a period of many years and thus some parts are crude. Related research has been done more elegantly by the present researchers [8, 10, 11]. The areas covered are 1) effects of catalysts, 2) rate of polymerization of 2-heptyl oxazoline with $HClO_4$ as a function of temperature and solvent, and 3) molecular weights of the polymers.

Effect of Catalyst

As is well known, the cyclic imino ethers are highly basic compared to other cationically polymerized monomers. They therefore destroy the Lewis acid catalysts [3]. However, oxygen acids, their esters, as well as iodides and bromides, are stable catalysts for the system. General data summarizing scanning work on yields of polymer covering various oxazolines and oxazines are given in Table 1.

BF_3 etherate and SbF_5 are consumed before polymerization is complete if higher M/I ratios are used. While the times of heating were long to insure maximum polymerization, most reactions were over in 1 to 2 hr. The only unusual point of Table 1 is that anhydrous $Mg(ClO_4)_2$ is an initiator for oxazoline polymerization.

Similarly, the effect of various oxygen and halogen acids, esters, and salts on the viscosity of polymers from phenyl oxazine is given in Table 2.

As can be seen, phenyl oxazine polymerized by a strong acid can achieve high molecular weights, $\eta_{sp}/c = 2.4$. Aliphatic oxazines cannot be prepared with viscosities greater than 1 at 130°C, no matter what catalyst is used [7].

Esters or anhydrides of weak acids, e.g., acetic acid, cannot initiate polymerization; the anion is too reactive and adds to the cationic end of the polymer. The same seems to be true for alkyl chlorides, though these do react very slowly. We have found that the use of chlorinated aliphatic solvents such as $CHCl_3$ or $C_2H_2Cl_4$ in monomer preparation can cause slow polymerization of the monomer when such a solution stands at room temperature for several months.

The effect of catalyst on the relative rates of polymerization and viscosities for 2-phenyl and 2-pentyl oxazine polymers is shown in Table 3.

The relative rates at 130°C for methyl tosylate and HI initiators are of the same order of magnitude as those found by Saegusa [10]. The major points are 1) the use of dimethyl sulfate or fuming H_2SO_4 lowers the viscosity by a factor of 2, showing that both ends of the sulfate group grow polymer molecules; and 2) the use of iodide or bromide initiators also lowers the viscosity tremendously, indicating some unknown chain transfer reaction which is greater for the more nucleophilic bromide.

The alkyl substituent allows more rapid polymerization by a factor of 2 to 3 over that of the aromatic substituent. Surprisingly, phenyl oxazoline seems to polymerize at about the same rate as phenyl oxazine.

Rates of Polymerization of 2-(n-Heptyl) Oxazoline

Polymerizations were done in bulk at 129.6°C using 2-(p-chloro-phenyl) oxazolinium perchlorate (PCOP) as initiator. A typical polymerization curve is shown in Fig. 1. The observed rate as a function of initiator concentration is shown in Fig. 2. This is about 2.5 to 3 times faster than that found for phenyl oxazoline in nitro-methane, a solvent which slowed the polymerization [12]. This corresponds to the relative rates shown in Table 3 for oxazines of about three to one.

TABLE 2. Acids, Esters, and Alkyl Halides as Polymerization Initiators

Monomer(M)	Initiator(I)	M/I	Time (hr)	Temp (°C)	% Polymer	η_{sp}/c [a]
Phenyl oxazine	$H_2SO_4(SO_3)_{0.3}$	200	4	130	100	0.45
	$CH_3C_6H_4SO_3H$	500	12	130	100	0.9
	$CH_3C_6H_4SO_3CH_3$	200	7	130	100	0.9
	$(CH_3)_2SO_4$	2,500	6	160	100	2.4
	$CH_3\overset{O}{\overset{\|}{C}}-OC_2H_5$	500	48	160	0	
	$(CH_3-\overset{O}{\overset{\|}{C}})_2O$	500	48	160	0	
	CH_3I	50	96	86	100	0.17
	C_3H_7Br	200	7	160	100	0.3
	$BrCH_2CH_2Br$	100	4	160	100	0.23
	$C_6H_5CH_2Cl$	200	40	160	0	
	$(C_6H_5CH_2)_3S^+ClO_4^-$	200	2.5	130	100	0.38

[a] All viscosities run at 1.5% solution in m-cresol at 30°C.

TABLE 3. Relative Polymerization Rates for Different Initiators

Monomer	Initiator	M/I	Time[a]	Temp (°C)	η_{sp}/c[c]
Phenyl oxazine	$HClO_4$[b]	200	3	130	0.8
	$HSO_3C_6H_4CH_3$[b]	200	6	130	0.9
	$CH_3SO_3C_6H_4CH_3$	200	6	130	0.9
	$H_2SO_4 \cdot (SO_3)_{0.3}$	200	4	130	0.4
	HI[b]	200	12	130	0.5
	CH_3I	200	8	160	0.45
	C_3H_7Br	200	7	160	0.3
Pentyl oxazine	$HClO_4$[b]	200	1.1	130	0.44
	$CH_3SO_3C_6H_4CH_3$	200	3	130	0.44
	$HSO_3C_6H_4CH_3$[b]	200	3-4	130	0.44
	$H_2SO_4(SO_3)_{0.3}$	200	1.5	130	0.24
	$(CH_3)_2SO_4$	200	1.5	130	0.25
Phenyl oxazoline	$HSO_3C_6H_4CH_3$[b]	200	6-7	130	0.91
	HI[b]	200	12-16	130	0.51

[a] Time in hours to estimated completion of polymerization. No further viscosity change.
[b] Used as the salt of phenyl oxazine.
[c] All viscosities run as 1.5% solution in m-cresol.

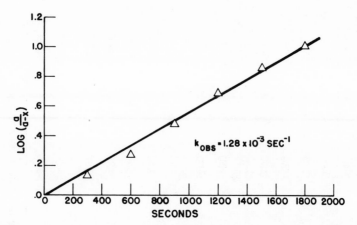

FIG. 1.   First-order plot for bulk polymerization of n-heptyl oxazoline at 129.6°C with 2-(p-chlorophenyl oxazolineium perchlorate as initiator.

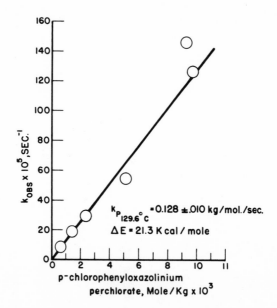

FIG. 2.   Determination of rate constant of polymerization for bulk polymerization of 2-(n-heptyl) oxazoline at 129.6°C.

The work at different temperatures was done in a solvent since the polymer crystallized below 130°C. Initially, dimethyl acetamide was chosen as the solvent. However, the rate constant of polymerization (first order in monomer and initiator) was solvent dependent as shown in Fig. 3.

(A wide variety of solvents has been investigated in the polymerization of phenyl oxazoline by Kagiya, Matsuda, and Hirata [12]. All polymerizations were done at 3 $\underline{M}$ concentration of monomer, and the relative rates of polymerization were compared as a function of polarity and nucleophilicity. For high dielectric constant solvents, the higher the nucleophilicity, the lower the rate.)

We found that the rate diminished rapidly by a factor of 2 up to 50% dilution and then seemed to remain constant. This cannot be explained in terms of competitive solvation and reaction of the active cation with a solvated monomer. It can be explained by saying that the monomer solvated cation attacks a free monomer twice as fast as a DMAC solvated cation, and at 50% solvent essentially all the cation is solvated with DMAC. This seems unlikely, though it may be true. Kagiya et al. [12] postulated a similar equilibrium.

We therefore chose to work in butyrolactone as it is a relatively nonnucleophilic, polar solvent. The rate constant at 50% solvent

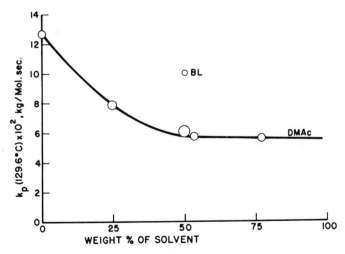

FIG. 3. Rate constant of polymerization as a function of monomer concentration in DMAC and butyrolactone (BL).

TABLE 4. Propagation Rate Constants (k_p) and Activation Parameters in the Polymerization of Heptyl Oxazoline with PCOP (50%
Monomer in γ-Butyrolactone)

T ($^\circ$C)	k_p (kg/mole sec)
115.0	0.0334
129.6	0.100
145.0	0.216
ΔE_p (kcal/mole)	21.3
A_p (kg/mole sec)	3.2×10^{10}

was 0.100 vs 0.127 kg/mole sec in bulk. In DMAC the rate constant
dropped by a factor of 2.3 from the bulk. The data for rate constants
as a function of temperature are given in Table 4.

The activation energy is about that found for most cyclic imino
ether polymerizations [10, 11]. However, our pre-exponential factor
is very high when compared to that of phenyl oxazine [11] though
lower than that for oxazoline [9]. As our lowest M/I ratio was 500,
complete initiation was achieved early in the polymerization and we
only observed the polymerization.

Chain Transfer and Repolymerization

Chain transfer has been noted for most oxazolines and oxazines.
Some extreme examples were found; benzyl oxazoline $(\eta_{sp}/c = 0.21$
at 130°C and M/I = 1000) [6] and 2-(acetoxymethyl) oxazoline
$(\eta_{sp}/c = 0.37$ at M/I = 1300 and $\eta_{sp}/c = 0.34$ at M/I = 500) [13].
For comparison, a normal n-alkyl oxazoline such as pentyl oxazoline will have $\eta_{sp}/c \approx 0.7$ at M/I = 560 [14] and one can easily get
reduced viscosities above 1 for 2-(n-alkyl) oxazolines at M/I > 2000.
(The mechanism for chain transfer in 2-phenyl oxazoline and oxazine
is still a puzzle.) We therefore decided to investigate the mechanism
and effect of chain transfer in alkyl oxazolines. Two sets of experiments were run. First, heptyl oxazoline was polymerized at 130°C

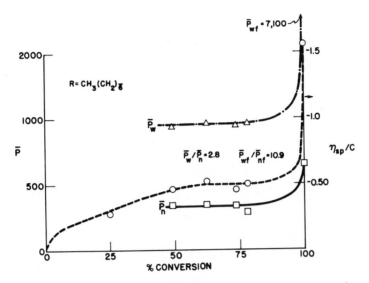

FIG. 4. $\overline{P}_n$, $\overline{P}_w$, and $\{\eta\}$ of polymers of 2-(n-heptyl oxazoline) as a function of conversion.

with M/I = 9810 (I = PCOP). Samples were removed at various times, and conversion, intrinsic viscosity, $\overline{P}_n$, and $\overline{P}_w$ were determined for each sample. The data are given in Fig. 4. From the time the conversion reached 50%, where enough polymer could be isolated for all the determinations, $[\eta]$, $\overline{P}_n$, and $\overline{P}_w$ remain constant at least to 78% conversion. At 100% conversion, 56 hr at 130°C, all three values rise considerably; the greatest rise is in $\overline{P}_w$ which goes from ~1000 to 7100. The constant value of $\overline{P}_n$ at ~300 and the relatively low and constant ratio of $\overline{P}_w$ to $\overline{P}_n$, 2.8, during the polymerization indicates that the major reaction controlling the degree of polymerization is chain transfer, and the chain transfer constant to a 2-(n-alkyl) oxazoline at 130°C is about 1/300.

The upturn in all the values at 100% conversion was puzzling. A series of experiments were therefore run with 2-pentyl oxazoline and 2-isobutyl oxazoline at various M/I concentrations (I = PCOP) from ~500 to ~40,000. All polymerizations were taken to 100% conversion (3.5 hr at 160°C and 0.5 hr at 180°C for 2-pentyl oxazoline except for M/I = 37,000, which was heated for 44 hr

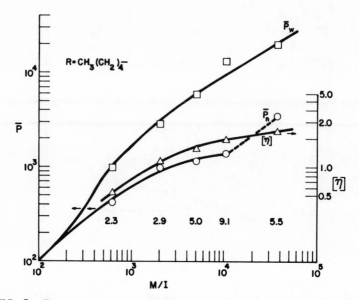

FIG. 5. Polymerization of 2-(n-pentyl oxazoline) to 100% conversion at various M/I ratios. I = PCOP.

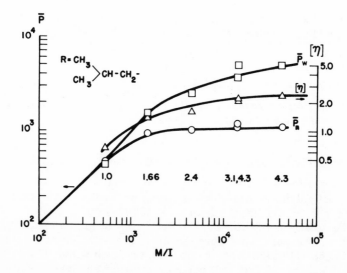

FIG. 6. Polymerization of 2-isobutyl oxazoline to 100% conversion at various M/I ratios. I = (PCOP).

at 160°C, and similar heating times for 2-isobutyl oxazoline). The data are presented in Figs. 5 and 6 and Tables 5 and 6.

In Fig. 5 and Table 5 it can be seen that even at M/I = 500, $\overline{P}_n$ is less than 500 while $\overline{P}_w$ is about 1000, with $\overline{P}_w/\overline{P}_n$ = 2.3. The ratio rises rapidly as M/I increases. Heating time and temperature obviously affect $\overline{P}_n$ greatly as shown by the 44-hr heating for M/I = 37,000 and the large $\overline{P}_w/\overline{P}_n$ value for the sample with M/I = 10,535. This last was heated to 180°C, above the melting point of the polymer.

The same effect is shown to a lesser degree by 2-isobutyl oxazoline, Fig. 6. At M/I = 500, $\overline{P}_n = \overline{P}_w$ = 500; living monodisperse polymer was obtained. The chain transfer constant is obviously much less than 1/500. As M/I increases, $\overline{P}_n$ becomes constant at about 1000, which means that the chain transfer constant is near 1/800 at 160°C. $\overline{P}_w$, however, increases steadily with increasing M/I as does $[\eta]$ to a lesser degree.

The above data make sense when the following reactions are postulated. The major site of chain transfer must be the carbon α to the 2-carbon of the ring. This is shown by the greater chain transfer on oxazoline polymers with activated α-methylene groups such as 2-benzyl [6] and 2-acetoxymethyl oxazoline [13]. Conversely, shielding the α-methylene group as in 2-isobutyl oxazoline lowers the chain transfer constant considerably. We can visualize the chain transfer reaction as

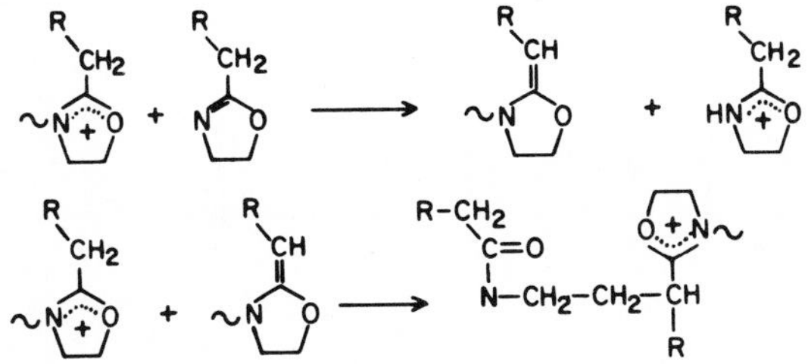

This was proposed earlier, but supporting evidence was not given [13]. The chain transferred polymer is now an enamine, though hindered, and can attack active sites such as quaternized oxazolines (polymer reactive ends), though apparently more slowly than does

TABLE 5. Molecular Weights and Viscosities as a Function of M/I (PCOP) for 2-Pentyl Oxazoline Polymerized at 160°C

| M/I | t (hr) | $\overline{P}_w$ | | $\langle \overline{r}^2 \rangle^{1/2}$ | $\overline{P}_n{}^d$ | $\overline{P}_w/\overline{P}_n$ | Viscosity parameters | | | |
		a	b				$[\eta]^c$	k'	$[\eta]^d$	k'
531	4[g]	1,000		274	425	2.3	0.53	0.55	0.34	0.52
2,030	4[g]	2,850		390	993	2.9	1.14	0.50	0.74	0.51
5,056	4[g]	5,850	6,050	570/620	1160	5.0	1.59	0.52	0.98	0.64
10,535	4[g]	12,910	13,130	720/910	1400	9.2	1.93	0.63	1.34	0.60
37,000	44	19,720			3550	5.5	2.36	0.86	1.40	1.10
5,750[e]	26	6,630		628	910	7.3	1.88	0.45		
f			1,550	437	930	1.67	1.19	0.17		

[a] Determination in acetic acid/0.5 M NaCl at 546 nm, 25°C.
[b] Determined in n-butanol at 546 nm, 25°C.
[c] Determined in m-cresol, 25°C.
[d] Determined in chlorobenzene, 25°C.
[e] Polymerized at 130°C.
[f] Initial M/I = 10,444, reaction stopped at 79% conversion. True M/I = 8270.
[g] Tubes kept at 160°C for 3½ hr, then raised to 180°C for ½ hr.

TABLE 6. Molecular Weights and Viscosities as a Function of M/I (PCOP) for 2-Isobutyl Oxazoline Polymerized at 160°C

M/I	t (hr)	$\bar{P}_w$ a	$\bar{P}_w$ b	$\langle \bar{r}_Z{}^2 \rangle^{1/2}$	$\bar{P}_n$ c	$\bar{P}_w/\bar{P}_n$	$[\eta]$ d
514	6	440		222	472	~1.0	0.66
1,581	6	1490		326	920	1.62	1.39
4,519	7.5	2600	2360	425/344	1030	2.41	1.59
13,940	7.5	3590	3690	484/490	1200	3.03	2.11
14,888	67		4720	/518	1100	3.88	2.20
41,180	25	5010	5260	523/537	1110	4.63	2.33

aDetermined in acetic acid at 25°C, 546 nm.
bDetermined in n-butanol at 25°C, 546 nm.
cDetermined in chlorobenzene at 37°C.
dDetermined in m-cresol at 25°C.

monomer. Thus polymerization is initially Poisson as shown here for 2-isobutyl oxazoline, Fig. 6, and by Kagiya and Matsuda for 2-phenyl oxazoline [15]. However, with alkyl oxazolines, chain transfer occurs readily, generating many polymer molecules per active center. These chain transferred molecules have a reactive, though hindered, enamine end. At the end of the polymerization, the chain transferred polymer molecules repolymerize on the active center, generating branches near its end. The molecule is probably star-shaped, with some extra branches along the chain.

The extent of repolymerization can be seen qualitatively in Tables 5 and 6. We can compare the last experiment in Table 5, M/I = 10,444 but polymerized only to 79% conversion, with that polymerized to 100% conversion. $\overline{P}_n$ rose by about 50% while $\overline{P}_w$ rose by about 840%. $\overline{P}_w/\overline{P}_n$ rose from 1.67 to 9.2. The individual determinations are accurate only to $\sim\pm20\%$, but the change is remarkable.

The star branching can also be noted in the Huggins constant, k', of the viscosity determinations. It is high, 0.5, even at M/I = 531, and rises steadily as M/I rises. However, for the incompletely polymerized sample, k' is only 0.17, showing a relatively linear molecule. There is probably some repolymerization during the whole polymerization, generating polymer with long branches.

Similarly, Table 6 compares two samples of about the same M/I, $\sim$14,000. Heating for 67 hr as compared to 7.5 hr did not change $\overline{P}_n$ but did raise $\overline{P}_w$ by about 25%. As the polymers were crystalline (T_m = 210°C [6]) and mobility is restricted, this is a significant change. $\overline{P}_w/\overline{P}_n$ rose, showing greater polydispersity. There was almost no change in viscosity at higher M/I values, showing that the average molecular volume in solution was almost independent of $\overline{P}_w$. The effect of steric factors on the repolymerization can be seen by comparing Tables 5 and 6. In spite of the greater chain transfer for the n-pentyl oxazoline, $\overline{P}_n$ is almost the same as for isobutyl oxazoline because of more repolymerization. However, due also to this, $\overline{P}_w$ is much higher at a given M/I for n-pentyl oxazoline.

The extent of repolymerization can be calculated easily from the rise in $\overline{P}_n$ and $\overline{P}_w$ if a chain transfer constant is assumed.

For example, take heptyl oxazoline in Fig. 3. $\overline{P}_n$ rises from ~ 300 to ~ 700 at 100% conversion, meaning that about 60% of the polymer molecules have repolymerized. Since there were about 35 chain transfers, the star molecules must contain about 21 chains each.

A similar effect, though much less marked, seems to be present for the phenyl derivatives. One can easily get very high viscosity polymers in this system. Table 7 gives some data on phenyl oxazine; we have similar results on phenyl oxazoline which are being analyzed further at present.

The Huggins constant is ~ 0.42. This is much higher than that found in truly linear molecules, where it is about 0.2 to 0.25, but lower than that found for pentyl oxazoline at 100% conversion, even at low M/I. Thus there is extensive branching. However, the molecule increases in size as M/I increases. It is obviously a longer molecule than those obtained from the alkyl oxazolines at similar M/I as shown by the higher $[\eta]$, and thus probably has less branching and longer branches.

The problem is: What is the mechanism of chain transfer? One tentative suggestion is that there is very slow cationic attack on the benzene rings, either of monomer or polymer. Another possibility is that the ring-opened ester, present in equilibrium with the salt, can have the elements of acid removed by monomer, leaving either a vinyl amide or a propenyl amide.

TABLE 7. Polymerization of 2-Phenyl Oxazine with Dimethyl Sulfate,[a] Effect on Viscosity

M/I	$\overline{P}_{n\ calc}$	η_{sp}/c[b]	$[\eta]$[c]	k'[d]
1000	500	0.9		
2500	1000	2.0	1.88	0.42
4700	2350	2.4	2.35	0.43

[a]Polymerized for 18 hr at 160°C.
[b]0.52% in m-cresol, 25°C.
[c]In m-cresol, 25°C.
[d]Huggins constant.

Theoretical Treatment of Repolymerization

An equation describing the final $\overline{P}_n$ and $\overline{P}_w$ can be derived readily if the following assumptions are made:

1. No repolymerization of polymer chains occurs until essentially all the monomer has been used up.
2. No chain transfer can occur on the dead polymer chains.
3. All active centers have an equal chance to repolymerize all the polymer molecules.
4. With no early repolymerization, $\overline{P}_n/\overline{P}_w \approx 2$.

We can make the following definitions:

$\overline{P}_{n(0)}$ = initial number average degree of polymerization resulting from chain transfer

x = number of chain transfers per active center
= $M/[I\overline{P}_{n(0)}]$

n = number of polymer molecules which have repolymerized on the active center by the end of the reaction

The following equations can be written for the star portion of the polymer. If i molecules have been added to the active end, each of average molecular length $\overline{P}_{n(0)}$, by standard molecular weight theory, with a most probable distribution:

$$\overline{P}_{n(i)} = (i + 1)\overline{P}_{n(0)} \tag{1}$$

$$\overline{P}_{w(i)} = (i + 2)\overline{P}_{n(0)} \tag{2}$$

We can sum over all i's. If the assumption is made that all active ends have had the same chance to add, i.e., the distribution of chains in a repolymerized active center is a Poisson distribution, the equations are easily solvable.

$$\overline{P}_{n(n)} = \frac{\displaystyle\sum_0^\infty (i + 1)P_{n(0)} f(i)}{\displaystyle\sum_0^\infty f(i)} \tag{3}$$

where

$$f(i) = \ell^{-n} \frac{n^i}{i!} \tag{4}$$

$$\overline{P}_{n(n)} = \ell^{-n} \overline{P}_{n(0)} \sum_0^\infty \frac{(i + 1)n^i}{i!} = (n + 1)\overline{P}_{n(0)} \tag{5}$$

$$\overline{P}_{w(n)} = \frac{\ell^{-2} \left[\overline{P}_{n(0)}\right]^2 \sum_0^\infty (i + 2)^2 \frac{n^i}{i!}}{\ell^{-2} \overline{P}_{n(0)} \sum_0^\infty (i + 1) \frac{n^i}{i!}} \tag{6}$$

Summing by standard methods:

$$\overline{P}_{w(n)} = \frac{n^2 + 5n + 4}{n + 1} \overline{P}_{n(0)} = (n + 4)\overline{P}_{n(0)} \tag{7}$$

The molecular weight of the total polymer is the average of the repolymerized star molecules and the single chain transferred molecules. It can be written as

$$\overline{P}_w = \frac{\Sigma[\overline{P}_w(i)]w(i)}{\Sigma w(i)} = \frac{\Sigma \overline{P}_w(i)\overline{P}_n(i)f(i)}{\Sigma \overline{P}_n(i)f(i)} \tag{8}$$

$$\overline{P}_n = \frac{\Sigma \overline{P}_n(i)f(i)}{\Sigma f(i)} \tag{9}$$

For each active center which has had n chain transfers and an average of x repolymerizations, there is one star molecule. With an average degree of polymerization described by Eqs. (5) and (7), and (x - n) chain transferred single molecules of average molecular weight $\overline{P}_{n(0)}$ and $\overline{P}_{w(0)} \approx 2 \overline{P}_{n(0)}$.

$$\overline{P}_n = \frac{(n + 1) + (x - n)}{(x - n) + 1} \overline{P}_{n(0)} = \frac{x + 1}{x - n + 1} \overline{P}_{n(0)} \tag{10}$$

Using Eq. (7) to calculate $\overline{P}_w$, we get

$$\overline{P}_w = \frac{(n^2 + 5n + 4) + 2(x - n)}{n + 1 + x - n}\ \overline{P}_{n(0)} = \frac{n^2 + 3n + 4 + 2x}{x + 1}\ \overline{P}_{n(0)} \quad (11)$$

The ratio $\overline{P}_w/\overline{P}_n$ is therefore

$$\frac{\overline{P}_w}{\overline{P}_n} = \frac{(n^2 + 3n + 4 + 2x)(x - n + 1)}{(x + 1)^2} \quad (12)$$

This can correlate the data for isobutyl oxazoline quite well using a chain transfer constant of 1/800. The equations work properly only when M/I is greater than three times the chain transfer constant, as the initial portion of the polymerization, before chain transfer has given an equilibrium distribution of molecular weights, is ignored. The results for 2-isobutyl oxazoline are given in Table 8, while those for 2-pentyl oxazoline are in Table 9.

One value did not fit; that where the sample M/I = 13,500 was heated for 67 hr. Here $\overline{P}_n$ was lower than that sample heated less time, showing that there was some random error.

TABLE 8. Comparison of Experimental and Theoretical Values of $\overline{P}_w/\overline{P}_n$ for Poly(N-isovaleryl Ethyleneimine)

M/I	x^a	n^b	$\overline{P}_w/\overline{P}_n$	
			Calc	Exptl
4,500	5.63	1.26	2.52	2.40
13,500[c]	16.9	5.64	3.30	3.04
40,000	50.0	14.0	4.86	4.62

[a]Calculated using a chain transfer constant of 1/800.
[b]Calculated from Eq. (9), $n = (x + 1)(1 - \overline{P}_{n(0)}/\overline{P}_n)$.
[c]Heated for 7 hr at 160°C.

TABLE 9. Comparison of Experimental and Theoretical Values of $\overline{P}_w/\overline{P}_n$ for Poly(N-hexanoyl Ethyleneimine)

M/I	x^a	n^b	$\overline{P}_w/\overline{P}_n$ Calc	$\overline{P}_w/\overline{P}_n$ Exptl
531	3.2	1.94	2.56	2.3
2,030	12.2	10.2	2.81	2.9
5,056	30.3	25.9	4.48	5.0
10,535	63.2	55.7	7.0	9.2
37,000	222.0	211.6	10.5	5.5

[a]Calculated using a chain transfer constant of 1/167.
[b]Calculated from Eq. (9).

A reasonable fit was found for polymers of n-pentyl oxazoline below M/C = 10,000, using a chain transfer constant of 150 to 200, Table 8. At the higher values of M/I, especially at 40,000, $\overline{P}_n$ determined by osmometry is too inaccurate to get good agreement.

For poly(N-octanoyl ethyleneimine) the agreement for the sample polymerized to 100% conversion is $\overline{P}_w/\overline{P}_{n\ exp}$ = 6.22, $\overline{P}_w/\overline{P}_n$ calc = 6.65 (chain transfer constant of 1/200) = 5.86 (chain transfer constant of 1/300). Overall, the data indicate that some repolymerization occurs with n-alkyl oxazolines before 100% conversion, but most happens at the end.

SUMMARY AND CONCLUSIONS

This paper covers several aspects of the polymerization of oxazolines and oxazines but concentrates on chain transfer. Results on the 2-phenyl derivatives imply that these polymers are branched at high M/I. Work in progress with T. Provder confirms this. Also, the iodide and bromide counterions participate in a chain transfer reaction, which may go through the ring-opened ester.

The 2-alkyl oxazines and oxazolines show clear evidence for extensive chain transfer. The chain transfer constant for n-alkyl oxazolines where n > 4 is in the neighborhood of 1/200 to 1/300. (It is larger for 2-methyl oxazoline, i.e., viscosities are lower for this polymer at comparable M/I as compared to other alkyl oxazolines.)

The chain transferred polymer has an end which is basically a substituted enamine ether and as such is quite nucleophilic. It thus can add back to the active center, regenerating another oxazolineium or oxazineium cation. However, as it is more hindered and less nucleophilic than a monomer, it reacts slowly. There is some repolymerization of chain transferred polymer molecules onto growing chains during the polymerization, giving long branches. However, most of the repolymerization occurs during the last few percent of the polymerization and thus tends to generate a star polymer. These may have as many as 100 to 200 branches to the star when M/I = 37,000.

With branching near the methylene carbon α to the ring, chain transfer and repolymerization are both reduced due to steric hinderance. The polymer molecular weight increases and $\overline{P}_w/\overline{P}_n$ decreases. One can easily obtain living polymers in this system.

ACKNOWLEDGMENTS

We wish to thank Allied Chemical Corp., Morristown, New Jersey, for permission to publish this work. We also thank Mr. E. Walsh for the molecular weight and viscosity determinations.

REFERENCES

[1] For Paper VIII of this series, see M. Litt, J. Herz, and E. Turi, in Block Copolymers (S. Aggarwal, ed.), Plenum, New York, 1970, p. 313.
[2] T. Kagiya, S. Narisawa, T. Manabe, and K. Fukui, J. Polym. Sci., Part B, 4, 441 (1966).
[3] D. A. Tomalia and D. P. Sheetz, J. Polym. Sci., Part A-1, 4, 2253 (1966).
[4] W. Seeliger and W. Thier, Angew. Chem., 78, 613 (1966).
[5] M. Litt. T. G. Bassiri, and A. Levy, Belgian Patents, 666,828 and 666,831 (1965).

[6]   T. G. Bassiri, A. Levy, and M. Litt, J. Polym. Sci., Part B, 5, 871 (1967).
[7]   A. Levy and M. Litt, Ibid., 5, 881 (1967).
[8]   T. Kagiya and T. Matsuda, J. Macromol. Sci.—Chem., A5, 1277 (1971); A6, 135 (1972).
[9]   T. Saegusa, H. Ikeda, and H. Fujii, Macromolecules, 6, 315 (1973).
[10] T. Saegusa, S. Kobayashi, and Y. Nagura, Ibid., 7, 264, 272 (1974).
[11] T. Kagiya, T. Matsuda, M. Kakato, and R. Hirata, J. Macromol. Sci.—Chem., A6, 1631 (1972).
[12] T. Kagiya, T. Matsuda and R. Hirata, Ibid., A6, 451 (1972).
[13] A. Levy and M. Litt, J. Polym. Sci., Part A-1, 6, 1883 (1968).
[14] A. Levy and M. Litt, Ibid., 6, 63 (1968).
[15] T. Kagiya and T. Matsuda, J. Macromol. Sci.—Chem., A5, 1265 (1971).

Oxonium Ion Ring-Opening Polymerization

M. P. DREYFUSS

Corporate Research
Research and Development Center
B. F. Goodrich Company
Brecksville, Ohio 44141

ABSTRACT

Oxonium ion ring-opening polymerization is reviewed with emphasis on unsolved problems. New findings in cationic ECH polymerization and in cyclic oligomer formation are given. Participation of the Cl of ECH in the cationic center is suggested.

The vigorous activity in the area of oxonium ion ring-opening polymerization in the past few years is reflected in the number of recent symposia [1-3] and reviews [4, 5]. This activity has contributed greatly to our knowledge and understanding of these polymerizations. It is the purpose of this paper to survey this progress while particularly emphasizing the gaps that still exist in our knowledge. When possible, recent progress in our own laboratories will be used to show how some of these gaps are being filled.

As the title of this paper indicates, the nature of the propagating species in cyclic ether ring-opening polymerization is now well established and accepted. For this reason and because we must initiate before we can propagate, we will first consider initiation.

INITIATION

Meerwein [6] first described the basic requirements necessary to generate oxonium ions in 1937. Many different methods and systems to accomplish the same end have been described since that time. While some are novel and different from Meerwein's work, an impressive number reduce to the basic principles embodied in Meerwein's work of the 1930s.

Initiation by trialkyl oxonium salts, shown in Eq. (1), is perhaps the most straightforward initiation reaction, since it is most nearly like the propagation reaction itself and directly gives the desired propagating species. This has been quite clearly shown to be the

$$R_3O^+ \; + \; \text{O} \longrightarrow R-\text{O+} \; + \; R_2O \tag{1}$$

case in four- and five-membered cyclic ethers [4, 5]. However, the situation is already more complex in the case of cyclic acetals where there is still considerable argument over the exact structure of the propagating species. Also, initiation by trialkyl oxonium salts may or may not be as straightforward as shown in Eq. (1) in the cases of the bridged cyclic ethers and especially of the 1,2-epoxides, where the occurrence of this alkylation reaction has yet to be demonstrated.

Related to the trialkyl oxonium salts are the carboxonium salts I and II. These salts behave as alkyl donors in a very analagous way.

I   II

Initiation with dioxolenium ion shown in Eq. (2) is just a little different.

$$\begin{array}{c} CH_2-O \\ | \quad\quad +\!\text{)}CH + \text{O} \longrightarrow H\overset{O}{\overset{\|}{C}}OCH_2CH_2-\text{O+} \\ CH_2-O \end{array} \tag{2}$$

The two alkyl groups of the dialkyl carboxonium ion are joined and hence the formate ester residue is not lost, but remains attached as an often convenient end group. The formate ester is useful for NMR end group analyses and can be removed by hydrolysis to give the often desired hydroxyl end group.

The acylium salts and the free acids seem relatively straight-forward at this point. But this may only be so because they have not been studied extensively. It is known that propagation beyond the first step, from species III and IV, is rather slow. Participation of the acyl oxygen in shielding the environment of the positive oxygen

$$
R-\overset{\overset{\text{O}}{\|}}{C}-\overset{+}{O} \qquad \overset{+}{O}H\cdots O
$$

III   IV

and the strong dietherate complexes that secondary oxonium ions form are probably the reasons for this slow reaction.

Initiation by the trityl carbenium ion—the triphenyl methyl carbenium ion—is now fairly well understood.

In all cases the first step is probably one of hydride ion abstraction to form triphenyl methane. What then? In the case of THF it appears to be the loss of a proton to complete a dehydrogenation, as shown in Eq. 3, and initiation is then by the proton dietherate.

$$
\emptyset_3 C^+ + \langle\!\!\!\diagup_O\!\!\!\diagdown\rangle \longrightarrow \left[\langle\!\!\!\diagup_{\overset{+}{O}}\!\!\!\diagdown\rangle\right] + \emptyset_3 CH
$$

THF

$$
\left[\langle\!\!\!\diagup_O\!\!\!\diagdown\rangle\right] + \overset{+}{O}H\cdots O \tag{3}
$$

In other cases, the loss of the hydride ion is followed by other reactions. For example, in the case of 1,3-dioxolane (Eq. 4),

$$
\overset{O}{\diagdown}\overset{O}{\diagup} + \phi_3 C^+ \longrightarrow \phi_3 CH + O\overset{+}{\diagdown}O \tag{4}
$$

a dioxolenium ion is formed and the initiation reduces to the use of
that species. The initiation chemistry using trityl ion initiation for
the 1,2-epoxides and the oxetanes has not been very extensively
studied. The loss of a proton to complete a dehydrogenation is
not very likely on the basis of such evidence as exists, but hydride
ion abstraction is probably still the important first step and has
been so shown in some cases. The next step or steps still await
study and definition.

The trialkylaluminum-water-promoter system was discovered
and extensively studied in Japan by Furukawa and Saegusa [7]. It
was for this system that Saegusa [5, 8] developed the elegant end
capping method for measuring the concentration of active ends—the
number of oxonium ions formed. However, this catalyst system is
still a complex system and little understood. The trialkyl-water
reaction is understood in broad terms, but the exact mixture of
products is not known. Since rather inefficient initiation based on
trialkylaluminum has been shown, a minor reaction product may
be responsible for initiation.

What is the initiation reaction? How are the oxonium ions
formed? What is the exact function and fate of the promoter?
What is the counterion? Is an aluminate ester involved? These
are important questions also because it is a very similar catalyst
indeed that constitutes the Vandenberg catalyst used to make high
molecular weight polyepichlorohydrin rubber. It is worth noting
that we use a trialkylaluminum-water catalyst with ECH promotion
to initiate cationic THF polymerization, and the Hercules (Vandenberg) process for high molecular weight polyepichlorohydrin rubber
also uses a trialkylaluminum-water catalyst. The latter is used
under conditions which result in what is generally agreed to be a
coordinate polymerization mechanism. A fruitful area of future
research should be the sorting out of the relationships in this
mystery and defining the chemistry of this system in a manner
that Kennedy is doing in the trialkylaluminum-initiated cationic
polymerization of olefins.

In regard to the effect of promoters, we might ask: Do we
really understand the chemistry in the Lewis acid-promoter
initiation? These initiating systems are frequently used, even
extensively studied, but our understanding of the initiator chemistry has not really progressed since Meerwein first studied
this chemistry some 40 years ago.

A relatively new class of initiators, the fluorosulfonates, was
discovered at the 3M Company by Smith and Hubin [9] in the early
1960s. These initiators include the esters and anhydrides of

trifluromethane sulfonic acid and fluorosulfonic acid. They have only recently been studied in any detail. Initiation with such an ester is shown by

$$CF_3SO_2OR + \underset{O}{\bigvee} \longrightarrow \underset{CF_3SO_3^-}{\bigvee_{+O-R}} \tag{5}$$

Both Saegusa [10] in Japan and Penczek [11] in Poland have gone a long way toward sorting out some interesting new chemistry. They are developing an understanding of the ion-ester equilibrium that is so important here:

$$\sim\sim CH_2-\underset{CF_3SO_3^-}{\overset{+}{\bigvee}} \rightleftharpoons \sim\sim CH_2-O(CH_2)_4-OSO_2CF_3 \tag{6}$$

The perchlorate counterion also appears to belong to this class of materials for which the formation of the corresponding ester must be considered to play an important role.

The use of the anhydrides presents an exciting new development since it allows the very simple preparation of dicationic systems. The direct interaction of, for example, trifluoromethane sulfonic anhydride with tetrahydrofuran yields a dicationically active polymer, as indicated by

$$(CF_3SO_2)_2O + 3\underset{O}{\bigvee} \longrightarrow \underset{CF_3SO_2O^-}{\bigvee_{+O}}-CH_2CH_2CH_2CH_2-\underset{CF_3SO_2O^-}{\bigvee_{O+}} \tag{7}$$

Polymers with active cations at both chain ends open up a host of possibilities for preparing block copolymers and polymers with functional end groups. Many of these possibilities had been recognized and were explored by Smith and Hubin [9], but some of the more elegant work in the area of multiblock copolymers, using quite different chemistry, has been carried out by Yamashita and co-workers [12]. This is certainly an area in which we can expect important new developments.

PROPAGATION

In many ways our knowledge about propagation is much better. We all seem to agree, at least with four- and five-membered cyclic ethers, that the growing species is an oxonium ion and propagation occurs, as shown in Eq. (8), by nucleophilic attack by an ether oxygen on an α-carbon atom:

$$
\begin{array}{c}
\overset{+}{O}R \\
CH_2 \\
O
\end{array}
\longrightarrow
RO \sim\sim\sim\sim\sim CH_2 - O+ \qquad (8)
$$

Studies in recent years have focused on the details of the propagation step. There have been several studies [13, 14] which sought to define more closely the role of the ion pair, solvent separated ion pair, and free ion equilibria in these systems. The results of these studies show that there is relatively little difference in the rates of propagation of free ions over ion pairs, and that these differences vary with the dielectric constant of the solvent. This suggests the importance of solvent separated ion pairs in higher dielectric solvents. The question remaining is the importance of monomer itself in separating the ion pairs, a question to which it is believed some groups are currently addressing themselves. Closely related to this problem is the ion-ester equilibrium occurring during the propagation (Eq. 6) involving certain counter ions. This is now becoming understood, as mentioned above. Again, here we see a significant effect of solvent polarity. The highly polar solvents are clearly found to favor the ionic form, while nonpolar solvents favor the ester.

All cyclic ethers do not propagate by a pure oxonium ion mechanism as indicated in Eq. (8). The polymerization of cyclic formals is perhaps the best example. Although there are those

who feel cyclic formals are no exception, others are equally certain
Eq. (8) is not applicable. Penczek [15] has recently summarized
these different points of view and presented his evidence supporting
a carboxonium active end that is complexed with and stabilized by
two polymer oxygen atoms. The resultant active end is Structure V.

$$\sim\sim\sim OCH_2CH_2O-CH_2\overset{\displaystyle \overset{CH_2\sim\sim}{O}}{\underset{\displaystyle \underset{CH_2\sim\sim}{O}}{+ \, CH_2}}$$

V

Of course, the actual propagation step must then be modified.
Thus, in the case of cyclic acetal polymerization, we may not
have the trialkyl oxonium ions that we have come to accept and
be comfortable with. Are there other cases?

We have no reason to doubt that trialkyl oxonium ion active
centers are a reasonable representation of the situation with the
larger single oxygen containing cyclic ethers. But the situation
in the cationic polymerization of 1,2-epoxides is a little less
clear. These materials can and do polymerize by cationic, anionic,
and coordinate mechanisms. We feel that the propagating species
in the cationic polymerizations of these epoxides is probably a
tertiary oxonium ion, but we have to consider that the oxirane
oxygen is one of the least basic, least nucleophilic, of the ether
oxygens. Thus one has to question the validity of the typical S_N2

propagation mechanism in this case. (See discussion of epichlorohydrin polymerization below.)

Another aspect of propagation is the concurrent formation of
cyclic oligomers in some cases. In the case of THF the monomer
is the most favored oligomer; it and polymer are the only products.
But the smaller ring systems do show cyclic oligomer formation
under conditions of cationic polymerization. In the case of ethylene
oxide, the formation of 1,4-dioxane is almost the predominant
reaction. Kern [16] has shown that cyclic tetramer is the predominant oligomer formed from propylene oxide, 1,2-butylene oxide,
and epichlorohydrin. Dimer and some higher cyclic oligomers were
also observed in smaller amounts, but no trimer. In their studies
of the cationic polymerization of oxiranes, Entelis and Korovina
[17] conclude that cyclic tetramer formation is an important

aspect of these polymerizations. Is cyclic oligomer formation a
competing side reaction to linear polymer formation? Or is it a
depolymerization reaction of linear polymer?

In his classic study of oxetane polymerization, Rose [18] showed
that cyclic tetramer formation was important in oxetane and in
3,3-dimethyloxetane polymerizations. The amount of cyclic tetramer observed was markedly dependent on temperature. This
aspect of oxetane polymerization was recently re-examined by
Dreyfuss [19]. The study has produced some interesting and
significant results. The disappearance of monomer was followed
by gas chromatography. At the same time, the formation of two
volatile products was observed. One corresponded to the cyclic
tetramer described by Rose and the other was shown to be the
cyclic trimer. There is no evidence as yet for cyclic dimer or
for cyclic oligomers larger than tetramer. In agreement with
Rose, polymerization temperature was found to be important.
Under the conditions used, little cyclic oligomer was formed
below about 25°C. As the temperature was raised, more oligomers were formed and the ratio of trimer to tetramer increased.

Counterion is very important. Dreyfuss finds that cyclic
oligomer formation is really an important reaction only when the
counterion is tetrafluoroborate. In his studies, Rose used BF_3
as initiator. Had he used another Lewis acid, he may never have
observed cyclic tetramer formation. Other counterions (such as
PF_6^-, SbF_6^-, $SbCl_6^-$) gave only very small amounts of cyclic
oligomer, but always there was evidence for both trimer and
tetramer.

The use of ethyl trifluoromethane sulfonate as initiator gave
significantly different results. Relatively much more trimer was
formed. The use of a nonpolar solvent such as benzene favored
trimer over tetramer. The effect of increasing polymerization
temperature in the benzene solvent case was to increase the
amount of cyclic trimer formed while the amount of tetramer
formed stayed more or less constant.

This appears to be an interesting approach to studying the
effect of the various counterions on the polymerization reaction.
Oligomer formation may be a more sensitive and more convenient
way of studying some of these differences.

As was just pointed out, counterions can have a very important
effect on the polymerization of cyclic ethers. All too often studies
are undertaken with little regard to counterions. Initiators (and

hence, counterions) are apparently chosen because of their availability or ease of preparation, rather than from any careful consideration of the properties of the counterion.

In general, the rates of oxonium ion polymerizations have been found to be quite independent of counterion. In recent years, great strides have been made, notably by Saegusa and co-workers [5], in the determination of the absolute rate constants and the thermodynamic parameters for a number of these systems. The ion-ester equilibrium mentioned above (Eq. 6) does have a marked influence on rate. The ester form is very much less reactive than the ionic form. Thus the amount of ester present greatly effects the overall rate of polymerization observed. The type of counterion used for a polymerization has a big influence on the transfer and termination reactions observed. Consequently, the counterion has an important influence on the molecular weight and molecular weight distribution of the polymers produced. These effects of counterions are really self-evident from oxonium ion chemistry. After all, the discovery and isolation of oxonium ion salts depended on the discovery and development of anions of very low nucleophilicity.

TRANSFER AND TERMINATION

The termination reaction that can occur with a counterion is well known:

$$\sim\!\!\sim CH_2CH_2O^+ \underset{BF_4^-}{\diagdown} \longrightarrow \sim\!\!\sim CH_2CH_2O\diagup\!\!\!\diagup_{CH_2F}^{CH_2\,CH_2}\!\!\!\diagdown_{CH_2}^{} + BF_3 \tag{9}$$

It is a termination reaction because generally BF_3 is not able to initiate a new chain by itself. On the other hand, in a case such as $SbCl_6^-$ the same reaction is also a transfer reaction. The $SbCl_5$ formed can initiate a new chain. But it probably takes two molecules of $SbCl_5$ to initiate a new chain. Thus since half a chain, that is, an active center, is lost by each such reaction, it is a combination of transfer and termination in this case. Transfer can and does occur with acyclic ethers:

$$\text{\small www}CH_2-O\overset{+}{\underset{\triangle}{\bigcirc}} \quad + \quad \overset{CH_3CH_2}{\underset{CH_3CH_2}{\diagup}}O \quad\longrightarrow\quad \text{\small ww}CH_2-O(CH_2)_4\text{-}O\overset{CH_2CH_3}{\underset{CH_2CH_3}{\diagup}}+ \qquad (10)$$

THF

$$\text{\small www}CH_2-O(CH_2)_4-OCH_2CH_3 \quad + \quad CH_3CH_2-O\overset{+}{\bigcirc}$$

Thus, obviously the polymer ether oxygens can and do also interact with the growing oxonium ion centers. Water and alcohols at low levels have also been reported to function as transfer agents. Long ago we reported that trimethylorthoformate is particularly effective as a transfer agent:

$$\text{\small www}CH_2-O\overset{+}{\bigcirc} \quad + \quad CH_3O-C\overset{OCH_3}{\underset{OCH_3}{\diagup}}H \quad\longrightarrow\quad$$

$$\text{\small www}CH_2-O(CH_2)_4-OCH_3 \quad + \quad HC\overset{OCH_3}{\underset{OCH_3}{\diagup}}+ \qquad (11)$$

THF

$$CH_3-O-\overset{\overset{O}{\|}}{C}-H \quad + \quad CH_3-O\overset{+}{\bigcirc}$$

In THF polymerization it can be used to reliably control molecular weight [4]. The problem is that one only gets the inert methoxy ether end groups. A material which would be as effective a transfer agent as trimethylorthoformate and which would give functional end groups would be most desirable indeed.

The discussion on transfer so far has dealt mostly with THF polymerization. Transfer and termination reactions seem to be much more important in the case of the bicyclic ethers and the 1,2-epoxides. But the reactions in these cases are little, if at all, understood. It seems clear that these monomers react differently than the apparently straightforward THF case. But how? And why?

Noncoordinate cationic polymerizations of 1,2-epoxides have not been reported to give high molecular weight polymers. In fact, molecular weights of 1000 or less seem to be the rule. Transfer reactions and reactions leading to cyclic oligomers appear to predominate. Our own work has made some significant progress

in controlling and perhaps even understanding these factors.
Although we have not yet been able to exercise complete control
over molecular weight, we have been able to polymerize epichlorohydrin cationically under certain conditions to considerably higher
molecular weights. Our data to date suggest that ECH may be a
somewhat unique monomer in this regard. This has led me to
suggest an involvement of the chlorine atom in stabilizing the
propagating end. This stabilization is thought to favor propagation
at the expense of transfer and termination reactions and of reactions
leading to cyclic oligomers. The chlorine atom that is invoked is the
one on the penultimate monomer unit and the result is the cyclic
intermediate VI. A modified oxonium ion is suggested. This ion

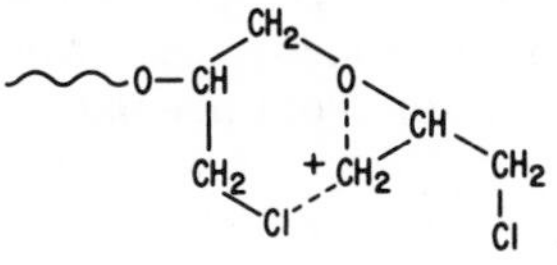

VI

is a six-membered ring active end in which the positive charge is
spread out. This is not a true oxonium ion propagating species,
but perhaps this is necessary for a successful cationic polymerization of an oxirane. A concerted growth step which directly forms
the next such species can be readily visualized and is depicted in
Structure VII. The two partial bonds break and a new active end

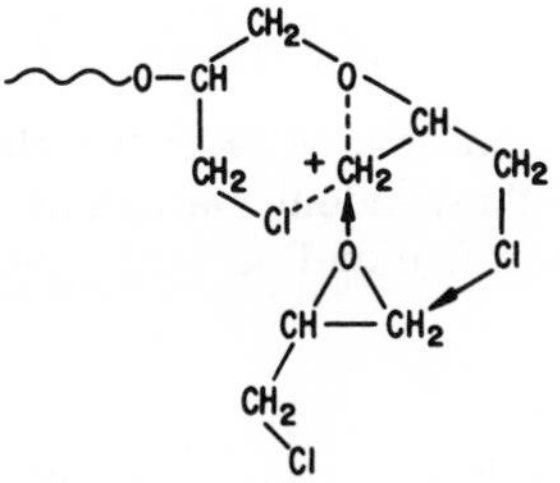

VII

just like the original one is formed. The process, of course, can
readily and rapidly continue. It is this process, then, that favors

propagation over transfer reactions and cyclic oligomer formation.
This picture was developed because under the same conditions other
epoxides do not polymerize as well nor lead to the same molecular
weight as does epichlorohydrin. Additional evidence for this picture
came from copolymerization studies. Copolymerization of other
cyclic ethers with ECH under the present conditions were found
to be generally very unsatisfactory, especially as compared to the
results of ECH homopolymerization. Evidently the propagation
pattern shown cannot be interrupted, even briefly. However, certain
epoxides do copolymerize very satisfactorily. Since these epoxides
were chosen on the basis of the scheme shown above, it has given
us greater confidence that there may be some validity to this suggestion. For example, a smooth and satisfactory copolymerization
was obtained with bischloromethyl oxetane and with phenyl glycidyl
ether. The comonomers were incorporated into the polymers. The
polymerizations did not suffer from premature termination, and the
molecular weights of the copolymers were in the same range as
observed for ECH homopolymerization. Both monomers can easily
form analogous stabilized active ends as shown in Structures VIII
and IX. It is interesting to note that alkyl glycidyl ethers were

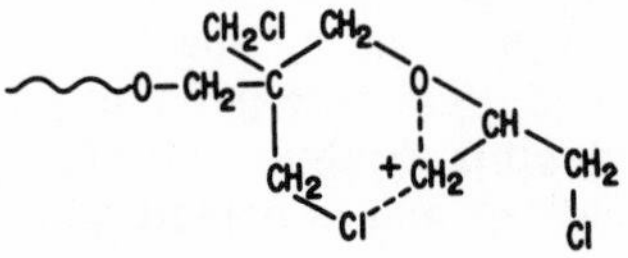

VIII 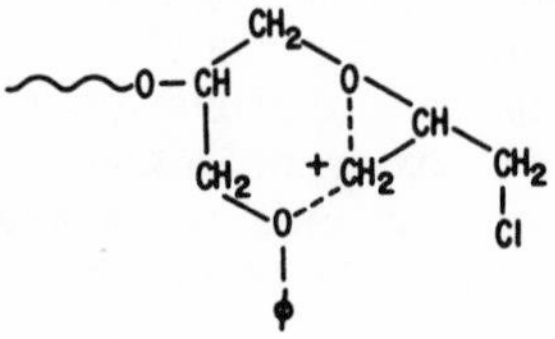   IX

unsatisfactory as comonomers. Apparently the nucleophilicity or
basicity of the atom on the pendent methylene of the epoxide must
be in a specific narrow range and be very much like a chlorine
atom.

THE POLYMERS

Monomers such as THF, BCMO, and oxetane give polymers of
varying properties depending on the molecular weight. But they
all give linear and readily crystallizable polymers. The bicyclic
ethers tend to give highly crystalline polymers having a trans

structure [20, 21]. Apparently the S_N2 propagation mechanism operates exclusively here to result in this trans product:

$$(12)$$

What about the corresponding cis isomer? Does the polymerization mechanism prevent our achieving this interesting isomer of the polymer shown in Eq. (12)? Or can we think of a way around this problem? Ionic polymerizations generally do not allow the preparation of isotactic and syndiotactic polymers. But it is interesting to point out here that isotactic polypropylene oxide is known, while the corresponding syndiotactic polymer is as yet unknown.

Copolymers can and are made by oxonium ion polymerizations. Mostly these are random copolymers. Some block copolymers have been made by some clever work of Saegusa [22] and by cotermination with anionic polymers [9, 12, 23]. But there has been relatively little work done on graft polymers using oxonium ion polymerization techniques. Alternating copolymers are another group that probably does not lend itself to a purely ionic polymerization process. Or can we find a way?

CONCLUSION

I have tried to point out and illustrate that we have many questions yet to be answered on all aspects of oxonium ion ring-opening polymerizations. I have indicated the type of progress we are making in some areas, notably on cyclic oligomer formation, which is very closely tied in with counterion effects. Also, in the area of 1,2-epoxide or oxirane polymerizations we may be making a small dent in understanding how to make higher molecular weights using a cationic mechanism.

ACKNOWLEDGMENTS

I wish to thank the B. F. Goodrich Company for their permission to publish this paper. The able experimental assistance of D. L. Hine

102 DREYFUSS

is greatly appreciated. I am very grateful for the prepublication
information furnished by Dr. P. Dreyfuss and appreciate, as always,
our many helpful discussions.

REFERENCES

[1]   Symposium on Polymerization of Heterocyclics, J. Macromol.
      Sci.—Chem., A6, 991-1199 (1972).
[2]   Symposium of Polymerization of Cyclic Ethers and Sulfides,
      Ibid., A7, 1359-1535 (1973).
[3]   Symposium on Cationic Polymerization (Rouen), Makromol.
      Chem., 175, 1017-1328 (1974).
[4]   P. Dreyfuss, J. Macromol. Sci.—Chem., A7, 1361 (1973).
[5]   T. Saegusa and S. Kobayashi, Progr. Polym. Sci. Japan, 6,
      107 (1973).
[6]   H. Meerwein, G. Hinz, P. Hofmann, E. Kroning, and E. Pfeil,
      J. Prakt. Chem., 147, 257 (1937).
[7]   J. Furukawa and T. Saegusa, Polymerization of Aldehydes
      and Oxides, Wiley-Interscience, New York, 1963.
[8]   T. Saegusa and S. Matsumoto, J. Polym. Sci., Part A-1, 6,
      1559 (1968).
[9]   S. Smith and A. J. Hubin, British Patent 1,120,304 (1968),
      (Chem. Abstr., 69, 68151e (1968)); S. Smith and A. J. Hubin,
      J. Macromol. Sci.—Chem., A7, 1399 (1973).
[10] S. Kobayashi, H. Danda, and T. Saegusa, Macromolecules, 7,
      415 (1974).
[11] K. Matyjaszewski, P. Kubisa, and S. Penczek, J. Polym.
      Sci., Polym. Chem. Ed., 12, 1333 (1974).
[12] Y. Yamashita, K. Nobutoki, Y. Nakamura, and M. Hirota,
      Macromolecules, 4, 548 (1971).
[13] J. M. Sangster and D. J. Worsfold, Ibid., 5, 229 (1972).
[14] E. J. Goethals, W. Drijvers, D. van Ooteghem, and A. M.
      Buyle, J. Macromol. Sci.—Chem., A7, 1375 (1973); E. J.
      Goethals, Makromol. Chem., 175, 1309 (1974).
[15] S. Penczek, Ibid., 175, 1217 (1974).
[16] R. J. Kern, J. Org. Chem., 33, 388 (1968).
[17] S. G. Entelis and G. V. Korovina, Makromol. Chem., 175,
      1253 (1974).
[18] J. B. Rose, J. Chem. Soc., 1956, 542.
[19] P. Dreyfuss, Private Communication.
[20] E. L. Wittbecker, H. K. Hall, and T. W. Campbell, J. Amer.
      Chem. Soc., 82, 1218 (1960).

[21] J. Kops and H. Spanggaard, J. Macromol. Sci.—Chem., A7, 1455
      (1973).
[22] T. Saegusa, S. Matsumoto, and Y. Hashimoto, Macromolecules,
      3, 377 (1970).
[23] G. Berger, M. Levy, and D. Vofsi, J. Polym. Sci., Part B, 4,
      183 (1966).

The Role of Donor-Acceptor Complexes in the Initiation of Ionic Polymerization

J. K. STILLE, N. OGUNI, D. C. CHUNG, R. F. TARVIN,
S. AOKI, and M. KAMACHI

Department of Chemistry
University of Iowa
Iowa City, Iowa 52242

ABSTRACT

Work carried out in the past few years aimed at elucidating
the mechanism of initiation of vinyl polymerization when a
donor and an acceptor molecule, one or both of which may
be vinyl monomers, is summarized. The emphasis of our
investigation has been on polymerizable ether donors and
strong electron acceptors which do not undergo polymerization, or the acceptor vinylidene cyanide. Alkyl vinyl ethers
were polymerized in the presence of tetracyanoquinodimethane
(TCNQ) and 2,3-dichloro-5,6-dicyano-p-benzoquinone (DDQ) in
polar solvents. Observation of the ESR spectrum of the DDQ
radical anion and the isolation of a 1:1 addition product of DDQ
and alkyl vinyl ether when the two are mixed in a 1:1 ratio and
quenched in alcohol support an initiation mechanism involving
a coupling reaction of the donor monomer (radical cation) and
the acceptor initiator (radical anion). The reaction of vinylidene cyanide (VC) with the vinyl ethers p-dioxene, dihydropyran,
ethyl vinyl ether, isopropyl vinyl ether, and ketene diethylacetal
in a variety of solvents at 25°C spontaneously afforded

poly(vinylidene cyanide), the cycloaddition products
7,7-dicyano-2,5-dioxo-bicyclo[4.2.0]octane, 8,8-dicyano-
2-oxo-bicyclo[4.2.0]octane, the 1,1-dicyano-2-alkoxycyclo-
butanes, and 1,1-diethoxy-2,2,4,4-tetracyanohexane,
respectively, and with the exception of p-dioxene, homopolymers of the vinyl ethers. In the presence of AIBN at
80°C, alternating copolymers were obtained in addition to
the homopolymers and cycloaddition products, supporting
the involvement of donor-acceptor complexes. The reaction
of styrene with VC spontaneously formed an alternating
copolymer in addition to the 1:2 head-to-head cycloaddition
product, 1,1,3,3-tetracyano-4-phenylcyclohexane. Mixing
VC with any one of the cyclic ethers tetrahydrofuran, oxetane,
2,2-dimethyloxirane, 2-chloromethyloxirane, and phenyloxirane
resulted in the polymerization of both the VC and the cyclic
ether to afford homopolymers of both. The cyclic ethers
trioxane, 3,3-bis(chloromethyl)oxetane, and oxirane initiated
the polymerization of VC, but did not undergo ring-opening
polymerizations themselves. Other ethers such as 1,3-dioxo-
lane, tetrahydropyran, and diethyl ether did not initiate the
polymerization of VC. In these polymerizations, VC and the
cyclic ethers polymerize via anionic and cationic propagation
reactions, respectively.

INTRODUCTION

The reaction of an electron donor with an electron acceptor, one
or both of which may be a vinyl monomer, often results in the polymerization of the donor, the acceptor, or both. Although donor-acceptor
complexes as well as their intermediate reaction products, radical
cations and radical anions, have been observed in these polymerizations, the initiation mechanisms are not well understood. Most
proposed mechanisms invoke a "T-class" reaction of the complex
in the first stages of the initiation step to give a radical cation
and a radical anion:

$$D + A \overset{K_e}{\rightleftharpoons} [D \to A] \rightleftharpoons D^{\ddot{.}}, A^{\ddot{-}} \rightleftharpoons D^{\ddot{+}} \text{ (solv)} + A^{\ddot{-}} \text{ (solv)}$$

Two types of spontaneous polymerization have been observed;
1) the homopolymerization of the donor monomer by a cationic
propagation and/or the homopolymerization of the acceptor monomer

by an anionic propagation, and 2) the alternating copolymerization of the donor and acceptor monomers by a free radical propagation. In addition, donor and acceptor monomers which form a complex and do not spontaneously homopolymerize may form an alternating copolymer by a radical propagation reaction. Which one of these types of polymerization reactions occurs has been postulated [1] to depend on the strength of the complex as measured by the equilibrium constant, K_{eq}.

SPONTANEOUS HOMOPOLYMERIZATION

Donor Monomer Homopolymerization via Nonmonomer Acceptors [2-5]

Although in the presence of tetracyanoethylene (TCNE) vinyl ethers form intermediate donor-acceptor complexes that ultimately lead to cyclobutane 2 + 2 cycloaddition products, the ethers polymerize in the presence of the acceptors tetracyanoquinodimethane (TCNQ) and 2,3-dichloro-5,6-dicyano-p-benzoquinone (DDQ), neither of which can undergo facile 2 + 2 cycloaddition.

High molecular weight polymers ($\overline{M}_n = 10^5$) are obtained from catalytic amounts of the acceptor via a cationic propagation mechanism. Immediately after mixing the vinyl ether and the TCNQ or DDQ acceptor, a color characteristic of the donor-acceptor complex was observed which eventually disappeared. The rate of initiation was dependent on the solvent, the temperature, and the vinyl ether; polar solvents and higher temperatures led to faster rates while the rates of initiation followed the order $t\text{-}C_4H_9 > i\text{-}C_3H_7 > C_2H_5$. The DDQ radical anion was observed by ESR using flow techniques. In certain cases the rates of formation and disappearance of the donor-acceptor complex as well as that of the radical anion could be obtained. When DDQ and methyl vinyl ether (1:1) were mixed in a flow system and then quenched in methanol, the 1:1 reaction product (1) was formed.

1

Thus the following mechanism of initiation was proposed, as illustrated with DDQ.

If this mechanism is correct, several important questions regarding it are unanswered. The role of the delocalized anion is not understood; whether it is destroyed early in the polymerization or serves as a counterion during propagation and ultimately is consumed, perhaps in an ion coupling termination reaction, is not known.

Quite unexpectedly, the highly polar electron donating monomer, ketene diethyl acetal, gives only low yields (<15%) of polymer in the presence of the electron acceptors TCNE, TCNQ, and DDQ. Contrary to earlier reports [6], polymer is obtained only in the polar solvent, acetonitrile, but not in toluene or methylene chloride. No pure cycloadduct could be isolated in the reaction with TCNE.

Why this monomer, which, with the same acceptors, forms a stronger donor-acceptor complex than the alkyl vinyl ethers, yet does not undergo polymerization, remains unexplained. The proposal [1] that monomers which form very strong complexes ($K_{eq} \geq 5$) do not undergo polymerization does not provide a satisfactory explanation.

Acceptor Monomer Homopolymerization via Nonmonomer Donor [5]

The polymerization of vinylidene cyanide (2), a strong electron acceptor monomer, takes place rapidly with an anionic catalyst, but only slowly in the presence of a radical initiator. Since p-dioxene (3) is an electron donor monomer that does not undergo homopolymerization, its reaction with vinylidene cyanide was studied.

High molecular weight poly(vinylidene cyanide) (4) and the cycloaddition product, dicyano-2,5-dioxo-bicyclo[4.2.0]octane (5), were obtained. Equal molar amounts of reactants gave a 2:1 ratio of polymer to cycloaddition product. Although no color characteristic of a donor-acceptor complex could be observed, the presence of a radical initiator led to the formation of poly(vinylidene cyanide), cycloadduct 5, and a 1:1 head-to-tail alternating copolymer (6, η_{inh} = 0.25, Table 1). The cycloadduct, 5, neither initiates the polymerization of vinylidene cyanide nor polymerizes in the presence of a free radical initiator. The structure of the copolymer was established by its solubility characteristics, its 1:1 composition regardless of the monomer feed ratio, and its ^{13}C NMR spectrum.

An initiation mechanism similar to that postulated for the vinyl ethers in the presence of TCNQ or DDQ (vide supra) can be written, but in this case is much less satisfactory.

Acceptor Monomer-Donor Monomer Homopolymerization [5]

The reaction of vinylidene cyanide, an electron acceptor monomer with the electron donor monomers, the vinyl ethers [7], ketene

TABLE 1. Copolymerization of Vinylidene Cyanide and p-Dioxene

2	3	4	5	6
Mole ratios		% Yields		
1	2	12	12	73
1	1	24	9	69
2.8	1	66	18	53

diethyl acetal, and dihydropyran in each case afforded cycloadduct, poly(vinylidene cyanide), and the corresponding polyether (Table 2).

Several features of these reactions are mechanistically note-worthy. The presence of light had little effect either on the yields of homopolymers or on the yields of cycloadducts. In general, an increase in the molar ratio of one of the two monomers resulted in an increased yield of the corresponding homopolymer. In the polymerization reactions of vinylidene cyanide and dihydropyran, higher conversions of vinylidene cyanide to polymer were observed in the more polar solvents, particularly in acetonitrile. Further, higher yields of cycloadducts were produced in the more polar solvents, but in a given solvent the yield of cycloadduct was rela-tively constant, regardless of the mole ratio of the donor and acceptor charged.

Under the reaction conditions, poly(vinylidene cyanide) (4) would not initiate the polymerization of the vinyl ethers, the poly(vinyl ethers) would not initiate the polymerization of vinylidene cyanide, and the cycloadduct would not initiate the polymerization of either vinylidene cyanide or the ether monomers. The observations [7] that the addition of trihydroxyethylamine or phosphorus pentoxide to the vinylidene cyanide-vinyl ether mixtures inhibited the polym-erizations of the vinyl ethers and vinylidene cyanide, respectively, suggest cationic and anionic propagation reactions of these monomers. Although no transient colors characteristic of the donor-acceptor complexes were observed, alternating copolymers 16 and 17 could be obtained from the reactions of vinylidene cyanide (2) and ethyl vinyl ether (7a) or dihydropyran (13), respectively, in the presence of a free radical initiator. Cycloadduct, 8 or 14, poly(vinylidene cyanide) (4), and poly(ethyl vinyl ether) (9a) were also produced along with the copolymer (Table 3). Neither cycloadduct 8 nor 14 polymerized in the presence of the free radical initiator, an anionic catalyst or a cationic catalyst.

16   17

The structures of the alternating copolymers (16 and 17) were established by their solubility characteristics, their 1:1 compositions regardless of the monomer feed ratios, and their ^{13}C NMR spectra.

In the case of polymer 17, a head-to-tail or a head-to-head structure could not be conclusively established by the ^{13}C NMR spectrum.

SPONTANEOUS ALTERNATING COPOLYMERIZATION [5,7]

Surprisingly, the reaction of vinylidene cyanide with styrene in the absence of a free radical initiator gave both a head-to-tail alternating copolymer (18) and the head-to-head cycloadduct (19) (Table 4). No homopolymer of either monomer was obtained. The structure of the copolymer was established from its solubility, its 1:1 composition regardless of monomer feed ratio, and its ^{13}C NMR spectrum.

The mechanisms of initiation of the polymerizations which take place when vinylidene cyanide and the vinyl ether donor monomers p-dioxene (3), ethyl vinyl ether (7a), i-propyl vinyl ether (7b), ketene diethyl acetal (10), and dihydropyran (13) are mixed in an inert solvent are certainly open to speculation. Tetracyanoethylene reacts with these monomers to give transient colors characteristic of the donor-acceptor complexes, yet in all cases only the cyclo-adduct is obtained* and no polymerization of the donor monomer is observed. The absence of color when vinylidene cyanide is the acceptor molecule does not necessarily imply the absence of a donor-acceptor complex, since it could be present in low concentrations as a result of a large difference in its relative rates of formation and disappearance. The observations that vinylidene cyanide and the vinyl ether donor monomers afforded alternating head-to-tail copolymers in the presence of free radical initiators implicates the presence of the donor-acceptor complex.

This series of reactions of vinylidene cyanide with donor vinyl monomers provides examples of the two modes of spontaneous polymerizations. The proposal [1] that spontaneous homopolymerization is observed when $K_{eq} \simeq 1$ to 5 and spontaneous alternating copolymerization occurs when $K_{eq} \simeq 0.1$ is generally observed. The estimated equilibrium constants for these complexes as obtained from the ionization potentials of the donor vinyl monomers and a calculated electron affinity for vinylidene cyanide show that a $K_{eq} \simeq 0.1$ gives the alternating copolymer (styrene, vinylidene

---

*The exception is ketene diethyl acetal, in which case only an impure unstable product is isolated.

TABLE 2. Vinylidene Cyanide–Vinyl Ether*

$$CH_2=C \overset{CN}{\underset{CN}{\big\backslash}} \; (\underline{2}) \; + \; CH_2=CH\!-\!OR \; (\underline{7}) \xrightarrow{C_6H_5CH_3} \text{(8)} + \left[CH_2-C(CN)_2\right]_n (\underline{4}) + \left[CH_2-CH(OR)\right]_n (\underline{9})$$

(8) = cyclobutane bearing CN, CN, OR

R = (a) C_2H_5; (b) $\underline{i}C_3H_7$

$\underline{2}$	$\underline{7}$	R	$\underline{8}$	$\underline{4}$	$\underline{9}$
3.3	1	a	27	52	84
2	1	b	13	52	81
1.2	1	a	17	75	69
1.5	1	b	9	71	67
4	1	a	33	85	53
4.1	1	b	20	86	43

$$CH_2=C \overset{CN}{\underset{CN}{\big\backslash}} \; (\underline{2}) \; + \; CH_2=C(OEt)_2 \; (\underline{10}) \xrightarrow{C_6H_5CH_3} (\underline{11}) + (\underline{4}) + \left[CH_2-C(OEt)_2\right]_n (\underline{12})$$

(11) = cyclohexane bearing EtO, OEt, CN, CN, NC, CN

$\underline{2}$	$\underline{10}$	$\underline{11}$	$\underline{4}$	$\underline{12}$
1.8	1	22	Trace	43
1	1.7	32	58	30

$$\underline{2} \;+\; \underline{13} \xrightarrow{\;C_6H_5CH_3\;} \underline{14} \;+\; \underline{4} \;+\; \underline{15}\,\big(\!+C_5H_8O\big)_n$$

$\underline{2}$	$\underline{13}$	$\underline{14}$	$\underline{4}$	$\underline{15}$
5	1	37	67	Trace
1	4.5	31	33	
1	1	26	52	
5.6	1	27	58	Trace

*Yield % with respect to reactant present in smaller molar amount.

TABLE 3. Copolymerization of Vinylidene Cyanide and Donor Monomers

Mole ratio			Product yield (%)[a]			
2	:	7a	8	4	9	16
1		3.3	Trace	Trace	83	38
1.2		1	Trace	60	80	30
3.85		1	Trace	72	58	42
2	:	13	14	4	-	17
1		2.7	12	12		73
1		1	9	24		69
3.3		1	18	66		53

[a]Based on the monomer present in the smaller mole ratio.

TABLE 4. Copolymerization of Vinylidene Cyanide and Styrene

$$\underline{2} + C_6H_5CH=CH_2 \longrightarrow \left[CH_2-\underset{CN}{\overset{CN}{C}}-CH_2-\underset{C_6H_5}{CH} \right]_n + \text{cyclohexane structure}$$

		18	19
Mole ratio			% Yield
1	1.37	49	33
1.45	1	53	23
2.17	1	78	20

cyanide) and a $K_{eq} \simeq 0.3$ affords homopolymer (ethyl vinyl ether, vinylidene cyanide).

A polymerization mechanism involving donor-acceptor complexes which satisfactorily accounts for the formation of cycloadduct and alternating copolymer in the presence of radical initiators does not satisfactorily explain or provide the details of the ionic homopolymerizations, however. The fact that both cationic and anionic propagation reactions take place in the same pot, whether they take place simultaneously or not, is quite astounding. Whether a four- or a six-membered cycloaddition product is obtained may depend either

on steric factors which prevent the closure to a four-membered ring or on the ability of the cation to be sufficiently delocalized, thus allowing insertion of a second vinylidene cyanide before closure.

VINYLIDENE CYANIDE POLYMERIZATION BY CYCLIC ETHERS [8]

These extraordinary results prompted us to explore the possibility that other monomers such as cyclic ethers, which are both electron donors and are susceptible to polymerization via cationic mechanisms, would also lead to the formation of homopolymers of each type of monomer (vinylidene cyanide and cyclic ether) when the two are mixed.

The reaction of vinylidene cyanide with certain polymerizable cyclic ethers in bulk or in toluene gave poly(vinylidene cyanide) (Table 5). The ethers which were studied can be divided into three categories: 1) those which induced the polymerization of vinylidene cyanide and polymerized themselves—tetrahydrofuran, oxetane, 2,2-dimethyloxirane, epichlorohydrin, and styrene oxide; 2) those which induced the polymerization of vinylidene cyanide but do not polymerize themselves—trioxane, 3,3-bischloromethyloxetane, and oxirane; and 3) those which do not polymerize vinylidene cyanide—1,3-dioxolane, tetrahydropyran, and diethyl ether.

TABLE 5. Polymerization of Vinylidene Cyanide (2) and Cyclic Ethers at 25°[a]

Cyclic ether	Mole ratio 2:Ether	Solvent	Insoluble fraction			Soluble fraction		
			% 4	% Conversion	$[\eta]$[d]	% Ether	% Conversion[a]	$[\eta]$[e]
Trioxane	1:1	CH_2Cl_2[b]	86	62			0	
	5:1	CH_2Cl_2[b]	86	32			0	
Tetrahydrofuran	1:10	Bulk[c]		Trace		86	25	1.28
	1:1	Bulk[c]	92	54	0.85	93	65	0.74
	5:1	Bulk[c]	93	36	1.36	95	100	0.27
Oxetane	1:5	Toluene	68	37	0.12	97	79	1.22
	1:1	Toluene	78	64	0.40	95	100	0.24
	5:1	Toluene	90	33	0.52	94	100	0.10
3,3-Bis(chloromethyl)oxetane	1:5	Toluene	86	34			0	
	1:1	Toluene	86	56			0	
	1:1	Acetonitrile	88	56			0	
2,2-Dimethyloxirane	1:5	Toluene	87	100	0.21	90	3	2.10
	1:1	Toluene	90	84	0.77	88	7	0.90
	5:1	Toluene	85	61	1.39	70	44	0.14

Monomer	Ratio	Solvent				
2-Chloromethyloxirane (epichlorohydrin)	1:10	Toluene	86	73		2
	1:1	Toluene	92	94	80	69
	10:1	Toluene	94	79		4
2-Phenyloxirane (styrene oxide)	1:5	Toluene	79	100	91	3[f]
	1:1	Toluene	87	100	83	45[f]
	5:1	Toluene	99	54	80	67[f]
Oxirane (ethylene oxide)	1:5	Toluene	87	16		1[g]
	1:1.5	Toluene	93	100		5[g]
	2.5:1	Toluene	92	84		2.5[g]
	1:10	Bulk	82	90	51[h]	1.5[g]

[a] Polymerizations were carried out for 96 hr, except with tetrahydrofuran, in which case the polymerizations were allowed to continue for 168 hr. Conversions are calculated on the basis of the respective monomer charged.

[b] Cyclic ether is insoluble in toluene.

[c] Polymerization of cyclic ether in toluene takes place only very slowly.

[d] Measured in dimethylformamide at 25°.

[e] Measured in benzene at 25°.

[f] A mixture of two diphenyl-1,4-dioxane isomers was obtained in each example.

[g] In each case, 1,4-dioxane was obtained from the reaction.

[h] Block copolymer as evidenced by its NMR spectrum and solubility in toluene. The NMR spectrum was identical with that of the two homopolymers.

Styrene oxide and ethylene oxide are the only two cyclic ethers in this series which also are known to afford polymer by anionic initiation and propagation mechanisms, and therefore the poly(ethylene oxide) and poly(styrene oxide) could have been produced anionically. In both cases, however, dioxane derivatives were isolated from the polymerization mixture containing polyether and poly(vinylidene cyanide); dioxane was obtained from the ethylene oxide polymerization and two of the diphenyl dioxanes (cis- and trans-2,5-diphenyl-1,4-dioxane) were obtained from the styrene oxide polymerization. These products are characteristic of the cationic polymerization of these monomers, and are not produced during the course of anionic polymerization.

When phosphorus pentoxide was present in the tetrahydrofuran or oxetane polymerizations, vinylidene cyanide did not polymerize; when pyridine was present, no polyethers were obtained. Kinetic studies showed a rapid rate of vinylidene cyanide polymerization $(K_2/K_{THF} \simeq 10^2)$.

In these polymerizations of vinylidene cyanide with the cyclic ethers, weak donors and a strong electron acceptor are involved, but no color, however transient, was observed. Attempts to initiate the polymerization of oxetane or tetrahydrofuran with the stronger electron acceptors, DDQ, TCNE, and TCNQ, produced only traces of polyethers. Thus a donor-acceptor mechanism is not consistent with these observations.

The observations that the rates of disappearance of the vinylidene cyanide in these polymerizations increased by increasing the amount of tetrahydrofuran initially charged, and that the molecular weight of each polymer increased as the ratio of the other monomer was decreased, indicate mutual initiation by the two monomers. Thus the initiator in the polymerization of vinylidene cyanide is either a cyclic ether or perhaps a complex of a cyclic ether. Clearly, anionic propagation in the polymerization of vinylidene cyanide and cationic propagation in the polymerization of the cyclic ethers are taking place. The studies of the rates of polymerization indicate, however, that the polymerization of vinylidene cyanide is rapid, and that significant polymerization of cyclic ethers does not take place until the propagation of vinylidene cyanide chains has ceased. Since the poly(vinylidene cyanide) precipitates in the early stages of the polymerization, the anion end may be buried in the polymer chain, fed by vinylidene cyanide monomer, and inaccessable to any terminating species such as a cationically propagating polyether.

A polar mechanism may be responsible for the initiation of polymerization. This mechanism relies on the rapid rate of propagation of vinylidene cyanide compared to the cyclization of the zwitterion.

Models show that cyclization of the dimer is not sterically possible, but cyclization could be accomplished after a second monomer addition. Cyclization at this stage may be precluded both by the rapid propagation rate and the fact that the anionic charge is delocalized over the nitrile, providing resonance stability and a diffuse charge. After the addition of subsequent vinylidene cyanide monomers, cyclization to ring sizes greater than nine becomes relatively more difficult. When the polymerization gets past the oligomer stage, the polymer precipitates, is immobile, and cannot cyclize readily. After precipitation, the chain end can be fed by vinylidene cyanide monomer by diffusion.

The steric bulk surrounding the oxonium ion end could effectively prevent attack of THF at carbon in a propagation step. Hydride transfer from monomer generates the free oxonium ion which is known to decompose to dihydrofuran and provide a proton initiator.

REFERENCES

[1] S. Iwatsuki and Y. Yamashita, Progr. Polym. Sci. Japan, 2, 1 (1971).
[2] S. Aoki, R. F. Tarvin, and J. K. Stille, Macromolecules, 3, 472 (1970).
[3] S. Aoki and J. K. Stille, Ibid., 3, 473 (1970).
[4] R. F. Tarvin, S. Aoki, and J. K. Stille, Ibid., 5, 663 (1972).

[5] J. K. Stille and D. C. Chung, Ibid., 8, 114 (1975).
[6] H. Noguchi and S. Kambara, J. Polym. Sci., Part B, 3, 271 (1965).
[7] H. Gilbert, F. F. Miller, S. J. Averill, E. J. Carlson, V. L. Folt, H. J. Heller, F. D. Stewart, R. F. Schmidt, and H. L. Trumbull, J. Amer. Chem. Soc., 78, 1669 (1956).
[8] N. Oguni, M. Kamachi, and J. K. Stille, Macromolecules, 6, 146 (1973); 7, 435 (1974).

High-Energy Phosphonium Compounds and Their Application to Polymer Synthesis

NOBORU YAMAZAKI and FUKUJI HIGASHI

Department of Polymer Science
Tokyo Institute of Technology
Ookayama, Meguro-ku, Tokyo 152, Japan

ABSTRACT

High-energy phosphonium compounds, the N-phosphonium salts of pyridines, were prepared by the oxidation of phosphorous acid and its esters with mercuric salts or halogens in pyridines, or by a hydrolysis-dehydration reaction of diphenyl and triaryl phosphites or phosphonites. These salts are very reactive to nucleophiles, activating carboxyl, amino, or hydroxyl compounds via the corresponding N-phosphonium salts to yield carboxylic amides and esters in high yields on further aminolysis, alcoholysis, and acidolysis. These reactions, especially the hydrolysis-dehydration reactions with phosphites, were successfully extended to the direct polycondensation reaction of dicarboxylic acids with diamines, of free α-amino acids or dipeptides, and of carbon dioxide and disulfide with diamines under mild conditions, yielding linear polymers of high molecular weight (polyamides, polypeptides, polyureas, and polythioureas).

INTRODUCTION

Adenosine triphosphate (ATP), a typical high-energy compound in living cells, plays an important role as an energy source in the production of lipids, proteins, and carbohydrates, and it is consumed and regenerated via a coupling to the phosphagen-ATP system [1]:

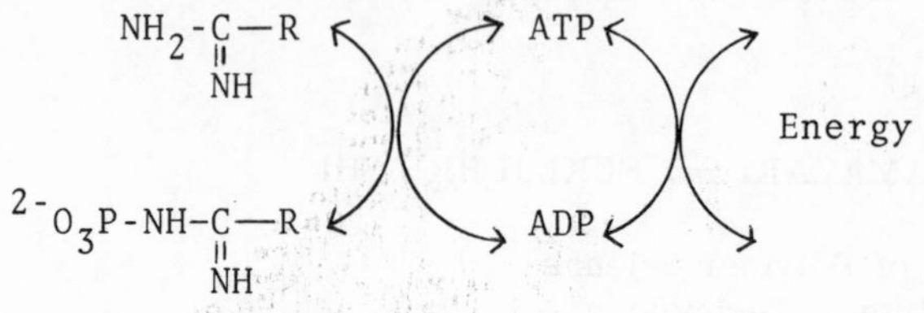

$$NH_2-\underset{\underset{NH}{\|}}{C}-R$$

$$^{2-}O_3P-NH-\underset{\underset{NH}{\|}}{C}-R$$

ATP

ADP

Energy

Phosphagen

$$\left(\begin{array}{l}Creatine\ ;\quad R=\ -\underset{\underset{CH_3}{|}}{N}-CH_2COOH\\[2em]Arginine\ ;\quad R=\ -NH-(CH_2)_4-\underset{\underset{NH_2}{|}}{CH}-COOH\end{array}\right.$$

Thus energies obtained from reactions such as the oxidative breakdown of glucose are transferred to (adenosine diphosphate) ADP, resulting in ATP, and the energies in ATP, in turn, are reserved in the phosphate bond of phosphagens.

During the course of studying chemical reactions via a process similar to the energy transfer in living cells described above, we have developed a new process like the phosphagen-ATP system which involved the oxidation of phosphorous acid and its esters with mercuric salts or halogens, or dephenoxylation of phosphites, giving rise to the high-energy phosphonium compounds, N-phosphonium salts of pyridines.

This paper describes studies on the reactions of the N-phosphonium salts of pyridines, and the application of the reactions to polymer synthesis.

RESULTS AND DISCUSSION

Reactions of the N-Phosphonium Salts of Pyridines

The N-phosphonium salts of pyridines given by the oxidation of phosphorous acid and its esters are very reactive to nucleophiles, activating carboxyl, amino, or hydroxyl compounds to yield the corresponding carboxylic amides and esters in high yields on further aminolysis, alcoholysis, and acidolysis [2].

These reactions were studied in terms of steric effect, acidity and basicity of carboxylic acids, amines, and tertiary amines such as pyridine by using phosphorous acid and its mono-, di-, and tri-esters, as proposed in Scheme 1 for the case of the reaction with diesters. These N-phosphonium salts (I-IV) were

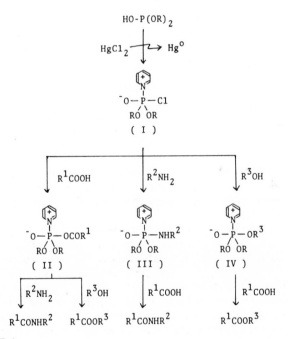

SCHEME 1.

separated and characterized on the bases of the IR spectra, acid-base titration, and their reactions.

These reactions were successfully used for the preparation of peptides and active esters of amino acids, with no detectable amounts of racemization (Table 1) [3].

We also showed that phosphites, especially diphenyl and triaryl phosphites, reacted nonoxidatively with carboxylic acids in the presence of pyridine to give the acyloxy-N-phosphonium salt of pyridine (V and VI in Scheme 2) accompanied by dephenoxylation,

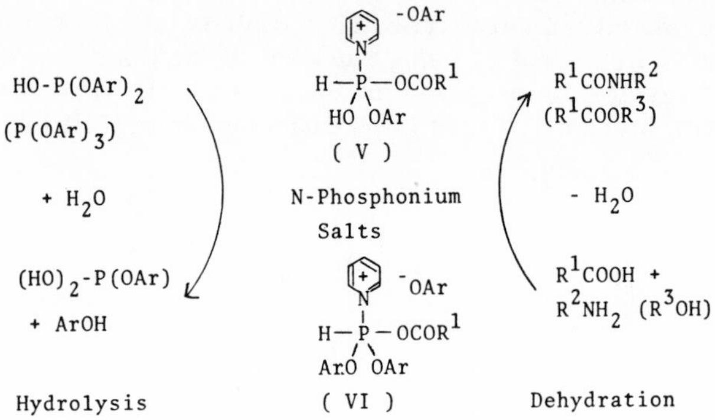

$$
\begin{array}{ccc}
\text{HO-P(OAr)}_2 &
\begin{array}{c}
\overset{+}{\text{N}} \quad {}^{-}\text{OAr} \\[2pt]
\text{H}-\overset{|}{\underset{\text{HO}}{\text{P}}}{\underset{\text{OAr}}{}}-\text{OCOR}^1 \\[2pt]
(\text{ V })
\end{array} &
\begin{array}{c}
\text{R}^1\text{CONHR}^2 \\
(\text{R}^1\text{COOR}^3)
\end{array} \\[20pt]
(\text{P(OAr)}_3) & & \\[10pt]
+ \ \text{H}_2\text{O} & \text{N-Phosphonium} & - \ \text{H}_2\text{O} \\
 & \text{Salts} & \\[10pt]
(\text{HO})_2\text{-P(OAr)} &
\begin{array}{c}
\overset{+}{\text{N}} \quad {}^{-}\text{OAr} \\[2pt]
\text{H}-\overset{|}{\underset{\text{ArO}}{\text{P}}}{\underset{\text{OAr}}{}}-\text{OCOR}^1 \\[2pt]
(\text{ VI })
\end{array} &
\begin{array}{c}
\text{R}^1\text{COOH} \ + \\
\text{R}^2\text{NH}_2 \ (\text{R}^3\text{OH})
\end{array} \\[10pt]
+ \ \text{ArOH} & & \\[6pt]
\text{Hydrolysis} & & \text{Dehydration}
\end{array}
$$

SCHEME 2.

which produced the corresponding amides and esters on aminolysis and alcoholysis. The N-phosphonium salts such as II were presumed from the stoichiometric relationship among phosphites, pyridine, and the carboxyl component [4].

In these reactions, hydrolysis of diphenyl and triaryl phosphites to monoaryl phosphites and phenol was coupled by dehydration between carboxylic acids and amines or alcohols to the corresponding amides and esters. Therefore, the reaction was generalized as a hydrolysis-dehydration reaction (Scheme 2).

The proposed concept of the hydrolysis-dehydration reaction using phosphites was shown to be applicable also to reactions with other phosphorous compounds, such as phosphonites, phosphinites, and phosphonates [5]. The aryl esters of these phosphorus compounds were effective for producing amides and esters, whereas alkyl esters were ineffective (Eqs. 1-3).

$$\underset{R}{\overset{R}{\diagup}}P-OR + R^1COOH + R^2NH_2 \ (R^3OH) \xrightarrow[Py]{} R^1CONHR^2(R^1COOR^3) + ROH$$

$$\underset{R}{\overset{R}{\diagup}}P-OH \tag{1}$$

$$R-P(OR)_2 + R^1COOH + R^2NH_2 \ (R^3OH) \xrightarrow[Py]{} R^1CONHR^2(R^1COOR^3) + ROH$$

$$R-P(OH)(OR) \tag{2}$$

$$R-\overset{O}{\overset{\|}{P}}(OR)_2 + R^1COOH + R^2NH_2 \ (R^3OH) \xrightarrow[Py]{} R^1CONHR^2(R^1COOR^3) + ROH$$

$$R-\underset{\underset{O}{\|}}{P}(OR)(OH) \tag{3}$$

TABLE 1. Peptide Synthesis by Means of Oxidation of Diphenyl Phosphite with Mercuric Chloride in Pyridine[a]

Peptide	Yield (%)	Method[b]
Z-Gly-Gly.OEt	84	A
	92	B
	95	C
Z-Phe-Gly.OEt	90	B
Z-Gly-Tyr.OEt	90	B
Z-α-Glu-Gly.OEt	70	B
Z-Glu(NH₂)-Gly.OEt	79	C
Z-Met-Gly.OEt	93	C

[a]The coupling reaction was carried out at 45°C for 12 hr.
[b]Method A: Activation of carboxyl components.
Method B: Activation of amino components.
Method C: Activation and coupling reaction in the presence of both components.

These reactions with phosphites, phosphinites and phosphonites were employed for the preparation of peptides and active esters of amino acids in good yields (Tables 2 and 3).

TABLE 2. Peptide Synthesis via the Hydrolysis-Dehydration Reaction with Phosphites and Phosphonite in Pyridine[a]

| | Yield (%) of peptide | | |
Peptide	HO-P(OPh)$_2$	P(OPh)$_3$	Et-P(OPh)$_2$
Z-Gly-Gly.OEt	91	92	92
Z-Phe-Gly.OEt	90	85	90
Z-Gly-Tyr.OEt	88	96	86
Z-Glu(NH$_2$)-Gly.OEt	85	-	78
Z-Met-Gly.OEt	91	95	93

[a]The reaction was carried out at 40°C for 6 hr using diphenyl phosphite (1 equiv), and for 12 hr using triphenyl phosphite (0.5 equiv) and diphenyl ethylphosphonite (1 equiv).

TABLE 3. Preparation of Active Esters via the Hydrolysis-Dehydration Reaction with Phosphites and Phosphonite in Pyridine[a]

| | Yield (%) of active ester | | |
Active ester	HO-P(OPh)$_2$	P(OPh)$_3$	Et-P(OPh)$_2$
Z-Gly-O-⬡-NO$_2$	77	83	74
Z-Gly-O-⬡-COOCH$_3$	69	70	75
Z-Gly-S-⬡	89	64	74
Z-Phe-O-⬡-NO$_2$	73	73	71
Z-Glu(NH$_2$)-O-⬡-NO$_2$	45	48	-

[a]The reaction was carried out at 40°C for 12 hr using diphenyl phosphite (1 equiv), triphenyl phosphite (0.5 equiv), and diphenyl ethylphosphonite (1 equiv).

Considering that the chemical reactivity of carboxylic acids is similar to that of carbonic acid, as is observed in amide and ester formation, we have attempted the substitution of carbon dioxide for carboxylic acids in the coupling reaction with amines by using phosphite in pyridine or imidazole, and found that ureas are, in fact, produced in good yields (Eq. 4) [6]. Similarly, carbon disulfide reacted with amines to yield the thioureas (Eq. 5):

$$CO_2 + 2RNH_2 + HO-P(OPh)_2 \xrightarrow{Py} RNHCONHR$$

$$+ (HO)_2-P(OPh) + PhOH \quad (4)$$

$$CS_2 + 2RNH_2 + HO-P(OPh)_2 \xrightarrow{Py} RNHCSNHR$$

$$+ HO-P(OPh)(SH) + PhOH \quad (5)$$

Based on the stoichiometric involvement of phosphites and pyridine in the reaction, the reaction was proposed to proceed via a carbamyl N-phosphonium salt of pyridine (Scheme 3).

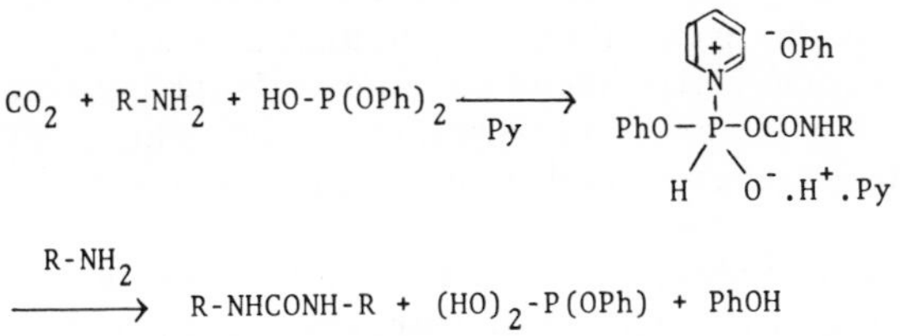

SCHEME 3.

In addition, we successfully applied the concept of the hydrolysis-dehydration reaction with phosphorus compounds to the reaction with sulfur compounds such as diaryl sulfites [7]:

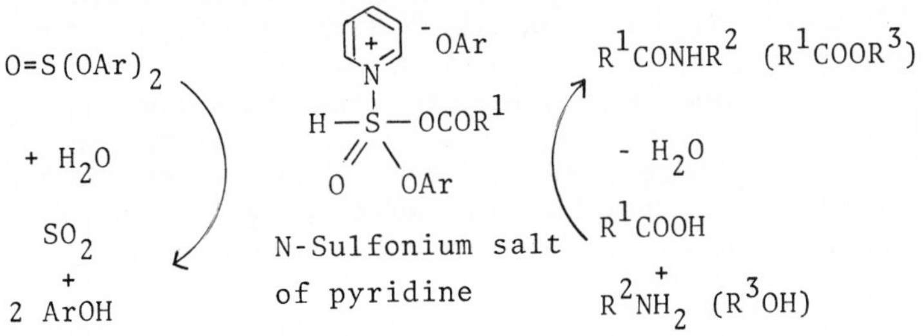

The Polycondensation Reactions

The reactions with phosphorus compounds were extended to the direct polycondensation reactions of dicarboxylic acids with diamines, of free α-amino acids or dipeptides, and of carbon dioxide and disulfide with diamines under mild conditions.

Polyamides

Direct polycondensations of aromatic diamines with dicarboxylic acids have generally been described as a poor route to high molecular linear polyamides. Recently, high molecular weight polyamides have been obtained with limited success by a melt polymerization of 4,4'-diaminodiphenylmethane with aliphatic dicarboxylic acids [8]. Surprisingly, by using the reactions via V and VI in Scheme 2, polyamides of high molecular weight were obtained directly from dicarboxylic acids and diamines in NMP solutions containing pyridine (Table 4) [9]. A combination of aromatic diamines with aliphatic dicarboxylic acids gave polymers of higher viscosity than an aliphatic diamine. On the other hand, 4,4'-diaminodiphenylsulfone yielded low viscous polymer, probably because of the lower basicity. Aromatic dicarboxylic acids, even with aromatic diamines and aromatic amino acids, did not form high viscous polymers. Isophthalic acid gave higher viscous polymer than terephthalic acid. This result led us to consider that higher solubility of polymer favors the polycondensation.

The difficulty of obtaining aromatic polyamides as shown above was overcome to a large extent by carrying out the polycondensation reaction in the presence of metal salts capable of improving the dissolution power of the polyamides [10].

As Table 5 shows, the addition of LiCl or $CaCl_2$ to the reaction mixture favored the polycondensation of p-aminobenzoic acid (p-ABA), giving poly-p-benzamide of high molecular weight in quantitative yield.

There are observed maxima of molecular weight of polymer at a concentration of about 4 wt% of LiCl or 8 wt% of $CaCl_2$ in the reaction mixture. Further addition retarded the reaction, and almost no polymer was obtained in the presence of more than 12 wt% LiCl or 20 wt% $CaCl_2$, where the reaction mixtures were deeply colored.

Considering that the presence of 2 wt% LiCl and 5 wt% $CaCl_2$, corresponding to an equivalent of phosphite and p-ABA, was very effective, the salts might participate in the reaction itself.

TABLE 4. Direct Synthesis of Polyamides by Using Phosphites and Phosphonite in NMP-Py Solution[a]

Dicarboxylic acid	Diamine	HO—P(OPh)$_2$		P(OPh)$_3$		Et—P(OPh)$_2$	
		Yield (%)	η_{inh}	Yield (%)	η_{inh}	Yield (%)	η_{inh}
HOOC-(CH$_2$)$_4$-COOH	NH$_2$-⬡-CH$_2$-⬡-NH$_2$	94	1.22	100	0.63	97	0.97
	NH$_2$-⬡-O-⬡-NH$_2$	100	0.95	100	0.74	97	1.65
	NH$_2$-⬡-SO$_2$-⬡-NH$_2$	86	0.28	78	0.18	62	0.22
	NH$_2$CH$_2$-⬡-CH$_2$NH$_2$	73	0.22	85	0.31	85	0.26
HOOC-(CH$_2$)$_8$-COOH	NH$_2$-⬡-CH$_2$-⬡-NH$_2$	97	1.45	100	1.07	96	1.12
	NH$_2$-⬡-O-⬡-NH$_2$	–	–	100	1.84	98	1.57
HOOC-⬡-COOH	NH$_2$-⬡-CH$_2$-⬡-NH$_2$	100	0.23	100	0.34	98	0.30
	NH$_2$-⬡-NH$_2$	100	0.12	–	–	–	–
HOOC-⬡(COOH)	NH$_2$-⬡-CH$_2$-⬡-NH$_2$	–	–	100	0.54	98	0.43
	NH$_2$-⬡-NH$_2$	84	0.26	–	–	–	–
	NH$_2$-⬡-COOH	98	0.16	100	0.22	97	0.21

[a][Monomer] = 0.25 mole/liter; [HO—P(OC$_6$H$_5$)$_2$] = [P(OC$_6$H$_5$)$_3$] = [C$_2$H$_5$—P(OC$_6$H$_5$)$_2$] = 1.0 mole/ mole of monomer; solvent = NMP/Py = 40/10(ml/ml); temperature = 100°C; time = 6 hr.

TABLE 5. Polycondensation Reaction of p-ABA in the Presence of Several Metal Salts[a]

	Polymer	
Metal salt	Yield (%)	η_{inh} [b]
LiCl	100	1.27
Li(AcAc)	0	-
CaCl$_2$	98	1.07
CaCl$_2$.2H$_2$O	71	0.04
KSCN	46	0.12
MgCl$_2$	100	0.31
ZnCl$_2$	97	0.20
None	100	0.22

[a][Monomer] = 0.4 mole/liter; [P(OC$_6$H$_5$)$_3$] = 1.0 mole/mole of monomer; [Metal salt] = 4 wt% in the solvent; solvent = NMP/Py = 40/10(ml/ml); temperature = 100°C; time = 6 hr.
[b]Measured in H$_2$SO$_4$ at 30°C.

Therefore, metal salts may contribute to the improvement of the dissolution power of the resulting polyamide and also to the depression of the side reaction owing to the formation of complexes between phenol derived from phosphite with metal salts, such as those of CaCl$_2$ with alcohols and phenols.

It was expected that the polycondensation reaction at high temperatures might favor the solubility of the resulting polymer, but be undesirable for the stability of the complexes of phenol with metal salts. As a consequence, an optimum of the reaction temperature might be observed in the polycondensation reaction.

An optimum of viscosity (η_{inh} = 1.71) was observed at a reaction temperature at around 80°C in the polycondensation of p-ABA. Above this temperature the viscosity decreased gradually with the temperature. Only low-viscosity polymer was obtained at a temperature of 60°C.

Solvents and the amount of pyridine in the NMP-pyridine mixed solvent affected the polycondensation of p-ABA in the presence of 4 wt% LiCl. Of the solvents tested, NMP was most effective and DMAc, in which the reaction mixture became light yellow, gave moderate results, whereas DMF largely retarded the reaction, probably because of a side reaction of DMF with LiCl at high temperatures, as indicated by deep colorization of the reaction mixture.

The viscosity of polymer varied with the amount of pyridine in the NMP-pyridine mixed solvent, showing the highest value in the solvent of relatively high pyridine content (40%) in spite of unfavorable results in pyridine alone. This result suggests that the solvent of this composition has a strong solvating power, as a combination of NMP and HMPA containing LiCl, each of which could not dissolve poly-p-benzamide, was a very powerful solvent.

Several wholly aromatic polyamides were prepared by using triphenyl phosphite in NMP-pyridine solution containing 4 wt% LiCl (Table 6). The combination of isophthalic acid with diamines gave a polymer of high viscosity, whereas terephthalic acid with pK_a values similar to those of isophthalic acid did not give high-viscosity polymers. The unfavorable results from terephthalic acid may be due to the lower solubility of polymers with a rigid structure. m-ABA of lower acidity than p-ABA did not yield a polymer with a high viscosity, although higher solubility was expected from polymer from m-ABA with a flexible structure.

Polypeptides

Though the preparation of polypeptides directly from free amino acids is very difficult because of their tendency to give cyclic dimers (diketopiperazines) by ordinary methods [11], we have succeeded in obtaining linear polypeptides with relatively high molecular weight by the direct polycondensation of α-amino acids through the use of diphenyl and triaryl phosphites in pyridine. We have also obtained polypeptides with ordered sequences by the indirect polycondensation of activated derivatives of peptides, such as their active esters by ordinary methods, directly from unactivated dipeptides (Table 7) [12].

The polycondensation of amino acids was affected significantly by the solvent. Interestingly, nonpolar solvents (such as n-hexane) and haloalkanes (such as chloroform) gave polymers of relatively higher viscosity than highly polar aprotic solvents (such as DMF) in spite of heterogeneity of the system in these nonpolar solvents (Table 8).

TABLE 6. Preparation of Aromatic Polyamides by Means of Triphenyl Phosphite in NMP-Pyridine Solution Containing 4 wt% LiCl[a]

Dicarboxylic acid[b]	Diamine	Polymer η_{inh}[c]
HOOC-⬡-COOH (3.72, 4.40)	NH$_2$-⬡-NH$_2$	1.14
	NH$_2$-⬡-O-⬡-NH$_2$	1.34
	NH$_2$-⬡-CH$_2$-⬡-NH$_2$	0.93
HOOC-⬡-COOH (3.54, 4.46)	NH$_2$-⬡-NH$_2$	0.19 (0.21)[d]
	NH$_2$-⬡-O-⬡-NH$_2$	0.32
	NH$_2$-⬡-CH$_2$-⬡-NH$_2$	0.33
	NH$_2$-⬡-COOH (3.07, 4.70)	1.32
	NH$_2$-⬡-COOH (2.28, 4.89)	0.43

[a][Monomer] = 0.6 mole/liter; [P(OC$_6$H$_5$)$_3$] = 1.0 mole/mole of monomer; solvent = NMP/Py = 20/15(ml/ml); temperature = 100°C; time = 3 hr.

[b]Values in parentheses are pK$_1$ and pK$_2$.

[c]Polymers were obtained in quantitative yields, and the viscosity was measured in H$_2$SO$_4$ at 30°C.

[d]CaCl$_2$ (8 wt%) was used in lieu of LiCl.

Polyureas and Polythioureas

The results of the preparation of polyureas under mild conditions (a pressure of less than 40 atm of carbon dioxide and a temperature around 40°C) and of polythioureas (using diphenyl phosphite in pyridine) are given in Table 9 [13]. Although the preparation of

TABLE 7. Direct Polycondensation of Amino Acids and Dipeptides by Using Phosphites and Phosphonite in Pyridine[a]

Amino acid and peptide	HO—P(OPh)$_2$		P(OPh)$_3$		Et—P(OPh)$_2$	
	Yield (%)	η_{inh}	Yield (%)	η_{inh}	Yield (%)	η_{inh}
Glycine	77	0.13	70	0.15	70	0.17
L-Alanine	45	0.12	100	0.14	73	0.18
L-Leucine	75	0.12	68	0.17	60	0.25
L-Phenylalanine	72	0.09	90	0.08	46	0.11
Glycylglycine	76	0.12	100	0.19	-	-
Glycyl-L-leucine	29	0.14	-		-	-

a[Monomer] = 1.25 mole/liter; [HO—P(OC$_6$H$_5$)$_2$] = 1.5 mole/mole of monomer; [P(OC$_6$H$_5$)$_3$] = [C$_2$H$_5$—P(OC$_6$H$_5$)$_2$] = 1.0 mole/mole of monomer; temperature = 40°C; time = 18 hr.

TABLE 8. Polycondensation of L-Leucine in Various Solvents[a]

Solvent	Yield (%)	η_{inh} [b]
n-Hexane	79	0.24
Dichloroethane	87	0.17
Chloroform	82	0.17
Benzene	78	0.16
Acetonitrile	82	0.16
Dioxane	87	0.13
Dimethoxyethane	90	0.13
Diisobutyl ketone	83	0.11
DMF	70	0.11
DMAc	70	0.11
NMP	65	0.10

[a][L-Leu] = 1.25 mole/liter; [Py] = 2.0 mole/mole of the phosphite; [HO—P(OC$_6$H$_5$)$_2$] = 1.5 mole/mole of monomer; temperature = 40°C; time = 18 hr.
[b]Measured in dichloroacetic acid at 30°C.

polyureas from carbon dioxide under drastic conditions (high temperatures and high pressures) or from carbon oxysulfide has been reported, neither preparative methods are operative under moderate conditions, nor have the methods for the synthesis of polythioureas from carbon disulfide been described.

As Table 9 shows, aromatic diamines, from which polymers with good solubility in pyridine were formed, gave polymers of higher molecular weight, whereas polymers from 4,4'-diaminodiphenyl-sulfone and p-phenylenediamine were insoluble even in HMPA and showed low viscosity in sulfuric acid. On the other hand, an aliphatic diamine with high basicity afforded a polymer of low

TABLE 9. Direct Polycondensation of Carbon Dioxide and Disulfide with Diamines by Using Diphenyl Phosphite in Pyridine[a]

Diamine	Polyurea				Polythiourea	
	Ordinary pressure		20 atm			
	Yield (%)	η_{inh}[b]	Yield (%)	η_{inh}[b]	Yield (%)	η_{inh}[d]
NH$_2$-⬡-CH$_2$-⬡-NH$_2$	100	0.32	100	2.24	100	0.18
NH$_2$-⬡-O-⬡-NH$_2$	98	0.22	100	0.51	100	0.25
[NH$_2$-⬡-O-⬡]$_2$ C=(CH$_3$)$_2$	95	0.44	100	1.87	78	0.15
NH$_2$-⬡-SO$_2$-⬡-NH$_2$	13	0.09[c]	86	0.14[c]	17	0.07
NH$_2$-⬡-NH$_2$	100	0.08[c]	100	0.09[c]	100	0.12[c]
NH$_2$CH$_2$-⬡-CH$_2$NH$_2$	33	0.08	46	0.13	0	-

[a][Monomer] = 0.26 mole/liter; [HO—P(OC$_6$H$_5$)$_2$] = 2.0 mole/mole of monomer; temperature = 40°C; time = 4 hr with CO$_2$ and 6 hr with CS$_2$.
[b]Measured in HMPA at 30°C.
[c]Measured in H$_2$SO$_4$ at 30°C.
[d]Measured in DMSO at 30°C.

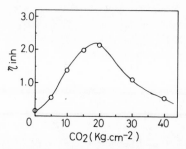

FIG. 1. Effect of the pressure of carbon dioxide upon the viscosity of polyurea from 4,4'-diaminodiphenylmethane at 40°C for 4 hr.

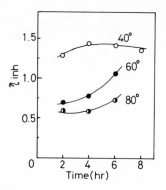

FIG. 2. Effect of the reaction temperature and time upon the viscosity of polyurea from carbon dioxide (10 atm) and 4,4'-diamino-diphenylmethane.

viscosity in low yield because of retardation by the formation of pyridine-insoluble and unreactive ammonium carbamate. These were also the results in the case of polythioureas.

The initial pressure of carbon dioxide and the reaction temperature affected the molecular weight of the resulting polyurea, a viscosity maxima being given by the reaction at around 40°C under a pressure of 20 atm of carbon dioxide. Above this pressure, the viscosity decreased with the pressure, dropping at 40 atm to one-fifth of that at 20 atm (see Figs. 1 and 2).

The unfavorable effects of both high temperature and high pressure upon the molecular weight may be caused by a depolymerization reaction or by an intermolecular or intramolecular exchange reaction between polymers, as observed in the Eqs. (6) and (7). These side reactions may be promoted by higher reaction temperatures and higher pressures of carbon dioxide, and also in the presence of carbon dioxide and/or diphenyl phosphite.

$$NH_2-R-NH_2 + NH_2CONH_2 \longrightarrow \frac{1}{n} \left(NH-R-NH-CO\right)_n + 2NH_3 \quad (6) \; [14]$$

$$R^1CONHR^2 + R^3NH_2 \xrightarrow[CO_2]{} R^1CONHR^3 + R^2NH_2 \quad (7) \; [15]$$

REFERENCES

[1] E. Baldwin, Dynamic Aspects of Biochemistry, Cambridge Univ. Press, New York, 1957.

[2] N. Yamazaki and F. Higashi, Bull. Chem. Soc. Japan, 46, 1235, 1239(1973); Tetrahedron Lett., 1972, 415; Synthesis, 1974, 436.

[3] N. Yamazaki and F. Higashi, Bull. Chem. Soc. Japan, 46, 3824(1974); 47, 170(1974); Synthesis, 1974, 495.

[4] N. Yamazaki and F. Higashi, Tetrahedron, 30, 1323(1974).

[5] N. Yamazaki, M. Niwano, J. Kawabata, and F. Higashi, Tetrahedron, 31, 665 (1975).

[6] N. Yamazaki, F. Higashi, and T. Iguchi, Tetrahedron Lett., 1974, 1191.

[7] N. Yamazaki, F. Higashi and M. Niwano, Tetrahedron, 30, 1319(1974).

[8] D. A. Holmer, O. A. Pickett, Jr., and J. H. Saunders, J. Polym. Sci., A-1, 10, 1547(1972).

[9] N. Yamazaki and F. Higashi, J. Polym. Sci., Polym. Lett. Ed., 12, 185(1974).

[10] N. Yamazaki, M. Matsumoto, and F. Higashi, Presented at the 23rd Symposium on Macromolecules of the Society of Polymer, Japan, October 1974.

[11] C. H. Bamford, A. Elliot, and W. E. Hanby, Synthetic Polypeptides, Academic, New York, 1956, p. 63.

[12] N. Yamazaki, F. Higashi, and J. Kawabata, Makromol. Chem., 175, 1825(1974).

[13] N. Yamazaki, F. Higashi, and T. Iguchi, J. Polym. Sci., Polym. Lett. Ed., 12, 517 (1974).

[14] H. Iijima, M. Asakura, and K. Kimoto, Kogyo Kagaku Zasshi, 68, 240(1965).

[15] Y. Otuji, N. Matsumura, and E. Imoto, Bull. Chem. Soc. Japan, 41, 1485(1968).

Carbamate Ions as Propagating Species in N-Carboxy Anhydride Polymerizations

CHARLES E. SEENEY and H. JAMES HARWOOD

Department of Polymer Science
The University of Akron
Akron, Ohio 44325

ABSTRACT

Polymerizations of N-carboxy anhydrides of L-phenylalanine, γ-benzyl-L-glutamate, O-carbobenzoxy-L-tyrosine, L-leucine, and sarcosine, as initiated by primary, secondary, and tertiary amines in N_2 or CO_2 atmospheres and in the presence or absence of NaH, indicate that they proceed via carbamate salt intermediates. This conclusion is supported by radiotracer studies as well as by NMR studies of the initial products of NCA-amine reaction mixtures.

The "activated monomer" mechanism of strong-base initiated polymerizations is discounted on the bases that polypeptides are not formed in aprotic tertiary amine-initiated systems (hydantoins and diketopiperazines are obtained instead) and that methoxyl end groups are detected in polypeptides initiated with ^{14}C-labeled $NaOCH_3$.

INTRODUCTION

There are many interesting aspects of the polymerization of N-carboxyl anhydrides [1-4], most of which are incompletely

understood. These include the mechanisms of stereoselection and
stereoelection [5-10], the influence of α-helical conformations [11]
of growing chains on the reaction, the influence of monomer adsorp-
tion by polymer [12-15] and the mechanism of the reaction.

The polymerization of N-carboxy anhydrides (NCA's) is believed
to proceed by either or two mechanisms, depending on whether the
reaction is initiated by weak or strong bases. Polymerizations
initiated by primary amines and some secondary amines are
believed to proceed by the addition of NCA's to chains containing
terminal amine units. Polymerizations initiated by strong bases,
such as tertiary amines, secondary amines, and $NaOCH_3$, are
thought to involve the addition of anions derived from NCA's to
chain ends containing N-acyl-NCA moieties, viz.:

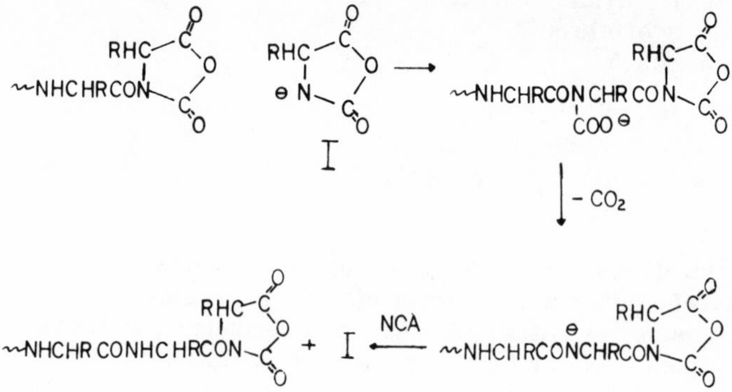

Evidence for the participation of species such as I in strong
base-initiated polymerizations is provided by the relative activities
of substituted pyridines as initiators [16], by the absence or low
amounts of strong base initiator fragments bonded to polypeptides
[17-19], by increases in molecular weight that occur when polymerization mixtures are concentrated [19-21], by the failure (in
some cases) of tertiary amines to initiate the polymerization of
N-substituted NCA's [16], and by the formation of 6-oxo-L-pipecolic
acid when the N-carboxy anhydride of δ-benzyl L-α-aminoadipate is
polymerized [22]. The "activated monomer" mechanism has been
the basis of a method for preparing block copolymers [23] and
silyl analogs of "activated monomers" have been described [24].

None of the evidence discussed above is unequivocal, unfortunately, and theoretical objections can be raised against certain

aspects of the "activated monomer" mechanism. The dimer II, which is generally proposed as the species first formed in such polymerizations, should be very prone to yield diketopiperazines (III) or hydantoin acetic acid (IV) derivatives. The behavior of

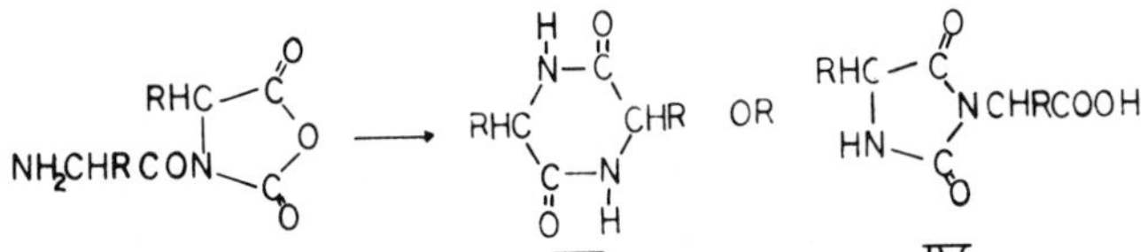

$$\text{II} \qquad \text{III} \qquad \text{IV}$$

substituted pyridines reflects their reactivity toward proton donors; these need not necessarily be NCA's. It is conceivable that species such as I can be formed by strong bases and that they could initiate polymerization, but reaction of I with an NCA should yield a carbamate ion which should not be particularly prone to react with NCA to yield I. The participation of "activated monomers" in initiation and cyclization reactions does not prove that they are participants in propagation reactions. There are even suggestions that "activated monomers" and their analogs are capable of rearranging to iso-cyantes, which subsequently yield diketopiperazines and hydantoins [24, 25].

From time to time, carbamate ions have also been suggested as being the propagating species in NCA polymerizations [26-30], viz.:

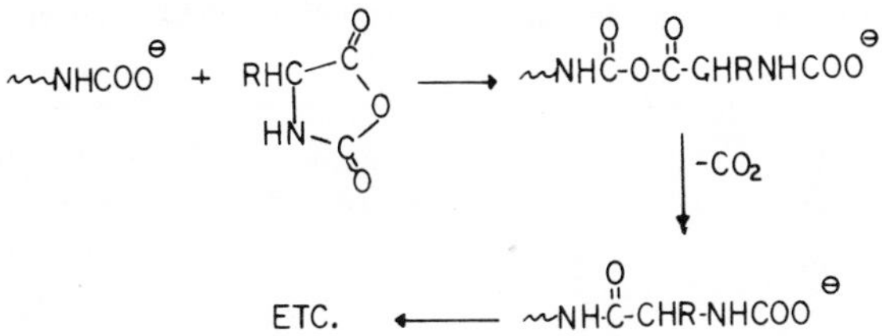

Mechanisms involving propagating carbamate ions were abandoned when radiotracer studies failed to detect initiator fragments in polymers initiated by carbamate salts, metal alcoholates, and some hindered amines [17-19]. Zilkha and co-workers [31, 32] showed, however, that alkoxide ions do add to NCA's, and it

would seem that the successful synthesis of depsipeptides [33]
would also require this. NCA polymerizations initiated by recently
developed catalysts are easily explained in terms of carbamate
intermediates [9, 34-38], and related polymerizations also seem
to involve such species [39]. The influence of CO_2 on NCA
polymerizations also suggests that carbamate ions are important
in such polymerizations.

The purpose of the present paper is to present results that lead
us to believe that both weak and strong base-initiated polymerizations
of N-carboxy anhydrides proceed by a common mechanism in which
the propagating species are carbamate ions rather than free amines
or N-acyl-NCA moieties.

EXPERIMENTAL

N-Carboxy Anhydrides

NCA's were prepared by adding liquid phosgene to amino acid
suspensions in dioxane at 60°C until clear, homogeneous solutions
were obtained. A slight excess of phosgene was added and the
solution was then flushed to remove excess phosgene and HCl. The
dioxane was then removed in vacuo at 60°C. The viscous residue
was taken up in an equal volume of ethyl acetate, and this solvent
was also removed in vacuo at 60°C. The solid residue was recrystallized repeatedly from ethyl acetate-hexane until the chloride content
[40] of the product was below 0.01% (0.05% in one instance). The
crystals were dried in vacuo at 40°C. The pure NCA's were stored
in evacuated containers at -30°C. The following melting points were
obtained for the NCA's prepared: O-carbobenzoxy-L-tyrosine,
104°C dec.; γ-benzyl-L-glutamate, 95°C; L-leucine 75°C dec.;
L-phenylalanine, 90°C dec.; L-sarcosine, 99 to 100°C dec.

Solvents

Dioxane and benzene were stored over 4A Molecular Sieves for
several weeks and then refluxed and distilled over NaH. Ethyl
acetate and hexane were dried over Molecular Sieves.

Initiators

Amines were stored over BaO and were distilled prior to use.
Solutions of amines in dioxane or benzene were stored under nitrogen

and over 4A Molecular Sieves. The sodium salts of amines were prepared by treating purified amines with NaH, followed by removal of excess amine in vacuo and washing of the salt with dry benzene. Sodium methoxide solutions were prepared by adding freshly cut sodium to anhydrous methanol (1 g/50 ml), followed by dilution of the resulting solution with fresh methanol and dry benzene. The resulting solution was stored under dry nitrogen.

Polymerizations

Polymerization experiments were conducted in glassware that had been heated at 125°C for 24 hr and that was assembled while being purged in a stream of dry nitrogen. The nitrogen was passed through 4A Molecular Sieves prior to use. Solutions of monomers in dioxane (2.0 to 2.5 wt%) were prepared under nitrogen using a glove bag and were charged with initiator. Reactions were allowed to proceed at ambient temperature under nitrogen. The course of reaction was monitored by IR spectroscopy. Polymerization mixtures were poured into ether to isolate the polymers. These were reprecipitated several times from dioxane into ether. When ^{14}C measurements were to be made, the polymers were reprecipitated as many as 10 times. Intrinsic viscosities of the polymers in dichloroacetic acid were determined using a wide bore Ubbelhode-type viscometer.

Radioactivity Measurements

Initiator and polymer activities were determined by scintillation counting using a Beckman β-Mate instrument and a dioxane-POP-POPOP cocktail. NSC Solubilizer was used when necessary to obtain homogeneous solutions. The ESR method was used to determine counting efficiency.

Nuclear Magnetic Resonance Studies

NMR studies of amine-NCA reaction mixtures in $CDCl_3$ were conducted with a Varian T-60 spectrometer. The general procedure was to add NCA to the amine solution and then record the spectrum of the mixture after 20 min. Polymer spectra were recorded using CF_3COOH as solvent. The ^{13}C-spectrum was obtained using a Varian CFT-20 spectrometer, courtesy of Dr. John Rieger, Varian Associates. This spectrum was recorded using CO_2-saturated $CDCl_3$ as a solvent and with an atmosphere of CO_2 above the

solution. (When air was allowed to enter the tube, a precipitate
began to form. This was probably the diketopiperazine derivative.)

RESULTS AND DISCUSSION

Our preference for a mechanism proceeding by carbamate ion
additions to NCA's is based on the following evidence.

1. PMR and CMR studies on the initial products of diethylamine-
NCA reaction mixtures indicate that carbamate salts are formed
and that they are reasonably stable in solution.

2. Tertiary amines fail to initiate NCA polymerizations in
carefully purified systems, but polymerization occurs immediately
upon the addition of small amounts of methanol or water. Similar
results have been obtained by others [41, 42].

3. Polymerizations initiated by 1°, 2°, or 3° amines at NCA/amine
ratios greater than 100 tend to terminate at low conversion when
conducted in N_2-swept systems, and large amounts of diketopiperazines and hydantoin derivatives are formed. However, polypeptides
are obtained in high yield in CO_2 saturated systems.

4. Addition of NaH to amine-initiated polymerization systems
causes an immediate rate enhancement and the formation of high
molecular weight polymers.

5. Radiotracer and spectroscopic studies indicate that methoxyl
groups are present in polypeptides obtained from $NaOCH_3$-initiated
polymerization reactions, in contrast to earlier reports [17, 18].

PMR and CMR Studies of Amine-NCA Reaction Mixtures

Figure 1 shows the 60 MHz PMR spectrum of L-phenylalanine
NCA and the spectra of reaction products derived from this NCA and
diethylamine. The spectra of the NCA-amine reaction products were
obtained by first recording the spectrum of the product derived from
a 3:1 amine/NCA mixture, followed by adding additional NCA to
obtain various amine/NCA ratios, and then recording their spectra.
The presence of two types of ethyl groups is clearly evident in the
spectra of the 1:3, 2:3, and 1:1 reaction products; these are present
in approximately equal amounts in the spectra of the 2:3 and 1:1
mixtures. Resonance is observed at 5.4 ppm in the spectra of all
the mixtures, and its intensity is approximately the same as that

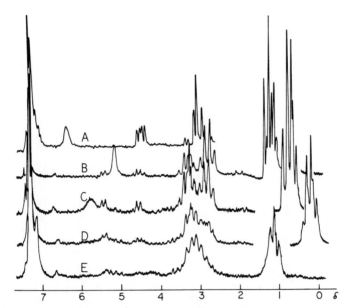

FIG. 1. 60 MHz PMR spectra of L-phenylalanine NCA—
diethylamine reaction mixtures. Anhydride/amine ratios are:
A, pure NCA; B, 1/3; C, 2/3; D, 1/1; and E, 2/1.

of methine proton resonance at ~4.5 ppm in the spectra of the 2:3
and 1:1 mixtures. We attribute this resonance to the NH proton
of carbamate salt end groups.

Figure 2 shows that the same results are obtained when
γ-benzyl-L-glutamate NCA-diethylamine reaction mixtures are
studied. Two types of ethyl groups, present in equal amounts,
are clearly indicated in the spectrum of the 2:1 reaction mixture.
One of these is lost when the reaction mixture is worked up and
the product is recrystallized. Also lost during workup is the
structure responsible for the resonance observed at 5.6 ppm.

These results are most easily explained if carbamate salts
result from the reactions of diethylamine with NCA's. Thus
reaction of an NCA with two molecules of diethylamine should
yield

$$Et_2N-\overset{\overset{\textstyle O}{\|}}{C}-CHR-NH-COO^{\ominus} Et_2NH_2^{\oplus}$$

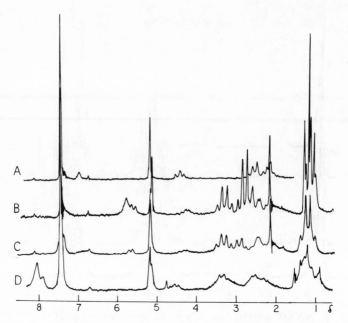

FIG. 2. 60 MHz PMR spectra of γ-benzyl-L-glutamate
NCA—diethylamine reaction mixtures. Anhydride/amine ratios
are: A, pure NCA; B, 1/1; C, 2/1; D, product isolated from C.

This structure contains two types of diethylamino moieties.
Subsequent reaction of this structure with additional NCA should
retain the integrity of these moieties:

$$Et_2N(COCHRNH)_n COCHRNHCOO^\ominus Et_2NH_2^\oplus$$

However, acidification of the carbamate salt during workup should
result in a product that would not contain the diethylamino moiety
responsible for resonance at ~2.9 and 1.2 ppm, but only the one
responsible for resonance at ~3.4 and 1.3 ppm. In addition, the
product would not resonate at ~5.6 ppm.

Additional support for the presence of carbamate salts in these
reaction mixtures was obtained by treating a 1:1 phenylalanine/
Et_2NH reaction mixture with benzyl chloride followed by isolation
of the reaction product and examination of its PMR spectrum.
The presence of the benzyl carbamate group is indicated by
resonances at 4.8, 5.3, and 7.6 ppm. The relative area of the
resonance due to phenyl protons of the phenylalanine unit is twice

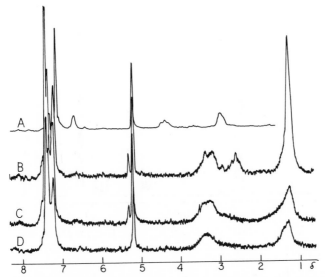

FIG. 3. 60 MHz PMR spectra of O-carbobenzoxy-L-tyrosine NCA and of reaction products in hexamethylenediamine. Anhydride/diamine ratios are: A, pure NCA; B, 1/1; C, 2/1; and D, 3/1.

that due to phenyl protons of the benzyl group, as would be expected for product derived from a 1:1 NCA/amine reaction mixture.

Figure 3 shows the 60 MHz PMR spectra of reaction products obtained from O-carbobenzoxy-L-tyrosine NCA and hexamethylene diamine. The resonance observed at 5.3 ppm decreases in relative intensity as the anhydride-amine ratio increases. This resonance is also attributed to a carbamate salt.

Figure 4 shows the 20 MHz CMR spectrum of a 1:1 phenylalanine NCA/Et$_2$NH reaction mixture. Its most significant features are two intense methyl carbon resonances (δc ~15 ppm), eight methylene carbon resonances (38 to 45 δc), two of which are very intense (attributed to methylene carbons of Et$_2$N moieties), eight aromatic carbon resonances (125 to 135 δc), several C-1 aromatic proton resonances at ~141 δc in addition to a resonance that may be due to a carbonyl carbon, and two additional carbonyl carbon resonances at 158 and 180 δc. The spectrum is consistent with the following average structure:

$$(CH_3CH_2)_2NCOCH-NH-CO-CH-COO^{\ominus}\overset{\oplus}{N}H_2(CH_2CH_3)_2$$
$$\underset{CH_2\phi}{|} \qquad \underset{CH_2\,\phi}{|}$$

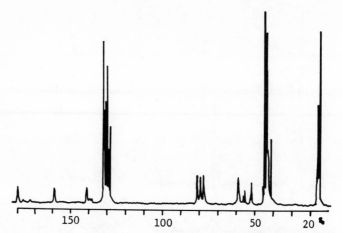

FIG. 4. 20 MHz CMR spectrum of a 1:1 L-phenylalanine—
diethylamine reaction mixture (CDCl₃ saturated with CO_2).

PMR studies on tertiary amine/NCA reaction mixtures were not
very instructive, except in the negative sense that no evidence
for the presence of "activated monomers" was obtained.

The Role of Carbon Dioxide

Other workers have noted that carbon dioxide enhances the
rates of NCA polymerizations, and it has been suggested that
this is due to the formation of acidic species that function as
catalysts. If propagation is via carbamate ions, an alternative
explanation is that CO_2 diminishes the tendency of carbamate
salts to decompose. The following equilibria can be expected
to occur in amine-initiated polymerizations:

$$\sim NHCOO^{\ominus} \ \overset{\oplus}{NH_2R_2} \rightleftharpoons \ \sim NHCOOH + R_2NH \rightleftharpoons \ \sim NH_2 + R_2NH + CO_2$$

A high concentration of CO_2 should force the equilibrium to the
left and thereby favor carbamate ion stability. This should lead
to enhanced rates, higher molecular weight polymers, and lower
yields of hydantoin and diketopiperazine byproducts. Such

by-products may result from reactions of terminal amine units with neighboring carbamoyl anhydride structures, viz.:

$$\sim NH\overset{O}{\overset{\|}{C}}\text{-}O\text{-}\overset{O}{\overset{\|}{C}}\text{-}CHR\text{-}NH\overset{O}{\overset{\|}{C}}\text{-}CHRNH_2 \longrightarrow \sim NHCOOH + \text{RHC}\overset{\overset{O}{\overset{\|}{C}-N}\overset{H}{}}{\underset{N-C}{\underset{H\ \ O}{}}}\text{CHR}$$

Our studies have shown that L-phenylalanine NCA shows a strong tendency to yield diketopiperazine and hydantoin derivatives in reactions initiated with Et_3N or Et_2NH in $CHCl_3$ or dioxane at room temperature when CO_2 is rapidly swept out of the reaction mixture, but that polypeptides are the principal products in CO_2 saturated systems.

Formation of a Derivative of the Carbamate Salt

To obtain chemical evidence for the presence of carbamate ions in NCA-secondary amine reaction mixtures, L-phenylalanine NCA was slowly added to a solution of diethylamine in dioxane at room temperature until the NCA/amine ratio was unity. The reaction mixture was then treated with benzyl chloride and an exothermic reaction was noted. The product isolated from this reaction mixture was washed several times with methanol. Its PMR spectrum is shown in Fig. 5. The presence of a benzyl ester is indicated by resonances at 7.5 and 4.6 δ. The relative intensities of the aromatic proton resonances due to the benzyl (7.5 δ) and phenyl (7.2 δ) groups are roughly 1:2, in keeping with the following overall structure:

$$\emptyset CH_2O\text{-}\overset{O}{\overset{\|}{C}}NH\overset{CH_2}{\underset{\emptyset}{\overset{|}{C}H}}\text{-}\overset{O}{\overset{\|}{C}}\text{-}NH\overset{CH_2}{\underset{\emptyset}{\overset{|}{C}H}}\text{-}\overset{O}{\overset{\|}{C}}\text{-}NEt_2$$

The Use of NaH in Initiator Systems

When we became convinced that NCA polymerizations proceed via carbamate salt intermediates, efforts were made to improve

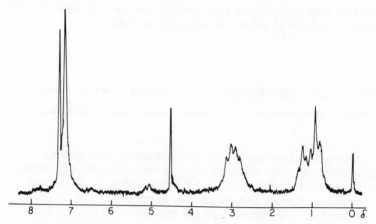

FIG. 5. 60 MHz PMR spectrum of the benzyl carbamate derived
from the reaction product of L-phenylalanine NCA and diethylamine.

the stability of such species in the polymerization systems. One
approach to accomplish this was to provide an aprotic counterion
for the carbamate ion. Sodium hydride was therefore added to various polymerization mixtures; the effects obtained were spectacular.
For example, when the polymerization of O-cbzo-L-tyrosine NCA
was initiated with hexamethylene diamine, a slow reaction yielding
mostly the diketopiperazine and low molecular weight polymers
ensued and terminated after 50% conversion. In contrast, a high
molecular weight polymer (DP $\sim$ 200) was obtained under the same
conditions when a product derived from the diamine and NaH was
used for initiation.

This behavior is also noted in tertiary amine-initiated reactions. Figure 6 (Curve A) shows the rate of disappearance of
O-carbobenzoxy-L-tyrosine NCA when initiated by Et_3N (anhydride/
amine = 100) in dioxane at room temperature. The reaction occurs
slowly, yields low molecular weight products, and terminates before
conversion is complete. However, addition of sodium hydride
(M/NaH = 2 to 5, although all NaH is probably not utilized) to the
reaction mixture after the limiting conversion has been obtained
causes a rapid reaction (Curve B) that yields high molecular weight
polymers, $[\eta]$ = 1.9. Identical behavior is observed in γ-benzyl-L-
glutamate NCA-triethylamine reactions (Fig. 7). It should be noted
that NaH alone does not initiate the polymerizations of NCA's. We
believe that products derived from NCA-tertiary amine reactions
react with NaH to yield the initiating species.

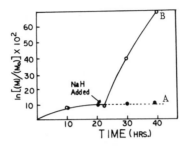

FIG. 6. Effect of adding NaH to a O-carbobenzoxy-L-tyrosine
NCA-triethylamine reaction mixture.

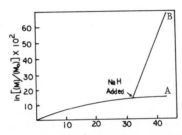

FIG. 7. Effect of adding NaH to a γ-benzyl-L-glutamate
NCA—triethylamine reaction mixture. The abscissa is time
in hours.

Detection of Methoxyl Groups in Polymers Initiated with NaOCH₃

The results obtained in the radiotracer studies conducted by
Goodman and Peggion [17-19] are inconsistent with an NCA
polymerization mechanism that would involve propagating carbamate
ions. It seemed important, therefore, to reinvestigate this aspect of
the reaction. Polymerizations of γ-benzyl-L-glutamate NCA,
sarcosine NCA, and L-leucine NCA in dioxane (first two) and DMSO
(leucine) were initiated by ^{14}C-labeled sodium methoxide-methanol
solutions [2.54×10^9 d/(min)(mole)]. The copolymerization of
O-carbobenzoxy-L-tyrosine NCA with γ-benzyl-L-glutamate in
dioxane was also initiated by the methoxide reagent. The polymers
obtained were carefully purified and were counted. The NMR
spectra of the polysarcosine (Figure 8) and polyleucine (Figure 9)
were also recorded. These clearly indicate the presence of

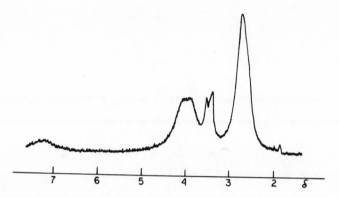

FIG. 8. 60 MHz PMR spectrum of poly(L-sarcosine) obtained
from $NaO^{14}CH_3$-$HO^{14}CH_3$ initiated polymerization in dioxane.
Solvent: CF_3COOH.

methoxyl protons at ~3.5δ. All the polymers were found to contain
appreciable quantities of methoxyl groups.

The polymers containing γ-benzyl-L-glutamate contained approximately one CH_3O group per molecule based on their radioactivity and
their molecular weights as estimated by intrinsic viscosity measurements. [Homopolymer: $\overline{M}_V$ ~ 91,000, activity = 1.98×10^4 d/(min)(g).
Copolymer: $\overline{M}_V$ ~ 108,000, activity = 1.24×10^4 d/(min)(g)]. However,
a control experiment in which inactive poly(γ-benzyl-L-glutamate)
was allowed to stand for 24 hr in the presence of $NaOCH_3$-CH_3OH
reagent yielded a polymer having an activity of 0.63×10^4 d/(min)(g).
It is apparent that transesterification of $NaOCH_3$ with benzyl ester
units on the polymer occurs to a significant extent and that experiments of this type are best done with NCA's that are devoid of
functional side groups. Low molecular weight (~600) polysarcosine
was obtained. Its radioactivity [1.18×10^5 d/(min)(g)] was lower
than expected for one CH_3O unit to be present per molecule, but
the NMR spectrum of the polymer clearly indicated the presence
of CH_3O-type protons.

The molecular weight of the polyleucine sample prepared has
not been determined, but its specific activity was 2.1×10^4 d/(min)(g).

Additional work needs to be done in this area, but there seems to
be little doubt that methoxide ion is capable of adding to NCA's to
generate carbamate or carboxylate ions, which can be the sites of
subsequent propagation.

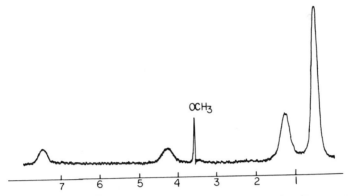

FIG. 9. 60 MHz PMR spectrum of poly(L-leucine) obtained from $NaO^{14}CH_3$-$HO^{14}CH_3$ initiated polymerization in DMSO. Solvent: CF_3COOH. The abscissa is in δ.

SUMMARY

On the basis of results presented here and after a careful reconsideration of previously reported results concerning NCA polymerizations, we believe that such polymerizations proceed via propagating carbamate ions. We suggest that if "activated monomers" are present in such polymerization systems, they are responsible for the formation of hydantoin and diketopiperazine derivatives.

High molecular weight polymers are believed to be obtained from tertiary amine-initiated polymerizations of NCA's simply because small amounts of protonic cocatalysts are present in systems that yield polymers and because tertiary amine salts of carbamic acids are relatively stable. In this connection, it is interesting to note that Mori and Iwatsuki [43] have recently found that combinations of tertiary amines with primary and secondary amines, mercaptans, and alcohols are very effective initiation systems for NCA polymerizations. Low molecular weight polymers are believed to result from primary amine-initiated polymerizations because of the tendency of salts of carbamic acids with such amines to decompose, yielding CO_2 and primary amines. The latter are more likely to participate in termination reactions than carbamate ions. Of course, metal salts of carbamic acids are very stable, and polymerizations involving such species are likely to be very successful. The considerable current activity along these lines indicates that others have also reached this conclusion.

ACKNOWLEDGMENT

The authors are grateful to the National Science Foundation for supporting this investigation.

REFERENCES

[1]   M. Szwarc, Adv. Polym. Sci., 4, 1 (1965).
[2]   V. V. Korshak, S. V. Rogozhin, V. A. Davankov, Y. A. Davidovich, and T. A. Makarova, Russ. Chem. Rev., 34, 329 (1965).
[3]   C. H. Bamford and H. Block, Poly(amino Acids), Polypeptides and Proteins (M. A. Stahmann, ed.), Univ. Wisconsin Press, Madison, 1962, p. 65.
[4]   C. H. Bamford, A. Elliott, and W. E. Hanby, Synthetic Polypeptides, Academic, New York, 1956.
[5]   T. Tsuruta, J. Polym. Sci., Macromol. Rev., 6, 179 (1972).
[6]   H. G. Buhrer, Chimia, 26, 501 (1972).
[7]   H. Buhrer and H. G. Elias, Makromol. Chem., 169, 145 (1973).
[8]   T. Akaike, T. Makino, S. Inoue, and T. Tsuruta, Biopolymers, 13, 129 (1974).
[9]   S. Yamashita, K. Waki, N. Yamawaki, and H. Tani, Macromolecules, 7, 410 (1974).
[10] Y. Iwakura, K. Uno, and M. Oya, J. Polym. Sci., Part A-1, 5, 2867 (1967).
[11] H. Weingarten, J. Amer. Chem. Soc., 80, 352 (1958).
[12] F. D. Williams and R. D. Brown, Makromol. Chem., 169, 191 (1973).
[13] F. D. Williams and R. D. Brown, Biopolymers, 12, 647 (1973).
[14] Y. Iwakura, K. Uno, and M. Oya, J. Polymer Sci., Part A-1, 6, 2165 (1968).
[15a] Y. Imanishi, K. Kugimiya, and T. Higashimura, Biopolymers, 12, 2643 (1973).
[15b] Y. Imanishi, K. Kugimiya, and T. Higashimura, Ibid., 13, 1205 (1974).
[16] C. H. Bamford and H. Block, J. Chem. Soc., 1961, 4989, 4992.

[17] M. Goodman and E. Peggion, Vysokomol. Soedin., Ser. A, 247 (1967); Polym. Sci. USSR, 9, 271 (1967).
[18] M. Goodman and J. Hutchison, J. Amer. Chem. Soc., 88, 3627 (1966).
[19] E. Peggion, M. Terbojevich, A. Cosaini, and C. Colombini, Ibid., 88, 3630 (1966).
[20] E. Peggion, E. Scoffone, A. Cosani, and A. Portolan, Biopolymers, 4, 695 (1966).
[21] M. Terbojevich, G. Pizziola, E. Peggion, A. Cosani, and E. Scoffone, J. Amer. Chem. Soc., 89, 2733 (1967).
[22] N. S. Choi and M. Goodman, Biopolymers, 11, 67 (1972).
[23] L. Reibel and G. Spach, Bull. Soc. Chim. Fr., 1972, 1025.
[24] H. R. Kricheldorf and G. Greber, Makromol. Chem., 104, 3131, 3168 (1971).
[25] R. Katakai, M. Oya, K. Uno, and Y. Iwakura, J. Org. Chem., 37, 327 (1972).
[26] T. Wieland, Angew. Chem., 63, 7(1951); 66, 507 (1954).
[27] M. Idelson and E. R. Blout, J. Amer. Chem. Soc., 79, 3948 (1957).
[28] W. Dieckmann and F. Breest, Ber., 39, 3052 (1906).
[29] M. Goodman and U. Aron, J. Amer. Chem. Soc., 65, 3627 (1965).
[30] M. Goodman and U. Aron, Biopolymers, 1, 500 (1963).
[31] Y. Avny, S. Migdal, and A. Zilkha, Eur. Polym. J., 2, 355 (1966).
[32] Y. Avny and A. Zilkha, Ibid., 2, 367 (1966).
[33] M. Goodman, C. Gilon, G. S. Kirshenbaum, and Y. Knobler, Isr. J. Chem., 10, 867 (1972).
[34] H. Fukushima, T. Makino, S. Inoue, and T. Tsuruta, J. Polym. Sci., Polym. Chem. Ed., 11, 695 (1973).
[35] T. Makino, S. Inoue, and T. Tsuruta, Makromol. Chem., 131, 147 (1970).
[36] T. Makino, S. Inoue, and T. Tsuruta, Ibid., 150, 137 (1971).
[37] S. Yamashita and H. Tani, Macromolecules, 7, 406 (1974).
[38] S. Freireich, D. Gertner, and A. Zilkha, Eur. Polym. J., 10, 439 (1974).
[39] H. G. Elias and H. G. Buhrer, Makromol. Chem., 140, 21 (1970).
[40] M. Idelson and E. R. Blout, J. Amer. Chem. Soc., 79, 3948 (1957).
[41] R. Ledger and F. H. C. Stewart, Aust. J. Chem., 19, 1729 (1966).
[42] H. R. Kricheldorf, Makromol. Chem., 172, 13 (1973).
[43] S. Mori and M. Iwatsuki, Kobunshi Kagaku, 30(333), 39, 365 (1973).

The Steric Course of the Cationic Polymerization of Vinyl and Related Monomers. The Counterion Effect

TOYOKI KUNITAKE

Department of Organic Synthesis*
Faculty of Engineering
Kyushu University
Fukuoka, 812 Japan

ABSTRACT

Polymerization with many triphenylmethyl salts was conducted for α-methylstyrene, isobutyl vinyl ether, t-butyl vinyl ether, and spiro[2,4]hepta-4,6-diene (SHD). The variation of polymer structure (the isotactic unit content for the first three monomer systems and the amount of the 1,4-addition structure for SHD) showed fairly simple correlations with the counteranion size. The results can be interpreted in terms of the tightness of the propagating ion pair within the framework of a theory of the cationic propagation which had been proposed. When the counteranion radius was greater than 3.5 Å, the counteranion exerts a parallel influence on the tightness of the growing ion pair without regard to the monomer structure. However, in the case of smaller counterions, the tightness appears to be determined by the relative sizes of counteranion and monomer. The penta-coordinated counteranions gave rise to the polymer

*Contribution No. 349.

structure which would arise from tighter ion pairs than expected from their sizes alone. The polymer structure was also affected by the initiator concentration in these cases. These results are attributed to peculiar characteristics of penta-coordinated anions.

INTRODUCTION

Among the major unsolved problems in cationic polymerization is the role played by the counterion. The counteranion in the propagating ion pair is, in most cases, derived from the catalyst, and the so-called catalyst effect observed in the cationic propagation must be largely related to the behavior of the counteranion. Unfortunately, however, the structure of the growing ion pair is usually not well defined, and it is as yet difficult to interpret the propagation data in a unified way in terms of the nature of the growing species.

The stable carbocation salts have been used increasingly as initiators of cationic polymerization [1, 2]. They include triphenylmethyl, tropylium, and xanthylium ions in combination with several counteranions. Their use in vinyl polymerization is advantageous in that the anion becomes a part of the growing species without changing its structure.

In the present report we describe the systematic preparation of triphenylmethyl salts and their use for the cationic polymerization of the following representative vinyl and related monomers:

$$CH_2\!\!=\!\!\overset{\displaystyle CH_3}{\underset{\displaystyle \bigcirc}{C}} \qquad CH_2\!\!=\!\!\overset{\displaystyle CH}{\underset{\displaystyle O}{\underset{\displaystyle iBu}{}}} \qquad CH_2\!\!=\!\!\overset{\displaystyle CH}{\underset{\displaystyle O}{\underset{\displaystyle tBu}{}}} \qquad \text{SHD}$$

αMS   IBVE   tBVE   SHD

Particular emphasis is placed on establishing the relation between the structures of counteranions and polymers, since the ambiguity which always accompanies the kinetic data is not present in this case.

RESULTS AND DISCUSSION

Initiation by Triphenylmethyl Salts
_____

Triphenylmethyl salts may exist in the following forms which are in equilibrium:

$$Ph_3CX + MX_n \underset{\rightleftharpoons}{K_1} Ph_3CX \cdot MX_n \underset{\rightleftharpoons}{K_2} Ph_3C^+MX_{n+1}^- \underset{\rightleftharpoons}{K_3} Ph_3C^+ + MX_{n+1}^- \quad (1)$$

contact and
agent-separated

There is a possibility that metal halides, MX_n, initiate polymerization by combination with H_2O and other contaminants. This possibility was denied from kinetic evidence in the $Ph_3C^+SnCl_5^-$-styrene-CH_2Cl_2 system [3]. Sambi et al. [4] and Ledwith [11] reached the same conclusion.

More direct evidence for initiation by the Ph_3C addition was obtained in the polymerization of isobutyl vinyl ether [5]. Table 1 shows the content of the triphenylmethyl group in the polymer as determined by UV spectroscopy. From a comparison with the number-average degree of polymerization, it is clear that 20 to 100% of the polymer chain is bonded to the triphenylmethyl group, depending on the polymerization condition. Therefore, the initiation is concluded to be the addition of the triphenylmethyl group by considering the facile chain transfer in the cationic polymerization:

$$Ph_3C^+MX_{n+1}^- + CH_2 = CH \longrightarrow Ph_3CCH_2\overset{+}{C}H \ MX_{n+1}^- \quad (2)$$

with the pendant O–i-Bu groups shown below each carbon.

NMR spectra of five triphenylmethyl salts ($Ph_3C^+MX_n^-$, $MX_n = BF_4$, $AlCl_4$, $AlBr_4$, $SnCl_5$, and $SbCl_6$) in ethylene dichloride or in acetonitrile showed a phenyl multiplet at around 8 ppm,

TABLE 1. Content of the Triphenylmethyl Group in Poly-IBVE

Initiator	Solvent	Number-average degree of polymerization[a] ($\overline{P}_n$)	Number of monomer unit per Ph_3C group incorporated	Number of chain transfer per initiation[b]
$Ph_3C^+ AlCl_4^-$	$MCH-CH_2Cl_2$ 8:3	227	530	1.3 ± 0.4
$Ph_3C^+ BF_4^-$	$MCH-CH_2Cl_2$ 8:3	240	326	0.4 ± 0.2
$Ph_3C^+ SnCl_5^-$	$MCH-CH_2Cl_2$ 8:3	134	237	0.8 ± 0.4
$Ph_3C^+ AlCl_4^-$	CH_2Cl_2	167	396	1.4 ± 0.5
$Ph_3C^+ BF_4^-$	CH_2Cl_2	175	410	1.3 ± 0.4
$Ph_3C^+ SbCl_6^-$	CH_2Cl_2	154	420	1.7 ± 0.8
$Ph_3C^+ SnCl_5^-$	CH_2Cl_2	130	170	0.4 ± 0.4
$Ph_3C^+ AlCl_4^-$	$CH_2Cl_2-CH_3CN$ 7:3	161	220	0.4 ± 0.3
$Ph_3C^+ BF_4^-$	$CH_2Cl_2-CH_3CN$ 7:3	142	320	1.3 ± 0.4
$Ph_3C^+ SbCl_6^-$	$CH_2Cl_2-CH_3CN$ 7:3	84	455	4.4 ± 0.8
$Ph_3C^+ SnCl_5^-$	$CH_2Cl_2-CH_3CN$ 7:3	57	57	0.1 ± 0.1

[a] Determined by vapor pressure osmometry.

[b] $\left\{ \left[\dfrac{\text{Monomer unit}}{Ph_3C} \right] \middle/ \overline{P}_n \right\} - 1$. The error was estimated from the reliability of determinations of molecular weight and the triphenylmethyl content.

which was identical with the spectrum of the triphenylmethyl cation in chlorosulfonic acid [6]. The phenyl proton due to triphenylmethyl halides was not detected. A similar spectrum was observed for $Ph_3C^+AlBr_4^-$ in a 6:4 mixture of ethylene dichloride and n-hexane. Therefore, it is assured that the triphenylmethyl species exist mostly in the ionic form in the polymerization system. On the other hand, Higashimura et al. determined the equilibrium constant between the ionic and nonionic forms of Ph_3CSnCl_5 (K = 0.5 to 0.8 $\underline{M}$) at the UV concentration ($\sim 10^{-4}$ $\underline{M}$).

From the above discussion, the MX_{n+1}^- anion in the triphenyl-methyl salts can be considered to become the counteranion of the growing chain without structural modification. This is a fundamental assumption in discussing the counterion effect in the propagation step. Direct evidence for this assumption is lacking. However, the fairly simple correlations described below provide strong circumstantial evidence.

Preparation of Triphenylmethyl Salts [7]

Although many triphenylmethyl salts have been reported in the literature, their systematic preparation has apparently not been attempted. Therefore, we carried out preparations of a number of the salts mostly from triphenylmethyl chloride or bromide and the corresponding metal halides by simply mixing and washing under nitrogen in appropriate solvents. The product was confirmed by elemental analysis and by IR and NMR spectroscopy.

The metal halides which have been successfully employed for the preparation for the first time include $GaCl_3$, $InCl_3$, $TlCl_3$, $ZrCl_4$, $TaCl_4$, $NbCl_5$, and $InBr_3$. Attempted preparations with the following metal halides were only partially successful: $HfCl_5$, $BiCl_4$, $BiBr_4$, $TeCl_5$, $SeCl_5$, $TiCl_4$, and ICl_3. In Table 2 the central metal atoms contained in the anion of triphenylmethyl salts ($Ph_3C^+MX_{n+1}^-$, X = F, Cl, and/or Br) are placed in their respective positions of the periodic table. The parentheses indicate isolation of impure salts, and underline indicates that pure triphenylmethyl salts were isolated when metal fluorides (but not chloride and bromide) are employed. It is clear from Table 2 that triphenylmethyl salts are most readily formed when the counteranion contains the IIIA group metal (B, Al, Ga, In, and Tl) and the 5th period metal (In, Sn, Zr, Sb, and Nb). The fluoride salts are more stable than their chloride and bromide counterparts, but the latter salts were usually employed in our study because of preparative convenience.

TABLE 2. Central Metals Contained in Triphenylmethyl Salts Prepared[a]

	II A	II B	III A	III B	IV A	IV B	V A	V B	VI A	VI B	VII A	VIII
2	(Be)[b]		B									
3			Al				P[c]					
4		(Zn)	Ga			(Ti)	As[c]		(Se)			Fe
5		(Cd)	In	(Y)[b]	Sn	Zr	Sb[c]	Nb[c]	(Te)		(I)	
6		(Hg)	Tl	(Ce)[b]		(Hf)	(Bi)	Ta[c]				
7						(Th)[b]						

[a]The parentheses indicate that impure salts were isolated. The underlines indicate that pure triphenylmethyl salts were isolated when metal fluorides (but not chloride and bromide) were employed.
[b]F. Fairbrother, J. Chem. Soc., 1945, 503.
[c]Preparation of triphenylmethyl metal fluorides was reported for these metals in D. W. A. Sharp and N. Shepperd, J. Chem. Soc., 1957, 674.

Polymerization of α-Methylstyrene [7,8]

The polymerization of α-methylstyrene was carried out in several solvent systems at -76°C by using Friedel-Crafts catalysts and the triphenylmethyl initiators mentioned above. The polymerization was generally slow, the polymerization period being more than 100 hr in 8:1 methylcyclohexane (MCH)-CH_2Cl_2.

The steric structure of the polymer was determined from the relative peak area of the three methyl signals in the NMR spectrum according to Brownstein et al. [9]. The probability of isotactic propagation σ as defined by Bovey was calculated from the fraction of the sydiotactic triad P_{mm}:

$$P_{mm} = (1 - \sigma)^2 \tag{3}$$

The plot of σ vs the triad fraction closely fitted the theoretical curve of one parameter σ for all the polymer samples, as in the previous cationic polymerizations of this monomer. Therefore, the penultimate effect is absent.

Figure 1 shows the variation of the σ value with initiator and solvent. There are several important features included in this figure. One is the solvent effect. In the abscissa the solvent systems are arranged in the order of increasing polarity. The σ value decreases with increasing polarity of the medium. Furthermore, the σ variation with initiator is greatest in the least polar solvent, and the difference diminishes as the polarity of the medium increases. In fact, the σ value was independent of the counteranion in 3:7 CH_3CN-CH_2Cl_2, indicating that the growing ion pair is too loose for a counteranion to affect the steric course of propagation. A third feature is that Lewis acids give rise to σ values which are smaller than those obtained with the corresponding triphenylmethyl salt (e.g., $SnCl_4$ vs $Ph_3C^+SnCl_5^-$). It is possible that counteranions derived from Lewis acids possess structures quite different from what is expected by simple combinations of Lewis acid and cocatalyst. These results point to the danger of discussing the propagating ion pair based on the data obtained with Lewis acids.

Figure 2 shows the variation of the σ values with the counterion radius for poly-α-methylstyrene obtained at -76°C in 8:2 MCH-CH_2Cl_2. They fall in the range of 0.1 to 0.2, except for a high value (0.281) for $ZrCl_5^-$. The counteranion radius was estimated, where necessary, as the sum of the length of the metal-halogen bond and the van der Waals radius of the halogen atom [5]. The σ value increases with

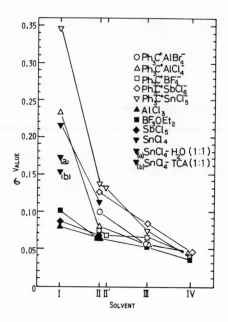

FIG. 1. Variation of the steric structure of poly-α-methylstyrene. Polymerization temperature: $-76°C$. Solvents: I, 10:2 MCH-CH_2Cl_2; II, 8:3 MCH-CH_2Cl_2; II', 7:3 MCH-CH_2Cl_2; III, CH_2Cl_2; IV, 7:3 CH_2Cl_2-CH_3CN.

increasing anion size from BF_4^- to ClO_4^- to BCl_4^-, and then decreases to a fairly constant value for counteranions of greater size. The penta-coordinated anions ($ZrCl_5^-$, $SnCl_5^-$, $SnBr_5^-$) gave σ values greater than expected from their sizes, and they are excluded from the correlation.

The σ value may well be correlated with other properties of the counteranion such as the first ionization potential or the electronegativity of the central metal atom. Figure 3 shows plots of σ vs electronegativity for metal chloride counteranions. There does not seem to exist any meaningful trend.

These data can be comfortably accommodated in our proposal on the steric course of propagation of vinyl and related monomers [10]. In this scheme the most stable conformation of the last two monomer units is assumed to be that shown by I. The monomer attack at the frontside (less hindered site) of the carbenium ion results in a syndiotactic placement, and the monomer attack at the backside in an isotactic placement. The relative ease of

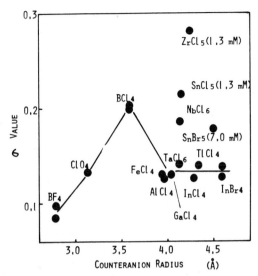

FIG. 2. Dependence of the steric structure of poly-α-methyl-styrene on counteranion size. Temperature: -76°C. Solvent: 8:2 MCH-CH_2Cl_2.

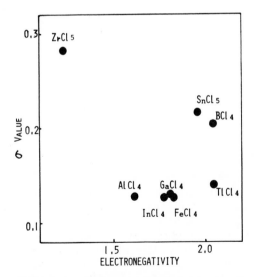

FIG. 3. Dependence of the steric structure of poly-α-methyl-styrene on the electronegativity of the central metal of the counter-anion. Polymerization condition: -76°C, 8:2 MCH-CH_2Cl_2.

monomer insertion at the frontside and the backside is influenced by
the tightness of the propagating ion pair and the difference in the
steric hindrance between the two modes of monomer attack. Therefore, the tightness is influenced only by counteranions for a given
monomer under a given polymerization condition. The decrease
of the σ value in polar media (Fig. 1) reflects a lowering of the
tightness, i.e., an increased frontside attack.

(4)

The syndiotactic placement at the frontside is favored in the
case of α-methylstyrene (small σ value) because the α-carbon is
disubstituted and the backside attack is subject to the enhanced
steric hindrance of the penultimate side chains.

Polymerizations of Isobutyl Vinyl Ether and t-Butyl Vinyl Ether

Vinyl ethers are one of the representative monomers in cationic
polymerization, and the variation of the steric structure of these
polymers has been studied to some extent. We selected two common
vinyl ethers and investigated the variation of the steric structure.

Isobutyl vinyl ether(IBVE) polymerizes quite readily with various
triphenylmethyl salts [5, 11, 12]. The steric structure of poly(isobutyl
vinyl ether) can be determined from the CH_2 dyad by means of CMR
spectroscopy [13]. Figure 4 shows examples of the FT-CMR spectrum
of IBVE polymer. In addition to the change of the CH_2 doublet, the

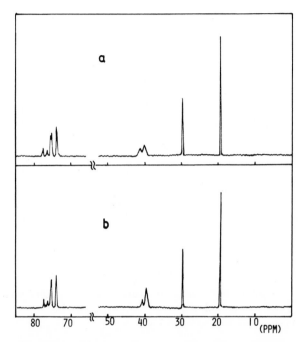

FIG. 4. ^{13}C NMR spectra of poly(isobutyl vinyl ether). (a) CDCl$_3$ solvent, 3600 scans, TMS reference. Polymerization conditions: -76°C, Ph$_3$C$^+$ClO$_4^-$, 5:5 C$_6$H$_{14}$-CH$_2$Cl$_2$. (b) CDCl$_3$ solvent, 5000 scans, TMS reference. Polymerization conditions: -76°C, Ph$_3$C$^+$BCl$_4^-$, 5:5 C$_6$H$_{14}$-CH$_2$Cl$_2$.

peaks of the methine and methylene carbons adjacent to oxygen are apparently affected by the steric structure. However, the assignment of the latter peaks has not been made.

The σ value was calculated from the relative peak area of the CH$_2$ dyad. It decreases with increasing polarity of the polymerization medium as in the polymerization of α-methylstyrene, reflecting loosening of the propagating ion pair. The variation of σ with counteranions is summarized in Fig. 5. The polymers were obtained at -76°C in 5:5 n-hexane—CH$_2$Cl$_2$. The σ value increases with increasing counterion sizes from BF$_4^-$ to BCl$_4^-$ via ClO$_4^-$, then decreases at AlCl$_4^-$, and again increases at GaCl$_4^-$ to TlCl$_4^-$. This trend is very similar to that of poly-α-methylstyrene, although it is less clear in the latter because of smaller changes of the σ value. Again the SnCl$_5^-$ counterion gave higher σ values.

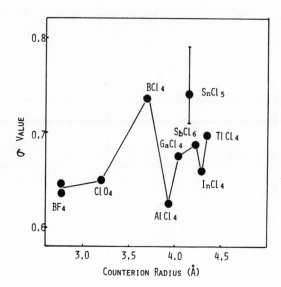

FIG. 5. Dependence of the steric structure of poly(isobutyl vinyl ether) on counteranion radii. Polymerization temperature: -76°C. Solvent: 5:5 C_6H_{14}-CH_2Cl_2.

Similarly, t-butyl vinyl ether (t-BVE) was polymerized with triphenylmethyl initiators [14]. Facile polymerization occurred. The polymer was converted to poly(vinyl alcohol) without separation [15], and the steric structure was determined from the splitting of the hydroxyl proton of poly(vinyl alcohol) in DMSO-d$_6$ [16].

$$-CH_2CH- \qquad\qquad -CH_2CH-$$
$$\overset{|}{O} \quad\xrightarrow{\text{HBr}}\quad \overset{|}{OH} \qquad\qquad\qquad (5)$$
$$\overset{|}{t\text{-Bu}}$$

The trends in the σ variation are rather similar between the two poly(vinyl ethers) for counteranions with radii greater than 3.5 Å (Fig. 6). However, the σ values for BF_4^- and ClO_4^- are greater than that for BCl_4^- in the case of poly-t-BVE, and the reverse is true for polyIBVE. These results suggest that the counterion effect on σ is not always independent of the monomer structure.

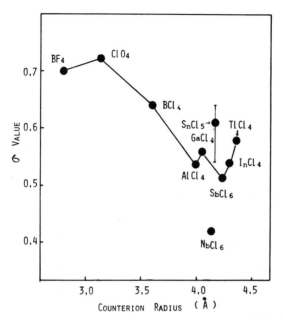

FIG. 6. Dependence of the steric structure of poly(t-butyl vinyl ether) on counteranion radii. Polymerization temperature: -76°C. Solvent: 7:3 CH_2Cl_2-toluene.

Polymerization of Spiro[2,4]hepta-4,6-diene

Cyclopentadiene polymerizes very readily with cationic initiators such as Lewis acids. Polycyclopentadiene formed under mild polymerization conditions is composed of the 1,2- and 1,4-addition structures. Their relative contents can be estimated by PMR spectroscopy [17]. Unfortunately, the structural variation with the polymerization condition was relatively small, and the counterion effect could not be studied in detail.

We recently found that spiro[2,4]hepta-4,6-diene (SHD) formed polymers very readily by cationic initiators [18]. This polymer is quite stable to autoxidation and heat compared with other cyclopentadiene polymers, and the polymer chain is presumed to be quite stiff from the viscometric data. On the other hand, spiro[4,4]nona-1,3-diene (SND) showed much less cationic reactivity [19].

$$\text{SHD} \longrightarrow \text{1,2-STRUCTURE} + \text{1,4-STRUCTURE} \qquad (6)$$

The structure (1,2- and 1,4-structures) of polySHD can be accurately determined from the olefinic proton peak. There was no indication of opening of the cyclopropyl ring. The content of the 1,4-structure decreases with increasing solvent polarity for a given initiator. This means that the 1,4-structure content can be a measure of the tightness of the growing ion pair. Figure 7 shows the dependence of the 1,4-structure content on the counterion size [20]. The structural variation is remarkable, considering that

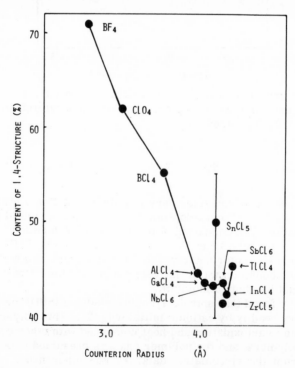

FIG. 7. Dependence of the structure of polyspiroheptadiene on counteranion radii. Polymerization temperature: -76°C. Solvent: 5:5 CH_2Cl_2-toluene.

the same polymerization medium is employed. The amount of the 1,4-structure decreases with increasing counterion radii from BF_4^- to $InCl_4^-$ and then increases at $TlCl_4^-$. It is noted that the structural change between $AlCl_4^-$ and $TlCl_4^-$ have some similarity to the variation of σ observed for other vinyl polymers. The penta-coordinated anions $SnCl_5^-$, $SnBr_5^-$, and $ZrCl_5^-$ again gave rise to the 1,4-structure contents, which are greater than expected from the correlation of Figure 7.

The influence of the polymerization conditions on the mode of propagation of SHD may be discussed using a propagation model related to the above-mentioned proposal. The structure of the propagating ion pair of SHD may be assumed to be

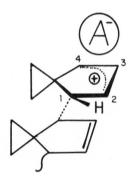

The counterion A^- is supposedly placed on the face of the five-membered ring which is remote from the penultimate unit. When the counteranion is only loosely associated with the growing cation, monomer molecule may react with the cation at the 2- and 4-positions from the less hindered side of the five-membered ring, avoiding the steric crowding of the penultimate unit. The formation of the 1,2-structure may be favored in this case because of the steric hindrance of the cyclopropyl group toward monomer attack at the 4-position. The difference in stability may also favor formation of the 1,2-structure.

As the tightness of the propagating ion pair increases, the reaction of the incoming monomer at the face of the five-membered ring opposite to the counterion (backside attack) will occur more readily. In this case the reaction at the 4-position is preferred because of the steric influence of the penultimate unit.

Influence of Initiator Concentration

In the previous sections the structural data obtained with penta-coordinated counteranions were omitted from the correlation because of their peculiar behavior. For example, in the polymerization of α-methylstyrene using $Ph_3C^+SnClBr_4^-$, the σ value was greater than expected from the counteranion size and, in addition, showed dependency on the initiator concentration [7].

Similar results are obtained with polySHD. When $Ph_3C^+SnCl_5^-$ was used as initiator, the content of the 1,4-structure decreased from 58 to 40% with an increase in the initiator concentration from 0.5 to 10 m$\underline{M}$ (-76°C, n-hexane—CH_2Cl_2). The polymer structure was independent of the initiator concentration (1,4-structure: 42%), when $Ph_3C^+InCl_4^-$ was used [20].

Figure 8 shows the structure changes of poly(vinyl ethers) with initiator concentrations. In the case of polyIBVE the same initiator dependence as mentioned above was observed: the σ value decreased with increasing concentrations of $Ph_3C^+SnCl_5^-$ initiator, and the concentration changes of $Ph_3C^+InCl_4^-$ and $Ph_3C^+SbCl_6^-$ did not affect the σ value. On the other hand, a somewhat different situation was found for poly-t-BVE. The σ value for this polymer increases with increasing concentrations of $Ph_3C^+SnCl_5^-$ and reaches a constant value. Again, the concentration change of $Ph_3C^+InCl_4^-$ did not affect the σ value. Interestingly, the σ value increased when t-BVE was polymerized by $Ph_3C^+SnCl_5^-$ (1 m$\underline{M}$) in the presence of 7 m$\underline{M}$ of $(n-Bu)_4N^+SnCl_5^-$.

Except for poly-t-BVE, the structures of polymers obtained with higher concentrations of $Ph_3C^+MX_5^-$ (saturation value) are quite close to those obtained with other initiators where no concentration effects are observed.

Counteranion Effects on Polymer Structure

According to the propagation scheme mentioned above, the counteranion effect is, in the main, related to the tightness of the propagating ion pair. Two major findings are obtained from the present research. One is the relatively simple correlation of the polymer structure with counteranion sizes. The other is the peculiarity of penta-coordinated counteranions.

The counteranions may be characterized by various properties in relation to the tightness of the propagating ion pair. When a counteranion is symmetrical or fully-coordinated, electrostatic

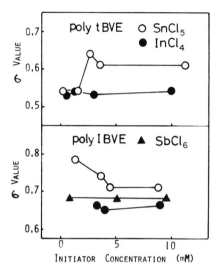

FIG. 8. Influence of the initiator concentration on the steric structure of poly(vinyl ethers). Polymerization conditions: for polytBVE, -76°C, 7:3 CH_2Cl_2-toluene; for polyIBVE, -76°C, 5:5 $CH_2Cl_2C_6H_{14}$.

and steric factors would suffice to describe its effect, provided that there are no specific interactions operating between counteranion and monomer. These two factors are primarily determined by the size of counteranions. The nature of the metal-halogen bond is influenced by the electronegativity or ionization potential. However, they are probably of secondary importance. In fact, the experimental results could not be correlated with these properties in any simple way. The specific solvation may alter the effective size of counteranions. This is again considered to be of minor importance. In anionic polymerization with alkali metals as countercations, Cs^+ is not specifically solvated in THF because of its size (radius, 1.69 Å) [21]. The counteranions employed in the present study are much larger than Cs^+, and the specific solvation is probably unimportant.

The trends shown in Figs. 2, 5, 6, and 7 are generally similar for counteranions larger than BCl_4^- in that general decreases of the σ values and the 1,4-structure are observed with increasing counteranion sizes. The variations between $AlCl_4^-$ and $InCl_4^-$ are fairly small, and detailed comparisons among counteranions and among polymers are not desirable. These data may be interpreted as showing that the tightness decreases due to weakened coulombic interaction for larger counteranions. The unmistakable increase in the tightness at $TlCl_4^-$ might be a reflection of the large ionization potential of Tl metal. It is of particular import that the common counterion effects were observed in this region.

On the other hand, the polymer structures depend on the counteranion size quite differently among polymers in the region of BF_4^-, ClO_4^-, and BCl_4^-. The monomer structure must affect the tightness (the relative ease of the frontside and backside attack) when counterions are small. This is readily seen by comparing the relative influences of these counteranions on the polymer structure. The σ value increases in the order of $BF_4^- < ClO_4^- < BCl_4^-$ for poly-α-methylstyrene and for polyIBVE. The order is $BF_4^- \lesssim ClO_4^- > BCl_4^-$ for poly-t-BVE, and the 1,4-structure of polySHD decreases in the order of $BF_4^- > ClO_4^- > BCl_4^-$. Steric crowding of the propagating terminal is apparently associated with the varying influences of these small counteranions. It may be postulated that the tightness of the ion pair decreases with increasing counterion sizes because of steric interference of counterion and side chain when the propagating end is crowded. The tightness increases with the increasing size of counteranions when crowding of the propagating chain end is reduced. In essence, the frontside attack is hindered most effectively when either (but not both) the counteranion or the side chain is bulky. However, this does not hold true if the counteranion size is greater than 3.5 to 4.0 Å. In this region the tightness becomes fairly insensitive to the sizes of counteranion and side chain.

The peculiar behavior of penta-coordinated anions may be attributed to their characteristic structures. It is known that the structure of penta-coordinated species (MX_5) is generally nonrigid, and one of the possible conformation—trigonal bipyramid—may readily convert to the other conformation—tetragonal pyramid [22].

$$X \leftarrow \overset{\overset{X}{|}}{\underset{\underset{X}{|}}{M}} {\overset{X}{\diagdown}} \rightleftarrows \overset{X}{\underset{X}{X} {\diagdown} M {\diagdown} X} \qquad (7)$$

This structural characteristic can give rise to an asymmetrical MX_5^- anion, and its effect could be different from what is expected from its size alone.

The tetragonal-pyramid conformation of the penta-coordinated anion possesses a vacant coordination site, in contrast to the fully coordinated tetra- and hexa-halide anions. The observed dependency of the polymer structure on the initiator concentration suggests the occurrence of aggregation of the penta-coordinated counteranions at high concentrations. The lack of the concentration effect for other anions is consistent with this view. The tightness (that is, the σ value and the 1,4-structure) in the penta-coordinated anion system was considerably greater than that of other anions at low initiator concentrations. The tetragonal-pyramid counteranion could yield an ion pair tighter than expected from other conformations. The tightness approaches a normal value with increasing initiator concentrations because of aggregation of counterion. The poly-t-BVE system is an exception, and the σ value increases with increasing concentrations of the $SnCl_5^-$ anion. This fact is hard to interpret.

CONCLUSION

Representative vinyl-type monomers for cationic polymerization were selected and subjected to polymerization by various triphenyl-methyl initiators. A fairly simple picture emerged concerning the counteranion effect on the steric (and geometrical) course of propagation, in spite of differences in the monomer structure.

The present study was limited to the structural change of polymers. It should be interesting to examine the implications of the present study on other polymerization characteristics. Further efforts must be directed to confirming the propagation model and to designing counteranions with desired characteristics.

ACKNOWLEDGMENT

The author appreciate the cooperation of the following co-workers during the course of the present study: Dr. Y. Matsuguma, Mr. S. Tsugawa, Mr. T. Ochiai, and Mr. K. Takarabe.

REFERENCES

[1] A. Ledwith, Makromol. Chem., 175, 1117 (1974).
[2] S. Penczek, Ibid., 175, 1217 (1974).

[3] T. Higashimura, T. Fukushima, and S. Okamura, J. Macromol.
     Sci.—Chem., A1, 683 (1967).

[4] M. S. Sambi and F. E. Treloar, J. Polym. Sci., Part B, 3, 445
     (1965); M. S. Sambi, Macromolecules, 3, 351 (1970).

[5] T. Kunitake, Y. Matsuguma, and C. Aso, Polym. J., Tokyo, 2,
     345 (1971).

[6] D. G. Farnum, J. Amer. Chem. Soc., 86, 934 (1964).

[7] T. Kunitake and S. Tsugawa, Macromolecules, In Press.

[8] Y. Matsuguma and T. Kunitake, Polym. J., Tokyo, 2, 353
     (1971).

[9] S. Brownstein, S. Bywater, and D. J. Worsfold, Makromol.
     Chem., 48, 127 (1961).

[10] T. Kunitake and C. Aso, J. Polym. Sci., Part A-1, 8, 665
     (1970).

[11] C. E. H. Bawn, C. Fitzsimmons, A. Ledwith, J. Penfold, D. C.
     Sherrington, and J. A. Weightman, Polymer, 12, 119 (1971).

[12] T. Kunitake, S. Tsugawa, and K. Takarabe, Manuscript in
     Preparation.

[13] K. Matsuzaki, H. Ito, T. Kawamura, and T. Uryu, J. Polym.
     Sci., Polym. Chem. Ed., 11, 971 (1972).

[14] T. Kunitake and K. Takarabe, Manuscript in Preparation.

[15] S. Okamura, T. Kodama, and T. Higashimura, Makromol.
     Chem., 53, 180 (1962).

[16] T. Moritani, and I. Kumura, Macromolecules, 5, 577 (1972).

[17] C. Aso, T. Kunitake, and Y. Ishimoto, J. Polym. Sci., Part A-1,
     6, 1163 (1968).

[18] O. Ohara, C. Aso, and T. Kunitake, J. Polym. Sci., Polym.
     Chem. Ed., 11, 1917 (1973).

[19] O. Ohara, C. Aso, and T. Kunitake, Polymer J., Tokyo, 5, 49
     (1973).

[20] T. Kunitake, T. Ochiai and O. Ohara, J. Polym. Sci., Polym.
     Chem. Ed., In Press.

[21] D. N. Bhattacharyya, J. Smid, and M. Szwarc, J. Phys. Chem.,
     69, 624 (1965).

[22] E. L. Muetterties and R. A. Schunn, Quart. Rev., 20, 245 (1966).

Recent Progress of Photo- and Radiation-Induced Ionic Polymerizations

M. IRIE, Y. YAMAMOTO, and K. HAYASHI*

The Institute of Scientific and Industrial Research
Yamadakami, Suita, Osaka, 565 Japan

ABSTRACT

Photoinduced ionic polymerizations of the monomers α-methylstyrene, cyclohexeneoxide, nitroethylene, and acrylonitrile were carried out in the presence of electron acceptor or donor molecules. These polymerizations are proved to be initiated by ions formed through the dissociation of the photoexcited electron donor-acceptor complex and to proceed by ionic mechanism.
The molecular weight distribution of the polymer and the light intensity dependency on the rate of polymerization indicate that free ionic and ion-pair propagations coexist in the cationic polymerization of α-methylstyrene.
Anionic polymerizations were observed for the nitroethylene-tetrahydrofuran and acrylonitrile-dimethylformamide systems.
Radiation-induced cationic polymerizations of styrene and α-methylstyrene were found to proceed by free cationic propagation. The effect of added electron acceptors in these polymerizations was investigated.

---

*The paper was presented by K. Hayashi.

PHOTOINDUCED IONIC POLYMERIZATION

α-Methylstyrene

   Although several photoinduced ionic polymerizations by charge
transfer interaction have been reported [1], the monomers used
have been limited to those containing hetero atoms, such as
N-vinylcarbazole, which have a very low ionization potential and
a high reactivity in cationic polymerization. We have found the
photoinduced cationic polymerization of α-methylstyrene [2],
which is a weak electron donor monomer and contains no hetero
atom, in the presence of tetracyanobenzene (TCNB). Extremely
well-dried α-methylstyrene polymerized in methylene chloride
by the photoillumination of the charge transfer band of the
α-methylstyrene—TCNB complex, though no polymer was
obtained in the dark. There was no polymerization after the light
was turned off. Water or triethylamine, which are typical inhibitors
of cationic polymerization, stopped the polymerization. This result
indicates that the photoinduced polymerization proceeds by a cationic
mechanism. This conclusion was further confirmed by a copolymerization with styrene at 0°C. The monomer reactivity ratios are r_1
(α-methylstyrene) = 2.0 and r_2 (styrene) = 0.11. These reactivity
ratios are close to those reported for cationic polymerization
initiated by $SnCl_4$ [3] (r_1 = 2.90, r_2 = 0.55), but different from those
of radical [4] (r_1 = 0.36, r_2 = 1.18 ± 0.04) and anionic [5] (r_1 = 0.08,
r_2 = 10.5 ± 0.5) copolymerizations. Both the effect of additives and
the reactivity ratios in copolymerization prove that the propagating
ends in the photopolymerization have a cationic nature.

   This polymerization is thought to be initiated through ions formed
by the dissociation of the excited electron donor-acceptor complex
as follows:

$$D + A \rightleftharpoons (D \cdot A) \qquad\qquad (D \cdot {}^+ \text{---} A \cdot {}_s^-) \xrightarrow{} \text{Polymer}$$

$$(D \cdot A) \xrightarrow{h\nu} (D \cdot A)^* \qquad\qquad D \cdot {}_s^+ + A \cdot {}_s^- \xrightarrow{} \text{Polymer}$$

$$A \xrightarrow{h\nu} A^* \xrightarrow{D} (D \cdot A)^*$$

$$(D \cdot A)^* \xrightarrow{S} (D \cdot {}_s^+ \text{---} A \cdot {}_s^-) \rightleftharpoons D \cdot {}_s^+ + A \cdot {}_s^-$$

where D, A, and S indicate α-methylstyrene, TCNB, and solvent, respectively.

We previously reported the fundamental processes involved in the formation of initiating species from the photoexcited complex by the use of various physicochemical methods, such as optical absorption and emission spectroscopies, laser flash photolysis, ESR, and photoconductivity [6, 7].

Figure 1 represents the cationic polymerization of α-methylstyrene in the presence of TCNB at -30°C. Curve A in Fig. 1 shows the yield of polymer obtained by illumination with wavelengths longer than 350 nm (with UV 31 filter), which involves both the charge-transfer band (363 nm) and the absorption band of TCNB (316 nm). Polymerization was also initiated by illumination with light of wavelength longer than 350 nm (with UV 35 filter), where only the charge-transfer absorption band exists, but the rate of polymerization was decreased by one-third (Curve B in Fig. 1). Illumination with light passed through a VY 42 filter, where the tail of the charge-transfer absorption band exists, scarcely initiates the polymerization, as shown in Curve C in Fig. 1 [8].

In ionic polymerization the dependence of the rate of polymerization, R_p, on the light intensity can give information on the form of propagating chain ends, whether they are free ions or ion pairs. When the chain ends are free, the dependence should be $R_p \propto I^{1/2}$. For ion pairs the dependence is expressed by $R_p \propto I$.

Figure 2 shows the dependence of the rate of polymerization on the

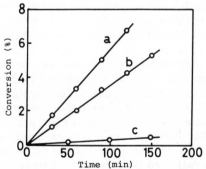

FIG. 1. Photopolymerization of α-methylstyrene (1.9 $\underline{M}$) in the presence of tetracyanobenzene (10^{-3} $\underline{M}$) in methylene chloride at -30°C. (a), (b), and (c): yields of polymer obtained by illumination with light passed through UV 31, UV 35, and VY 42 filters, respectively.

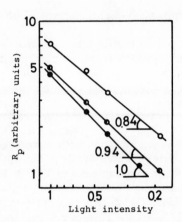

FIG. 2. Light intensity dependence of the rate of polymerization at various temperatures: (O) -74°C, (◐) -35°C, and (●) -15°C. [α-MeSt] = 1.9 $\underline{M}$, [PMDA] = 4 × 10⁻⁴ $\underline{M}$. Solvent: CH_2Cl_2.

light intensity at three temperatures [9]. Hereafter we use pyromellitic dianhydride (PMDA) as electron acceptor instead of TCNB. The dependence decreased from first order at -15°C to 0.84 at -74°C. The result suggests that free growing chain ends also contribute to the propagation process at the lower temperature.

The molecular weight distributions of the polymers obtained at four temperatures measured by GPC (gel permeation chromatography) are shown in Fig. 3. The polymer obtained at -74°C has a bimodal distribution, which clearly indicates that two chain ends with different structures propagate independently. The height of the polymer peak at the lower molecular weight decreased to a great extent with an increase of the temperature, though the peak positions are the same. On the other hand, the other peak increased both in its intensity and elution volume counts with increasing temperature. The deviation from first-order dependency at the lower temperature shown in Fig. 2 indicates the lower molecular weight peak arises from free propagating ends. The other peak is presumably caused by the ion-pair ends. The relative contributions of these two mechanisms, free ion and ion pair, are considered to depend on the polarity of the systems. Upon lowering the temperature the polarity of the system is expected to increase, and the facilitates ionic dissociation of ion pairs into free ions [10].

The relationship between the propagation mechanism and the polarity was also revealed by the effect of the monomer concentration.

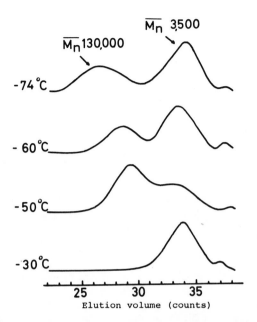

FIG. 3. Molecular weight distribution of poly-α-methylstyrene formed by GPC method. Polymerization was accomplished with a monomer concentration of 0.57 $\underline{M}$ at various temperatures.

When the concentration of α-methylstyrene in the system is decreased, so that the system is made more polar, the light intensity dependence also approached 0.5, as shown in Fig. 4. The molecular weight distribution curves shown in Fig. 5 indicate that the relative intensity of the peak at lower molecular weight, due to free ionic polymerization, increases with decreasing monomer concentration.

The rate of polymerization increased when the monomer concentration and temperature were lowered. This result is attributable to the facts that an increase in the polarity of the solution facilitates the dissociation of the ion pair to free ions, and the free ionic propagation has a higher rate than the ion pair one.

The light intensity dependencies of the photocurrents during photoillumination, which corresponds to the concentration of free ions, was always 0.5 as shown in Fig. 6. The photocurrent increased with decreasing temperature as expected. It also increased with a decrease in the concentration of α-methylstyrene

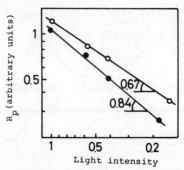

FIG. 4. Light intensity dependence of the rate of polymerization
at various monomer concentrations. (O) [α-MeSt] = 0.53 M, (●)
[α-MeSt] = 1.9 M, [PMDA] = 4 × 10⁻⁴ M, Solvent: CH₂Cl₂.
Temperature: -74°C.

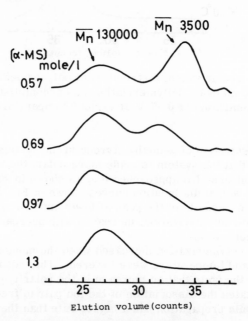

FIG. 5. Molecular weight distribution of poly-α-methylstyrene
formed by GPC method. Polymerization was done at -74°C for
various monomer concentrations.

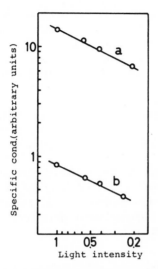

FIG. 6. Light intensity dependence of photocurrent. [α-MeSt] = 1.9 M̲, [PMDA] = 4 × 10⁻⁴ M̲. (a) -74°C, (b) -15°C.

in the system, and this results in an increase of the polarity of the system. The steady-state yield of the free ions at monomer concentrations of 0.53 M̲ is four times higher than that at 1.9 M̲.

On the assumption that the mobility of ion is 5×10^{-4} $cm^2/(V)(sec)$ [11], the rate constant of free ionic propagation is obtained tentatively to be $(3.5 \pm 1.8) \times 10^3$ $M^{-1}sec^{-1}$ at -74°C by the use of the above photocurrent value and the rate of the polymerization measured by dilatometry. This value is one order less than that of free ions obtained in radiation-induced ionic polymerization [12], even when it is extraporated to room temperatures by the use of an activation energy of 4 kcal/mole. A part of the decrease is possibly due to the solvation of the propagating chain ends by the polar solvent during the photopolymerization.

Cyclohexene Oxide [13]

Not only π-electron-donor monomers, such as styrene or α-methylstyrene, but also n-electron-donor monomers, such as cyclohexene oxide, are expected to be photoionized when complexed with suitable acceptors.

Photopolymerization of cyclohexene oxide was observed in the presence of TCNB or PMDA, while no polymer was obtained in the dark. The molecular weight of the polymer formed in the presence of PMDA at -78°C was 477,000, which is as large as that obtained by γ-irradiation. Small amounts of triethylamine inhibited the polymerization, which indicates that this polymerization proceeds by a cationic mechanism. The yield of polymer obtained in the presence of PMDA is three times that obtained in the presence of TCNB.

The dependence of the rate of polymerization of cyclohexene oxide on the wavelength of illuminating light in the presence of PMDA are shown in Fig. 7. The photopolymerization of cyclohexene oxide induced by the light exciting the acceptor band is much faster than that of α-methylstyrene. Hardly any poly-(cyclohexene oxide) is obtained by illumination with light of wavelengths larger than 390 nm.

Nitroethylene [14]

Anionic as well as cationic polymerization is expected to occur in photoexcited charge transfer systems, since laser photolysis studies have proved that cations and anions are formed simultaneously from the excited complex [7]. The photoinduced anionic

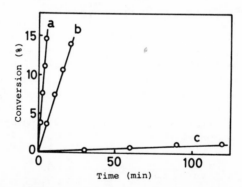

FIG. 7.   Polymerization of cyclohexene oxide in the presence of pyromellitic dianhydride at -78°C. (a), (b), and (c): yields of polymer obtained by illumination with light passed through Toshiba UV 31, UV 35, and UV 39 filters, respectively. Concentrations of cyclohexene oxide and pyromellitic dianhydride were 2.5 and 1×10^{-3} M respectively.

polymerization of nitroethylene and acrylonitrile was carried out with the purpose of extending the scope of photoinduced ionic polymerization to anionic systems. It is well known that nitroethylene has a high electron affinity, that it is readily polymerized by anionic catalysts, and that tetrahydrofuran is a suitable solvent for anionic polymerization as well as an n-electron donor.

The spectrum of the mixture of nitroethylene and tetrahydrofuran has a weak charge transfer band around 450 nm, which indicates the existence of a charge-transfer interaction between these compounds. Photoexcitation of the charge-transfer band by the light passed through a VY-45 filter ($\lambda > 450$ nm) gives rise to polymerization of nitroethylene as shown in Fig. 8(b). No polymer was obtained in the dark. In this system there was neither an induction period nor postpolymerization after turning off the light. Elemental analysis and the IR spectrum proved that the polymer obtained is poly(nitroethylene). The molecular weight was determined by viscometry to be 17,000.

The light which covers both the charge-transfer band and the acceptor band gives a high polymer yield, although hardly any polymer was obtained with light of wavelength longer than 520 nm, as shown in Figs. 8(a) and 8(c). In addition, no polymer was obtained in the absence of tetrahydrofuran. These results imply that the polymerization is initiated by the photoexcited charge-transfer interaction between nitroethylene and tetrahydrofuran.

A trace of hydrogen chloride (5×10^{-3} $\underline{M}$) inhibited the polymerization completely, which suggests that this reaction proceeds by an anionic mechanism. This was confirmed by a

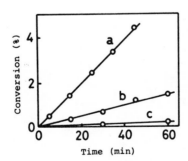

FIG. 8. Photopolymerization of nitroethylene in tetrahydrofuran at 4°C. Time-conversion curves were attained by the light passed through UV-39 filter (a), VY-45 (b), and VO-52 (c). Concentration of nitroethylene in tetrahydrofuran was 3.0 $\underline{M}$.

copolymerization with acrylonitrile, for which the light was passed
through a VY-45 filter to avoid a direct radical polymerization of
the acrylonitrile. The content of nitroethylene in the copolymer is
shown as a function of monomer composition in Fig. 9. The monomer
reactivity ratios were estimated to be r_1 (nitroethylene) = 25 ± 10
and r_2 (acrylonitrile) = 0.24 ± 0.20, which are similar to those for the
radiation-induced anionic copolymerization in tetrahydrofuran at
-78°C (r_1 = 63 ± 15, r_2 = 0.01 ± 0.01) [15].

Acrylonitrile [16]

The intermolecular interaction between acrylonitrile and DMF
in the ground state is very weak, because the absorption spectrum
of DMF was scarcely shifted to longer wavelength by the addition
of acrylonitrile. The excitation of the DMF band by the light passed
through Toshiba UV-29 filter ($\lambda > 290$ nm) induced the polymerization
of acrylonitrile. The light hardly excites the acrylonitrile band.
The polymerization was completely inhibited by the addition of
small amount of hydrogen chloride or a trace of moisture. In
addition, preilluminated DMF cannot initiate the polymerization.
These results indicate that the polymerization is initiated by the
excited complex between acrylonitrile and DMF and proceeds by
anionic mechanism.

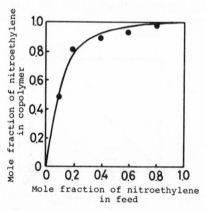

FIG. 9. Photoinduced copolymerization of nitroethylene and
acrylonitrile in tetrahydrofuran at 4°C. The concentration of the
mixture of nitroethylene and acrylonitrile in tetrahydrofuran was
3.0 M. r_1 (nitroethylene) = 25 ± 10; r_2 (acrylonitrile) = 0.24 ± 0.20.

The mechanism was further confirmed by copolymerization with methacrylonitrile at -30°C, and the monomer reactivity ratio agreed with the value reported for anionic polymerization initiated by NaNH$_2$ [17].

The dependence of the rate of polymerization on the light intensity was measured at -30°C to clarify the structure of the propagating chain ends, as shown in Fig. 10. The slope is found to be 0.63. The dependence of close to 0.5 indicates that almost all propagating chain ends are free at -30°C in this system. The free ion mechanism in this very polar solvent agrees well with that proposed for the case of the cationic polymerization of α-methylstyrene.

Simultaneous Cationic and Anionic Polymerizations [18]

Simultaneous polymerization is expected when two monomers can form an excited complex by photoirradiation, and the donor monomer can be polymerized cationically and the acceptor monomer anionically. In this study, cyclohexene oxide and nitroethylene were used as the electron-donor and -acceptor monomers, respectively.

Nitroethylene has an absorption tail around 470 nm. The addition of cyclohexene oxide to nitroethylene causes a small red shift in the absorption, which indicates a weak contact-type charge-transfer interaction between these monomers.

The excitation of the charge-transfer band by light passing through the Toshiba UV-39 filter induces the polymerization of both monomers at 4°C. Polymerization was not observed in the

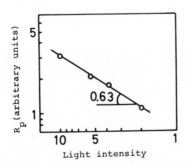

FIG. 10. Light intensity dependence of the rate of photopolymerization of acrylonitrile in DMF at -30°C. The concentration of acrylonitrile was 3.0 M.

absence of either nitroethylene or cyclohexene oxide or in the dark. There was neither an induction period nor postpolymerization after turning off the light.

Figure 11 shows the dependence of the rate of polymerization on the feed composition. The maximum rate was attained at almost equimolar monomer composition, and the rate decreased markedly on deviating from this composition. The fact that the highest rate of polymerization occurs at equimolar composition seems to be due to the formation of the highest concentration of initiating active species by photoirradiation of the 1:1 EDA complex. Fractionation of the polymer obtained indicates that it is a mixture of the homopolymers of cyclohexene oxide and nitroethylene and a copolymer, probably a block copolymer.

RADIATION-INDUCED IONIC POLYMERIZATION

It has been found that the radiation-induced polymerizations of styrene and α-methylstyrene are considerably enhanced by the use of the extreme drying technique [19, 20]. Evidence that those polymerizations are due mainly to a cationic propagation mechanism comes both from scavenger studies and from the determination of reactivity ratios in copolymerization experiments. k_p values in bulk were determined to be 10^6 $\underline{M}^{-1}sec^{-1}$ for both monomers by

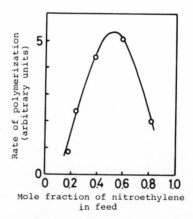

FIG. 11. Dependence of the rate of photopolymerization of mixture of cyclohexene oxide and nitroethylene on the composition of the two monomers in the feed at 4°C.

the electrical conductivity method [21]. The large k_p values are explained as a characteristic ion-dipole reaction for the free ionic propagation [22, 23]. The propagation reaction of styrene in free cationic process in bulk is characterized by a lower activation energy as compared with that in other mechanisms [24]. A decrease of k_p value owing to the solvation by polar solvent such as CH_2Cl_2 to growing cation was observed preliminary [25]. The radiation-induced cationic polymerization of superdried isobutyl vinyl ether (IBVE) in bulk gave a half-power dose rate dependence of the rate of polymerization [26]. A propagation reaction of this monomer was found to be characterized by a high activation energy of 9.6 kcal/mole. In order to elucidate the reason for the consider-able difference in the activation energies for the propagation of styrene and IBVE, radiation-induced copolymerization of these monomers was carried out in bulk at various temperatures [27]. It is found that in the reactivity of cation, styrene is greater than IBVE; however, the relation is reversed in relative reactivity of monomer. The extraordinary high activation energy (E_p) of IBVE is presumably explained in terms of the solvation of IBVE to cation and resonance stabilization of the growing IBVE cation as follows.

$$-CH_2-\overset{+}{C}H \longleftrightarrow -CH_2-CH$$
$$\hspace{1.2cm} | \hspace{3.2cm} |$$
$$\hspace{1.2cm} O \hspace{3.2cm} \overset{+}{O}$$
$$\hspace{1.2cm} | \hspace{3.2cm} |$$
$$\hspace{1.2cm} R \hspace{3.2cm} R$$

On the other hand, radiation-induced anionic polymerization was achieved by using well-purified nitroethylene [28] and acrylonitrile [29].

Recently the formation of an ion-pair by ionizing radiation was found for electron-donor acceptor (EDA) systems by the use of the optical absorption technique in glassy matrices at low temperature [30]. By the addition of ethylbenzene the yield of radical anion of pyromellitic anhydride (PMDA) increased and the absorption band of this intermediate shifted from 664 to 668 nm. These results are explained in terms of the formation of ion pairs through the ionization of the EDA complex of PMDA and ethylbenzene. The rate of polymerization in the radiation-induced ionic polymerization of α-methylstyrene in methylene chloride was increased by the addition of an electron acceptor such as PMDA [31]. Figure 12 represents the increase of molecular weight of polymer formed

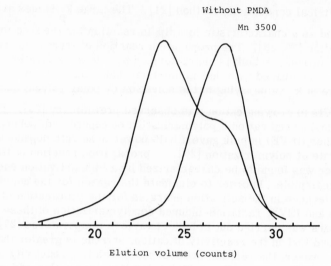

Without PMDA

Mn 3500

Elution volume (counts)

FIG. 12. Molecular weight distribution of poly-α-methylstyrene obtained by ionizing radiation; effect of added PMDA (3×10^{-3} $\underline{M}$) at -78°C in CH_2Cl_2.

caused by the addition of PMDA in this polymerization. These results indicate that the polymerization mechanism changes from free cationic propagation to ion-pair as discussed previous for the photopolymerization of this monomer.

REFERENCES

[1] For example: (a) S. Tazuke, Adv. Polym. Sci., 6, 321 (1969); (b) M. Yamamoto, S. Nishimoto, M. Ohoka, and Y. Nishijima, Macromolecule, 3, 706 (1970); (c) K. Tada, Y. Shirota, and H. Mikawa, J. Polym. Sci., Polym. Chem. Ed., 11, 2961 (1973).
[2] M. Irie, S. Tomimoto, and K. Hayashi, J. Polym. Sci., Part B, 8, 585 (1970).
[3] E. B. Lindvig, A. P. Gantmakker, and S. S. Medvedev, Dokl. Akad. Nauk SSSR, 119, 90 (1950).
[4] A. V. Golubera, N. F. Usmanova, and A. A. Vansheidt, J. Polym. Sci., 52, 63 (1961).
[5] A. V. Tobolsky and R. J. Boudreau, Ibid., 51, S53 (1961).

[6] M. Irie, S. Tomimoto, and K. Hayashi, J. Phys. Chem., 76, 1419 (1972).
[7] M. Irie, M. Masuhara, K. Hayashi, and N. Mataga, Ibid., 78, 341 (1974).
[8] M. Irie, S. Tomimoto, and K. Hayashi, J. Polym. Sci., Polym. Chem. Ed., 10, 3235 (1972).
[9] M. Irie, S. Irie, and K. Hayashi, To Be Published.
[10] M. Szwarc, Carbanions, Living Polymers and Electron Transfer Processes, Wiley-Interscience, New York, 1968, p. 416.
[11] M. Katayama, H. Yamazaki, and Y. Ozawa, Bull. Chem. Soc. Japan, 42, 2410 (1969).
[12] F. Williams, K. Hayashi, K. Ueno, K. Hayashi, and S. Okamura, Trans. Faraday Soc., 63, 1501 (1969).
[13] M. Irie, S. Tomimoto, and K. Hayashi, J. Polym. Sci., Polym. Chem. Ed., 10, 3243 (1972).
[14] M. Irie, S. Tomimoto, and K. Hayashi, J. Polym. Sci., Polym. Lett. Ed., 10, 699 (1972).
[15] H. Yamaoka, P. Uchida, K. Hayashi, and S. Okamura, Kobunshi Kagaku, 24, 79 (1967).
[16] M. Irie, S. Sasaoka, and K. Hayashi, To Be Published.
[17] F. Dawans and G. Smets, Makromol. Chem., 59, 163 (1963).
[18] M. Irie, S. Tomimoto, and K. Hayashi, J. Polym. Sci., Polym. Chem. Ed., 11, 1859 (1973).
[19] K. Ueno, Ff. Williams, K. Hayashi, and S. Okamura, Trans. Faraday Soc., 63, 1478 (1967).
[20] R. C. Potter, R. H. Bretton, and D. J. Metz, J. Polymer Sci., Part A-1, 4, 2295 (1966).
[21] Ka. Hayashi, Y. Yamazawa, T. Takagaki, F. Williams, K. Hayashi, and S. Okamura, Trans. Faraday Soc., 63, 1489 (1967).
[22] F. Williams, Ka. Hayashi, K. Ueno, K. Hayashi, and S. Okamura, Ibid., 63, 1501 (1969).
[23] K. Hayashi, Actions Chim. Biol. Radiat., 15, 145 (1971).
[24] K. Hayashi, K. Hayashi, and S. Okamura, Polym. J., 4, 426 (1973).
[25] T. Hibi, K. Hayashi, M. Irie, and K. Hayashi, To Be Published.
[26] K. Hayashi, K. Hayashi, and S. Okamura, J. Polym. Sci., Part A-1, 9, 2305 (1971).
[27] K. Hayashi, K. Hayashi, and S. Okamura, Polym. J., 4, 495 (1973).
[28] H. Yamaoka, F. Williams, and K. Hayashi, Trans. Faraday Soc., 63, 376 (1967).

[29] S. Iiyama, S. Abe, and K. Namba, Kobunshi Kagaku, 30, 134
      (1973).
[30] M. Irie, S. Irie, Y. Yamamoto, and K. Hayashi, J. Phys.
      Chem., 79, 699 (1975).
[31] Y. Yamamoto, M. Irie, and K. Hayashi, 15th Annual Meeting
      of the Polymer Society, Japan, Tokyo, June 1974.

Block and Bigraft Copolymers by Carbocation Polymerization*

JOSEPH P. KENNEDY, EARL G. MELBY,[†] and ALAIN VIDAL[‡]

Institute of Polymer Science
The University of Akron
Akron, Ohio 44325

ABSTRACT

The recent observation that the rate of methylation of t-BuCl
by Me_3Al is several orders faster than that of t-BuBr, which
in turn is much faster than that of t-BuI, provides the basis
of a new synthetic method for the preparation of block and
bigraft copolymers. This presentation concerns the synthesis
and characterization of the first well-defined block and bigraft
copolymers produced by the carbenium ion mechanism i.e.,
poly(styrene-b-isobutylene), poly[(ethylene-co-propylene)-
g-styrene-g-isobutylene)], and poly[(ethylene-co-propylene)-
g-styrene-g-α-methylstyrene)]. The synthesis of the block
copolymers involved three key steps: 1) the synthesis of a
chlorobrominated alkane initiator, 2) the selective initiation
of styrene polymerization by the chlorobrominated alkane
in conjunction with alkylaluminum halide under conditions
of no chain transfer, and 3) selective initiation of isobutylene
polymerization by the polystyrene-Br in conjunction with
alkylaluminum halide. Selective solubility, DSC, and GPC

---

*In the authors' absence, the paper was delivered by Dr. Patricia
Dreyfuss.
†Present address: Union Carbide Corp., Bound Brook, New Jersey
08805.
‡Present address: Cent. Rech. Phys. Chim. Surf. Solides,
Mulhouse, France.

data indicate pure block copolymer. The effect of temperature on the intrinsic viscosity of poly(styrene-b-isobutylene)
exhibited peculiarities characteristic of block copolymers.
The bigraft was synthesized by chlorobrominating an
ethylene-propylene copolymer and selectively initiating
first the polymerization of styrene and subsequently that
of α-methylstyrene or isobutylene by the chlorobrominated
copolymer plus an alkylaluminum compound.

INTRODUCTION

The recent discovery that alkylaluminum/alkyl halide combinations, e.g., $Et_2AlCl/t-BuCl$, are excellent initiator systems for the
cationic polymerization of a large variety of olefins [1, 2] has given
rise to vigorous fundamental and development studies in the United
States [3, 4] and in Europe [5, 6]. In particular, the $Et_2AlCl/t-BuCl$
system initiates the rapid polymerization of styrene to a very high
molecular weight product and, importantly, the polymerization
proceeds in the absence of chain transfer. The latter circumstance
led to the development of novel graft copolymers [7] and, more
recently, to the development of block and bigraft copolymers, the
subject of this presentation.

The discovery of the principle that led to the first carbenium
ion synthesis of block and bigraft copolymers was made during
our fundamental studies concerning the effect of the nature of
the tertiary butyl halide on the rate of methylation by Me_3Al, i.e.,

$$Me_3Al + t-BuX \rightarrow t-BuMe + Me_2AlX$$

where X = Cl, Br, and I [8].

Significantly, the rate of methylation of t-BuCl by Me_3Al in
methyl chloride diluent was found to be many orders of magnitude
larger than that of t-BuBr. Table 1 shows the findings.

SYNTHESIS AND CHARACTERIZATION OF POLY(STYRENE-b-ISOBUTYLENE)

Synthesis

It was theorized that the huge difference in methylation rates
shown in Table 1 could be exploited in the synthesis of a block

TABLE 1. Rates of Alkylation of t-Butyl Halides with Trimethyl-aluminum at -40°C

| t-BuX | $k[\text{liter}/(\text{mole})(\text{sec})] \times 10^6$ | | | |
	MeCl	MeBr	MeI	Cyclopentane
t-BuCl	$>10^9$ [a]	4100	2000	1.2
t-BuBr	13000	1800	· 100	0.7 [b]
t-BuI	75 [c]	15 [c]	3.2 [c]	0.005 [b]

[a]Calculated from the observed rate at -80°C, assuming an activation energy of ~10 kcal/mole.
[b]Calculated from the measured rates in the temperature range -0 to +50°.
[c]Calculated from the measured rates in the temperature range -20 to +20°.

copolymer, for example, poly(styrene-b-isobutylene). The plan was to synthesize a small molecule containing both a tertiary chloride and a tertiary bromide, and using it in conjunction with an alkyl-aluminum compound to initiate the polymerization of a monomer first only from the chloride site and subsequently to initiate the polymerization of another monomer from the bromine site. The concept using styrene and isobutylene as the two monomers may be schematized as follows:

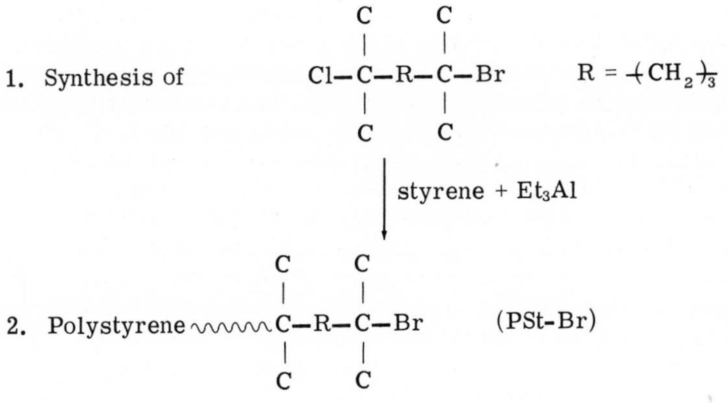

1. Synthesis of

$$
\begin{array}{ccc}
& C & C \\
& | & | \\
Cl-&C-R-&C-Br \\
& | & | \\
& C & C
\end{array}
\qquad R = \left(CH_2 \right)_3
$$

styrene + Et$_3$Al

2. Polystyrene〜〜〜〜

$$
\begin{array}{cc}
C & C \\
| & | \\
C-R-&C-Br \\
| & | \\
C & C
\end{array}
\qquad \text{(PSt-Br)}
$$

3. Remove unreacted chlorobromo initiator

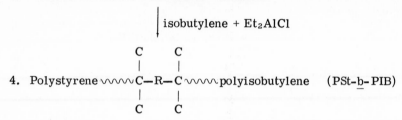

4. Polystyrene〜〜〜C—R—C〜〜〜polyisobutylene   (PSt-b-PIB)

For this plan to succeed it was essential to work out conditions
under which sequential initiation can be achieved. Model studies and
other experiments indicated that 2-chloro-6-bromo-2,6-dimethyl-
heptane (ClBrDMH) in conjunction with Et_3Al or Et_2AlCl was suitable
for this purpose. For example, about 50% of the chlorine in this
chlorobrominated alkane can be substituted before any bromine
loss using Et_3Al in CH_3Cl at -70°C.

Fundamental studies had shown previously that the polymerization
of styrene coinitiated, for example, by Et_3Al proceeds essentially
free of chain transfer. This background information was needed for
the "clean" synthesis of polystyrene possessing a terminal tertiary
bromine (PSt-Br).

Again, model studies indicated that the addition of ClBrDMH to a
charge of styrene and Et_3Al in C_2H_5Cl at -80°C readily gave PSt-Br
of $\overline{M}_n$ = 20,000, suitable for the subsequent isobutylene polymerization step.

Prior to the isobutylene polymerization, unreacted ClBrDMH had
to be removed. Residual ClBrDMH initiator, in conjunction with the
subsequently added alkylaluminum compounds, might give homopolyisobutylene and thus contaminate the block copolymer. Thus unreacted
ClBrDMH was removed by dissolving the PSt-Br in CH_2Cl_2, filtering
(to remove some aluminum oxides which arose during the quenching
of the styrene polymerization with cold methanol), and pouring the
CH_2Cl_2 solution into an excess of methanol. The precipitated
PSt-Br was washed several times with methanol and dried.

The final isobutylene polymerization step was carried out by
dissolving PSt-Br in a mixture of methylene chloride/hexane at
-55°C, adding isobutylene, and ultimately the Et_2AlCl coinitiator.
Since more aggressive conditions are necessary to achieve fast
initiation of isobutylene from the tertiary bromine site than from
the tertiary chlorine site, we used a stronger Lewis acid Et_2AlCl
(than Et_3Al) and higher temperatures -55°C (rather than -70°C).
Table 2 shows some representative examples of isobutylene
polymerizations using PSt-Br and Et_2AlCl.

TABLE 2. Isobutylene Polymerizations Initiated by PSt-Br/Et$_2$AlCl^{a,b}

	Reaction conditions						Products^c			
C$_4$H$_8$ (g)	PSt-Br (g)	Et$_2$AlCl ($\underline{M} \times 10^{-2}$)	Temperature (°C)	Time (min)	Conversion (%)	MEK insoluble (PSt-b-PIB, PIB)	MEK + pentane soluble (PSt-b-PIB)	MEK + heptane soluble (PSt-b-PIB)	Pentane or heptane insoluble (PSt)
2.1	0.92	1.4	-45	45	95	59% PSt content = 15%	38% PSt content = 66% $\overline{M}_n$ = 38,000	-	3% PSt content = 100%
31.5	7.8	1.4	-55	30	38	59% PSt content = 20%	27% PSt content = 70% $\overline{M}_n$ = 34,000	12% PSt content = 79% $\overline{M}_n$ = 35,000	2% PSt content = 100%
2.1	0.50	4.2	-65	30	43	16% PSt content = 17%	38% PSt content = 21% $\overline{M}_n$ = 42,000	44% PSt content = 47% $\overline{M}_n$ = 55,000	2% PSt content = 100%

^{a}PSt-Br synthesis conditions: To a solution of styrene, 0.10 mole, in ethyl chloride, 80 ml, and Et$_3$Al, 4.8×10^{-3} mole, introduce 2-bromo-6-chloro-2,6-dimethylheptane, 8.0×10^{-4} mole, at -80°C; quench with methanol after 5 min; yield 7.7 g (74%), $\overline{M}_n$ = 20,000.

^{b}Solvent for isobutylene polymerization, v/v: CH$_2$Cl$_2$/hexane = 65/25.

^{c}Percent on basis of final polymer yield. Experimental error: Experiments 1 and 3 = 10%, Experiment 2 = 4%.

Concurrently with polymerization experiments, blank experiments were also run to monitor the purity of the reagents used. Thus monomer, solvent, and alkylaluminum coinitiator were combined in the same proportion as used for polymerization; however, no initiator was added. After quenching the polymerization and the blank experiments with methanol, the absence of any polymer in the blank was an indication of the satisfactory purity level of the chemicals used.

Since the possibility for chain transfer in isobutylene polymerization initiated by the PSt-Br/Et₂AlCl system exists, the possibility for homopolyisobutylene formation also arises. Consequently, a selective extraction procedure was developed to separate the pure poly(styrene-b-isobutylene) PSt-b-PIB from the crude product, i.e., that contaminated by homopolyisobutylene.

Scheme 1 illustrates our selective extraction procedure together with the yields (wt%) and composition (wt% by NMR) of the fractions obtained from the polymer prepared at -55°C. Methyl ethyl ketone, MEK, a nonsolvent for polyisobutylene, dissolves polystyrene and PSt-b-PIB, rich in polystyrene. The MEK-insoluble fraction contains homopolyisobutylene along with PSt-b-PIB of lower polystyrene content. The subsequent extraction of the MEK-soluble material with pentane and heptane, nonsolvents for polystyrene, resulted in soluble fractions containing pure PSt-b-PIB. The fact that only insignificant quantities (2 to 3%) of homopolystyrene were recovered demonstrates the substantial absence of chain transfer in the synthesis of PSt-Br and leads to the expected high levels of terminal t-bromine.

Since the higher molecular weight block copolymer was found to be insoluble in MEK, we were not successful in isolating homopolyisobutylenes and thus determine the total quantity of PSt-b-PIB formed. However, it can be seen from Table 2 that the MEK-insoluble fraction is smallest for the product obtained at -65°C, demonstrating the presence of a significantly lower amount of homopolyisobutylene. This is consistent with the fact that chain transfer is reduced at lower temperatures.

Gel permeation chromatography studies confirmed our extraction results.

Characterization

Solvent fractionation coupled with NMR, IR, and GPC studies established the existence of PSt-b-PIB. In addition to these

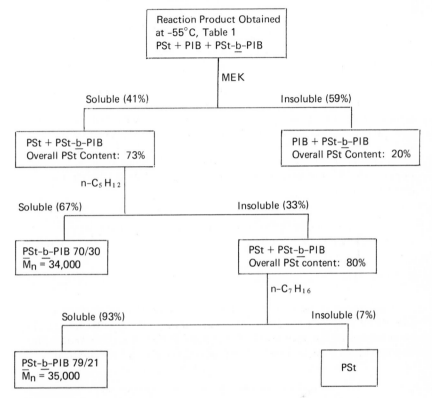

SCHEME 1. Extraction procedure used to obtain pure PSt-b-PIB.

analytical techniques, we have investigated the solubility and film behavior, glass transition temperature, and intrinsic viscosity of our block copolymer.

The PSt-b-PIB formed cloudy solutions in n-pentane (a solvent only for polyisobutylene) and in MEK (a good solvent only for polystyrene). In cyclohexane at room temperature slightly hazy solutions were obtained which, however, became clear when heated above the theta temperature of polystyrene (35°C), the temperature level beyond which cyclohexane becomes a good solvent for polystyrene. In contrast, the block copolymer formed visually clear solutions in toluene, benzene, and CCl₄, good solvents for both polystyrene and polyisobutylene.

Films cast from solutions of PSt-b-PIB in benzene were homogeneous and partially transparent. Films cast from cyclohexane
were striped, presumably due to phase separations since cyclohexane
is a poor solvent for polystyrene below 35°C.

The copolymer exhibited two T_g's (by DSC) at 369 and 199°K,
characteristic of polystyrene and polyisobutylene, respectively.

The intrinsic viscosity, $[\eta]$, of PSt-b-PIB as a function of
temperature in the range from 15 to 55°C is illustrated in Fig. 1.
The viscosity increases rapidly to a maximum followed by a
sudden decrease. After a minimum, $[\eta]$ again begins to increase.
This abnormal temperature dependence of $[\eta]$ has been observed
previously [9] and was shown to be characteristic of block and
graft copolymers.

BIGRAFT COPOLYMERS: SYNTHESIS, CHARACTERIZATION, AND PHYSICAL PROPERTIES

Synthesis

It occurred to us that the principle of selective initiation, as
described above in regard to the synthesis of PSt-b-PIB, could also
be extended to the synthesis of bigraft copolymers. The basic plan
was to select a suitable backbone polymer, chlorobrominate it to

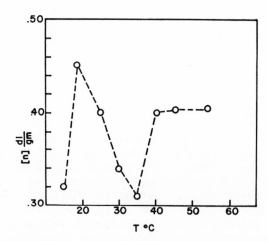

FIG. 1. The effect of temperature on the intrinsic viscosity
of poly(styrene-b-isobutylene) in toluene.

produce a chain containing randomly distributed tertiary or allylic
—Cl or —Br functions, and to use this species in conjunction with
a suitable alkylaluminum compound for the selective sequential
initiation of two monomers.

For a variety of reasons, such as ready availability, relatively
simple microarchitecture, desirable solubility characteristics,
and presence of tertiary and allylic positions, we decided to use
Nordel-1440 for the backbone polymer. Nordel-1440 is a copolymer
of ethylene and propylene containing a few percent 1,4-hexadiene.
For the first monomer we chose styrene and for the second
monomer α-methylstyrene or isobutylene.

The following scheme further illustrates the experimental
procedure and provides some detailed information as to the
chemicals used, copolymers contained, grafting efficiencies,
molecular weights, etc:

1. The Backbone Polymer (Nordel):

```
+C—C+C—C+C—C-wwwwwwC—C wwww
     |   |          |
     C   C          C
     |                  |
     C              C
     ||             ||
     C              C
     |                  |
     C              C

                              t-BuOCl
                          ——————————————>
```

```
~(C-C+C-C+C-C wwwwww C-C wwww
       |   |            |
       C   C            C
           |                |
           C            C
           ||           ||
           C            C
           |                |
         C-Cl           C

                              t-BuOBr
                          ——————————————>
```

```
ww+C-C+C-C+C-C wwwww C-C wwww
       |   |            |
       C   C            C
           |                |
           C            C
           ||           ||
           C            C
           |                |
         C-Cl          C-Br
```

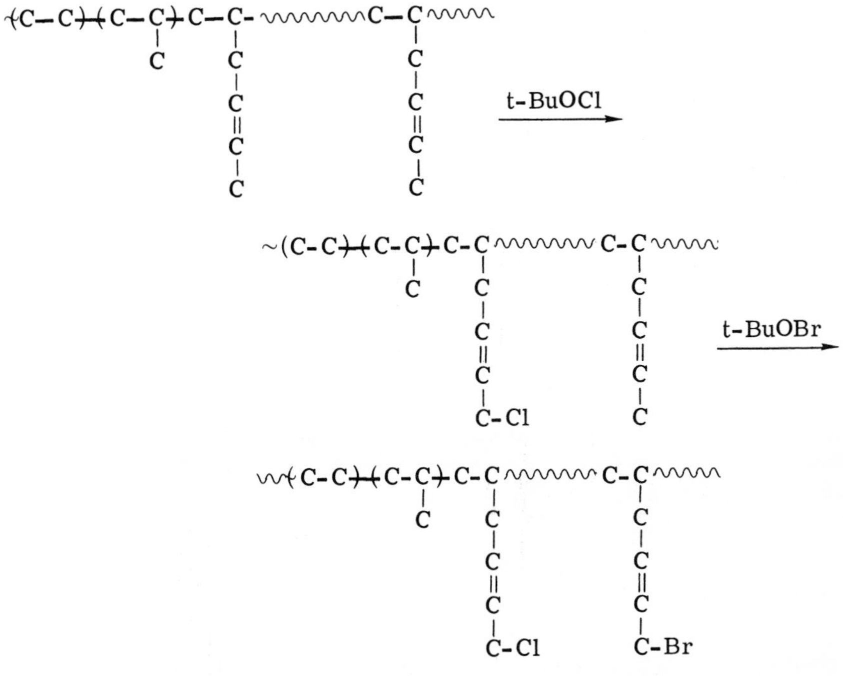

(ClBrNordel)

$\overline{M}_n$ = 57,000, Cl = 1.25%, Br = 5.88%.

2. The Monograft: Grafting of Styrene:

$$\text{ClBrNordel} \xrightarrow[\text{Et}_2\text{AlCl}, -30°\text{C}]{\text{styrene,}} \xrightarrow{\text{acetone extractions}}$$

~(C—C)—(C—C)—C—C www C—C www
       | | |
       C   C   C
            | |
            C   C
            ‖   ‖
            C   C
            | |
            C   C-Br
            |
        polystyrene

(BrNordel-g-polystyrene)

$\overline{M}_n$ = 98,000, G.E. = 40.5%, $\overline{M}_n$ homopolystyrene = 37,400.

3. The Bigraft: Grafting of α-Methylstyrene:

$$\text{BrNordel-g-polystyrene} \xrightarrow[\text{Et}_2\text{AlCl}, -20°\text{C}]{\alpha\text{-methylstyrene}} \xrightarrow{\text{3-pentanone extractions}}$$

~(C—C)—(C—C)—C—C www C—C www
       | | |
       C   C   C
            | |
            C   C
            ‖   ‖
            C   C
            | |
            C   C
            | |
      polystyrene   poly-α-methylstyrene

(Nordel-g-polystyrene-g-poly-α-methylstyrene)

$\overline{M}_n$ = 189,000, G.E. = 11.7%.

Syntheses have also been performed in which the second monomer was isobutylene. This procedure yielded Nordel-g-polystyrene-g-polyisobutylene, $\overline{M}_n$ = 190,000.

A large amount of research effort was necessary to find suitable conditions for all the steps of these syntheses. For example, the allylic chlorobromination of Nordel had to be worked out. We preferred to use pendant allylic halogens as initiators because preliminary and other information indicated high allylic initiation activity coupled with selectivity. Also, the purification of first the monograft and subsequently the bigraft copolymers had to be worked out. We found acetone extractions to be satisfactory for the removal of the homopolystyrene formed via chain transfer in the first grafting step. Pertinent background information was available to guide us in accomplishing this purification step from previous work in this laboratory [10]. Thus a series of BrNordel-g-polystyrenes have been prepared and characterized. These materials are interesting thermoplastic elastomers whose properties are, obviously, quite similar to EPR-g-polystyrenes previously synthesized in our laboratories [10].

The purification of the bigrafts was a difficult undertaking. A large number of solubility and other studies was carried out before a satisfactory solution was found. A key finding was that cold 3-pentanone dissolves only homopoly-α-methylstyrene but not the Nordel-g-polystyrene-g-poly-α-methylstyrene bigraft. Interestingly, hot 3-pentanone can also be used to separate homo-polyisobutylene in the Nordel-g-polystyrene-g-polyisobutylene synthesis. Whereas homopolyisobutylene is insoluble in boiling 3-pentanone, the bigraft dissolves in this solvent.

Another fractionation technique was to gel the bigraft (by treatment with sulfur monochloride) and extract all the soluble (i.e., nonbigraft) fraction.

The compositions of pure bigrafts were determined spectro-scopically (IR, NMR); number-average molecular weights were established by osmometry and molecular weight distributions by GPC.

Table 3 shows some representative data.

Characterization and Physical Properties

At this time we wish to report some results in regard to our characterization and physical property studies carried out with various bigrafts.

The glass transition temperatures were determined by DSC. As expected, the Nordel-g-polystyrene-g-poly-α-methylstyrene and the Nordel-g-polystyrene-g-polyisobutylene both exhibited T_g's, indicating the presence of molecularly incompatible domains.

Stress-strain and permanent set data was obtained with micro-dumbbells prepared of solution cast films. Figure 2 shows some representative results.

TABLE 3. The Synthesis and Characterization of Bigraft Copolymers

Backbone (g) (initiator)	Monomer (M) Coinitiator (M) Solvents (v/v) T (°C)	Conversion (%)	Homopolymer[a]	Graft	Grafting efficiency (%)	Branches/backbone
I. Nordel-g-PSt-g-PαMeSt						
1. The Monograft Nordel-g-PSt						
ClBr-Nordel ~ 4 g $\overline{M}_n = 57,000$	Styrene (1.5) AlEt$_2$Cl (1.32×10^{-2}) n-Heptane/EtCl (85/15) -30	5.45	$\overline{M}_n = 37,400$	Br-Nordel-g-PSt $\overline{M}_n = 98,000$ PSt content = 17.6%	40.6	1.1
2. The Bigraft Nordel-g-PSt-g-PαMeSt						
Br-Nordel-g-PSt ~ 4 g $\overline{M}_n = 98,000$	α-Methylstyrene (1.5) AlEt$_2$Cl (1.32×10^{-2}) n-Heptane/EtCl (80/20) -20	19	$\overline{M}_n = 51,500$	Nordel-g-PSt-g-PαMeSt $\overline{M}_n = 189,000$ PSt + PαMeSt content = 46.2%	11.7	1.8 (relative to PαMeSt grafts)

II. Nordel-g-PSt-g-PIB

1. The Monograft Nordel-g-PSt

ClBr-Nordel ~ 4 g	Styrene (1.5)	16.6	$\overline{M}_n$ = 36,500	Br-Nordel-g-PSt	30	1.16
	AlEt₂Cl (1.32 × 10⁻²)			$\overline{M}_n$ = 99,500		
$\overline{M}_n$ = 57,000	n-Heptane/EtCl (80/20)			PSt content = 31.5%		
	-30					

2. The Bigraft Nordel-g-PSt-g-PIB

Br-Nordel-g-PSt ~ 4 g	Isobutylene (1.5)	44.05	$\overline{M}_n$ = 99,300	Nordel-g-PSt-g-PIB	3.7	1.0 (relative to PIB grafts)
	AlEt₂Cl (1.32 × 10⁻²)			$\overline{M}_n$ = 190,000		
$\overline{M}_n$ = 99,500	n-Heptane/EtCl (70/30)			PSt + PIB content = 64.05%		
	-30					

a Molecular weights of extracted homopolymers.

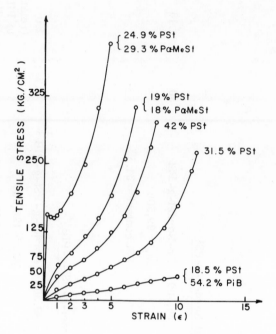

FIG. 2. Stress-strain properties of representative monograft
and bigraft copolymers. The numbers indicate the composition
of the branches of the graft.

ACKNOWLEDGMENTS

Acknowledgment is made to the donors of the Petroleum Research
Fund, administered by the American Chemical Society, and to the
National Science Foundation for partial support of this research.

REFERENCES

[1] J. P. Kennedy, Belgian Patent 663,319 (April 30, 1965).
[2] J. P. Kennedy and F. P. Baldwin, Belgian Patent 663,320
     (April 30, 1965).
[3] J. P. Kennedy and S. Rengachary, Adv. Polym. Sci., 14, 1
     (1974).
[4] J. P. Kennedy, Makromol. Chem., 175, 1101 (1974).

[5] M. Baccaredda, M. Bruzzone, S. Cesca, M. DiMaina, G. Ferraris, P. Giusti, P. L. Magagnini, and A. Priola, Chim. Ind., Milan, 55, 109 (1973).

[6] A. Priola, S. Cesca, and G. Ferraris, Makromol. Chem., 160, 41 (1972).

[7] J. P. Kennedy, J. J. Charles, and D. L. Davidson, in Recent Advances in Polymer Blends, Grafts and Blocks (L. H. Sperling, ed.), Plenum, New York, 1974, p. 157.

[8] J. P. Kennedy, N. V. Desai, and S. Sivaram, J. Amer. Chem. Soc., 95, 6386 (1973).

[9] A. Dondos, Makromol. Chem., 99, 275 (1966).

[10] J. P. Kennedy and R. R. Smith, in Recent Advances in Polymer Blends, Grafts and Blocks (L. H. Sperling, ed.), Plenum, New York, 1974, p. 303.

Cationic Polymerization with Expansion in Volume

WILLIAM J. BAILEY

Department of Chemistry
University of Maryland
College Park, Maryland 20742

ABSTRACT

It is demonstrated that polymerization of bicyclic monomers in which two rings are opened for every new bond formed in the polymer chain will produce either very small changes in volume or significant expansion.

For a number of industrial applications, such as strain-free composites, potting resins, high gloss coatings, and binders for solid propellants, it appears to be highly desirable to have monomers that will polymerize with near zero shrinkage. For other applications, such as precision castings, high strength adhesives, prestressed plastics, rock-cracking materials, and dental fillings, it appears highly desirable to have monomers that would undergo positive expansion on polymerization. A number of examples from other fields would lead one to predict that such an expansion in volume during polymerization would be highly useful. For example, since water undergoes an expansion in volume of about 4% when it freezes, ice will adhere to almost any surface, including Teflon, which it does not even wet, by what is called micromechanical inter-action. Alloys that expand on solidification are used commercially to make precision type, and amalgams that expand are used to fill teeth.

Many composites involving high strength fibers in a polymeric matrix fail because of either poor adhesion between the matrix and the fibers or because of voids and microcracks in the matrix. Both of these problems are at least partially related to the fact that when available materials polymerize or cure, a pronounced shrinkage takes place. In bulk plastics some of these stresses can be relieved with a total shrinkage in the overall dimensions of the article. However, in a composite the reinforcing material, which has a high modulus, will often not permit appreciable shrinkage in the overall dimensions of the molded object, and as a result enormous stresses are built up in the composite. These stresses can be relieved either by an adhesive failure, in which the matrix pulls away from the reinforcing fiber, or by a cohesive failure, in which a void or a microcrack is formed.

For these reasons a research program was initiated to find mono-mers that undergo either zero shrinkage or expansion on polymeri-zation. Shrinkage that occurs during polymerization arises from a number of factors. One of the most important, however, is the fact that the monomer molecules are located at a van der Waals' distance from one another while in the corresponding polymer the monomeric units move to within a covalent distance of each other. The atoms are thus much closer to one another in the polymer than they were in the original monomer. In some cases the polymer actually packs better than the monomer. This would be true in going from an amorphous or liquid monomer to a semicrystalline polymer to pro-duce a greater than normal shrinkage. On the other hand, in going from a crystalline monomer to an amorphous polymer which may have a great deal of free volume, the shrinkage could be much less than normal.

In a condensation polymerization, one would expect shrinkage, since a small molecule is eliminated during the formation of the new bond. One can get large shrinkages during such condensation polym-erization depending on the size of the molecule eliminated. Table 1 lists some typical calculated shrinkages during condensation polym-erization. Obviously, the larger the molecule split out during the polymerization, the larger the shrinkage will be. During addition polymerization, even though there is no small molecule eliminated, there is still a fairly large shrinkage in many cases, since the atoms are much closer to one another in the polymer than they were in the monomer. Table 2 lists the shrinkage that occurs in a number of very common monomers. The shrinkage is related, of course, to a

TABLE 1. Calculated Shrinkage During Condensation Polymerization

Monomer A	d_4^{20}	Monomer B	d_4^{20}	Polymer (d_4^{20})	Shrinkage (%)
Adipic acid	1.360	$NH_2(CH_2)_6NH_2$	0.863	1.14	22
Adipic acid	1.360	$O=C=N(CH_2)_6-N=C=O$	1.0528	1.14	29
Dimethyl adipate	1.0625	$NH_2(CH_2)_6NH_2$	0.863	1.14	31
Dibutyl adipate	0.965	$NH_2(CH_2)_6NH_2$	0.863	1.14	53
Dioctyl adipate	0.913	$NH_2(CH_2)_6NH_2$	0.863	1.14	66

TABLE 2. Calculated Shrinkages for Addition Polymerization [1]

Monomer	Shrinkage (%)
Ethylene	66.0
Vinyl chloride	34.4
Acrylonitrile	31.0
Methyl methacrylate	21.2
Vinyl acetate	20.9
Styrene	14.5
Diallyl phthalate	11.8
N-Vinylcarbazole	7.5
1-Vinylpyrene	6.0

large extent to how many monomer molecules undergo polymerization per unit volume. For example, styrene, which has approximately four times the molecular weight of ethylene, gives a shrinkage during polymerization which is approximately one-quarter of that of ethylene.

In ring-opening polymerization, the shrinkage may be less than in either of the two cases just discussed. Not only is a small molecule not eliminated during the polymerization reaction, but for every bond that undergoes a change from a van der Waals' distance to a covalent distance, another bond goes from a covalent distance to a near van der Waals' distance. Table 3 lists the shrinkage that would occur during the ring-opening polymerization of a number of cyclic monomers. It is obvious from this table that the shrinkage is related to the number of rings being opened per unit volume as well as to the size of the ring. By analogy with the molecular volume of ethylene, one might at first glance expect the shrinkage of ethylene oxide to be approximately 42%. Thus the reduced shrinkage must be at least partly due to the opening of the three-membered ring.

In a program designed to find monomers that would give near zero shrinkage or even expand on polymerization, a fruitful approach appeared to be the study of bicyclic compounds. In such bicyclic monomers, for every bond that undergoes a shift from a van der Waals' distance to a covalent distance, at least two bonds go from a covalent distance to a near van der Waals' distance. In a study of densities

TABLE 3. Calculated Shrinkage during Ring-Opening Polymerization [2, 3]

Monomer	Shrinkage (%)
Ethylene oxide	23
Propylene oxide	17
Cyclopentene	15
Cyclopentane	12
Styrene oxide	9
Cyclooctene	5
Bisphenol-A digylcidyl ether, diethylaminopropylamine	5
Cyclododecatriene	3
5-Oxa-1,2-dithiacycloheptane	3
Dimethylsilane oxide cyclic tetramer	2

of polycyclic hydrocarbons compared to the densities of the corresponding linear polymers which have the same empirical formula, one can become convinced that this approach would indeed give monomers that expand on polymerization.

In the hypothetical case shown, cyclopentene, which undergoes a 15% shrinkage when it is converted to the polycyclopentenomer, would undergo a 20% shrinkage when converted to its saturated dimer. If a catalyst were available for the polymerization of the saturated dimer to the same polycyclopentomer, a 6% expansion would be predicted. Similarly, if the cyclopentene is converted to adamantane, a 26% shrinkage would result, and if adamantane could somehow be converted to the same cyclopentenomer, a 17% expansion in volume would be calculated. Unfortunately, the catalysts which would make these conversions possible are not known.

For the reasons just discussed, a search was made for oxygen analogs of a number of bicyclic monomers, since catalysts were available for the cationic polymerization of cyclic ethers and cyclic esters. The monomer that appeared particularly attractive was the spiro ortho ester, 1,4,6-trioxaspiro[4,4]nonane, which was first reported by

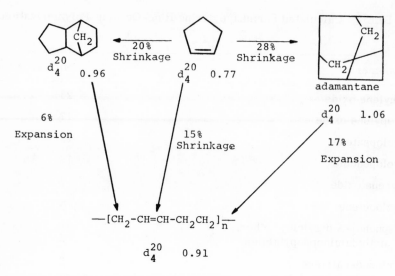

Bodenbenner [4] in 1959. Although there are a number of Japanese patents [5-8] which describe the copolymerization of this monomer with trioxane to produce copolymers with improved thermal stability, no description of the homopolymerization of this monomer appears in the literature. Bodenbenner reported that when the monomer was allowed to stand at room temperature with boron trifluoride, it gave a viscous oil but which was not characterized as to either structure or molecular weight. This spiro ortho ester was synthesized in a 33% yield [9].

$$CH_2\!\!-\!\!CH_2 \;\; + \;\; CH_2\!\!-\!\!CH_2 \qquad \xrightarrow[\text{13\% Shrinkage}]{BF_3} \qquad CH_2\!\!-\!\!CH_2 \qquad O\!\!-\!\!CH_2$$

$$d_4^{20} \;\; 0.869 \qquad d_4^{20} \;\; 1.11 \qquad\qquad\qquad\qquad d_4^{20} \;\; 1.16$$

By a comparison of the densities of the materials involved, it is obvious that the spiro compound is a very compact monomer. When this monomer was polymerized in the dilatometer, placed in a constant temperature bath of 25° C in the presence of boron trifluoride, polymerization occurred over a 24-hr period. During the polymerization the meniscus remained essentially at a constant level and indicated a slight increase in volume of 0.1%. Purification of the polymer by reprecipitation gave a 95% yield of a viscous liquid with a molecular weight of about 25,000. Although the polymer was difficult to purify, its density indicated that a slight shrinkage occurred during the polymerization (less than 1%).

The mechanism of the polymerization undoubtedly involves a oxonium ion:

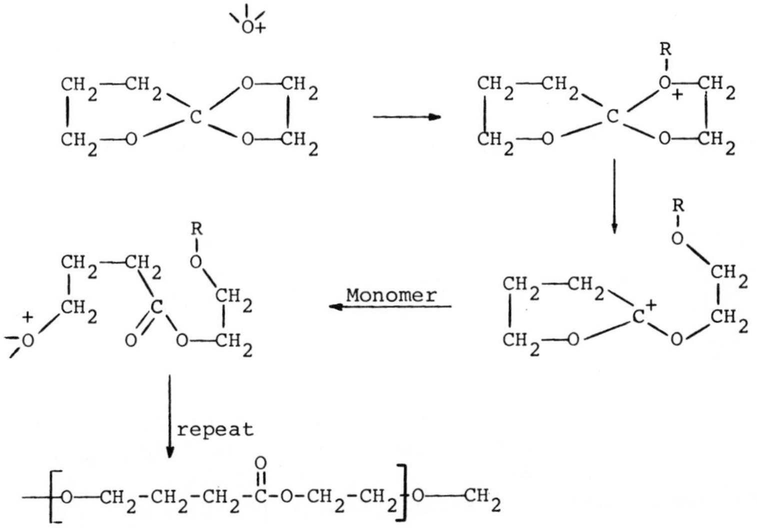

$$\text{Monomer}$$

repeat

Since the monomer contains two different types of oxygen atoms, attack can occur at the other oxygen:

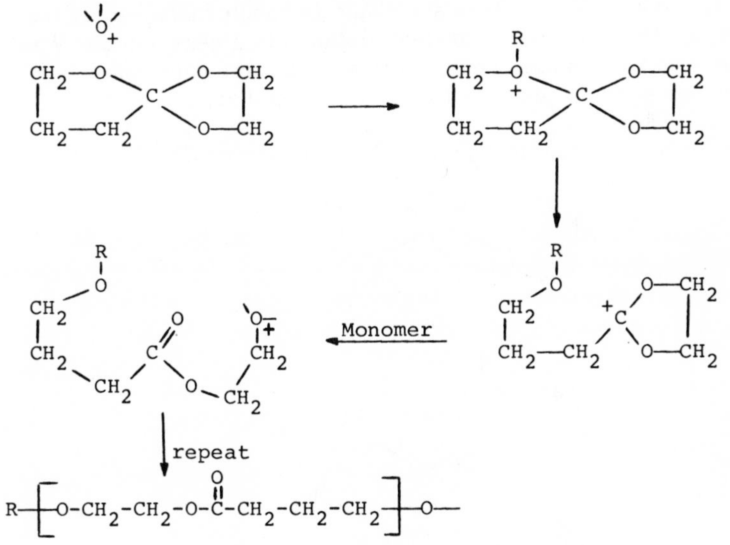

$$\text{Monomer}$$

repeat

On this basis one would predict the formation of a polyester-ether containing both head-to-tail and head-to-head units. This fact was verified by the synthesis of the head-to-tail polymer by a direct esterification procedure:

$$HO-CH_2-CH_2-O-CH_2\,CH_2\,CH_2\,\overset{\overset{\displaystyle O}{\|}}{C}-OH \longrightarrow$$

$$\left[O-CH_2-CH_2-O-CH_2-CH_2-CH_2-\overset{\overset{\displaystyle O}{\|}}{C} \right]_x$$

Comparison of the NMR spectra of the two polymers gave strong evidence for the presence of a head-to-head units (10 to 20%) in the ring-opened polymer.

The reason for this low shrinkage during polymerization can be rationalized by comparing the original monomer with the final ether-containing polyester. There are two processes which would lead to some contraction; one bond goes from a van der Waals' distance to a covalent distance, and one bond goes from a single bond to a double bond. This shrinkage is counterbalanced by the two bonds that go from a covalent distance to a near van der Waals' distance in the final polymer. In this particular case, these processes seem to just about cancel one another.

Interesting analogs of this spiro ortho ester can be prepared by the same general technique. For instance, a crystalline monomer from phenyl glycidyl ether can be prepared by the following reaction in a 50% yield [10]:

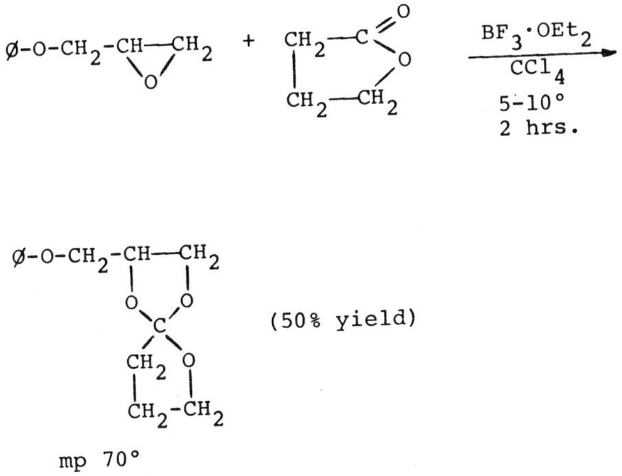

$$\emptyset\text{-O-CH}_2\text{-CH}\overset{\displaystyle\diagdown_{\text{O}}\diagup}{}\text{CH}_2 \quad + \quad \begin{array}{c}\text{CH}_2\text{---C}\diagup^{\text{O}}_{\diagdown\text{O}} \\ | \qquad | \\ \text{CH}_2\text{---CH}_2\end{array} \quad \xrightarrow[\substack{\text{CCl}_4 \\ 5\text{-}10° \\ 2 \text{ hrs.}}]{\text{BF}_3\cdot\text{OEt}_2}$$

$$\begin{array}{c}\emptyset\text{-O-CH}_2\text{-CH---CH}_2 \\ | \qquad | \\ \text{O} \quad \text{O} \\ \diagdown_{\text{C}}\diagup \\ | \\ \text{CH}_2 \quad \text{O} \\ | \qquad | \\ \text{CH}_2\text{-CH}_2\end{array} \qquad (50\% \text{ yield})$$

mp 70°

When this monomer is polymerized at 30° C with boron trifluoride etherate, a polymer results with high molecular weight. From the density of the polymer compared with that of the monomer, one would calculate that a 12% expansion in volume had occurred. When the same process is carried out at 75° C, a 1% shrinkage in volume is calculated. Since the monomer is crystalline, it packs very well, but the polymer is amorphous and does not pack well. Therefore, in addition to the volume change that would occur during polymerization, a large additional change in volume occurs when the crystalline monomer is converted to an amorphous material. Several other cases in which a crystalline monomer is converted to an amorphous polymer with a large increase in volume are discussed later.

Still another variation in this series is the bifunctional monomer that can be prepared from the hydroquinone diglycidyl ether:

$$BF_3 \cdot OEt_2 \qquad 5-10°$$
$$CCl_4 \qquad 1.5 \text{ hrs.}$$

mp 176° (30% yield)

On polymerization this monomer will give a highly cross-linked material, but it can also be used to vulcanize the linear ether-containing polyester obtained from the previously discussed spiro ortho ester.

If one desires a higher melting polymer from an spiro ortho ester, the introduction of cyclohexane rings gives a fairly large increase in the softening point. The introduction of one cyclohexane ring produces a monomer that gives a substantial increase in volume upon polymerization (3%). The introduction of two cyclohexane rings gives a material with a softening point of over 100° C, but there is essentially no change in volume.

Although the adduct between an epoxy resin and the butyrolactone did not give a material that could be isolated in the pure state, it was possible to use this reaction to prepare prepolymers which had a large shrinkage during the initial portion of the reaction when the material was liquid. When this prepolymer was further polymerized with boron trifluoride, a cross-linked material resulted with essentially no change in volume near the last part of the polymerization where the material becomes viscous and gels. This technique should allow the production of strain-free materials at a reasonable cost.

Another very interesting class of compounds appeared to be the spiro ortho carbonates. Sakai, Kobayashi, and Ishii [11] recently described a method for synthesizing ortho carbonates using tin compounds with carbon disulfide. Using their method, we were able to synthesize a series of spiro ortho carbonates by the following set of reactions. This method worked well for 1,2-, 1,3-, or 1,4-glycols to produce crystalline monomers [12].

Since the spiro ortho carbonate was a highly crystalline material, initial polymerization studies were carried out above its melting point at 142° C. Although the polymerization could be carried out with a variety of cationic catalysts, such as boron trifluoride gas, boron trifluoride etherate, and aluminum chloride, boron trifluoride etherate proved to be the most convenient.

Thus, when the polymerization of molten spiro ortho carbonate was carried out in bulk with boron trifluoride etherate at 142° C, a quantitative yield of polymer was obtained after several hours.

m.p. 141°

[When polymerization was carried out at higher temperatures, the evolution of a gas (CO_2) was observed.] The polymer was purified by dissolution in chloroform followed by extraction of the solution with water. The structure of the polymeric material was proven not only by elemental analysis, but also by NMR and IR spectra. The polymer had an intrinsic viscosity of 0.26 in chloroform at 25° C. Although the relationship between molecular weight and intrinsic viscosity is unknown for this series of polymers, a reasonable assumption of the constants would indicate a molecular weight in excess of 100,000. These results would indicate that the strain inherent in the ortho carbonate structure provides a strong driving force for the polymerization. A very similar polymerization could be carried out at 100° C by addition of catalyst to the solid monomer.

When the polymerization was carried out in a dilatometer in which the bath was held at a constant temperature (142° C), the meniscus, instead of falling as is the usual case during polymerization, actually rose quite substantially. A calculation of the extent of change in volume indicates an expansion in excess of 2%. This compares very favorably with the very slight increase (0.14%) in volume reported earlier for the polymerization of a spiro ortho ester. This example, then, represents the first reported case in which a substantial amount of expansion in volume occurs during polymerization.

An even more remarkable relationship was discovered when the densities of the monomer and polymer were determined as a function of temperature. Table 4 lists the densities of the two materials at 25, 100, 130, and 142° C.

TABLE 4. Calculation of Expansion During Polymerization

Temperature (°C)	Density of monomer (g/cc)	Density of polymer (g/cc)	Expansion in volume (%)
25	1.31	1.20	9
100	1.30	1.14	14
130	1.30	1.11	17
142	1.12	1.10	2

The density of the amorphous liquid polycarbonate varied quite regularly and smoothly with changes in temperature from 1.20 g/cc at 25° C to 1.10 g/cc at 142° C. The density of the monomer, however, changed quite abruptly when it went from the molten monomer at 142° C to the crystalline monomer at temperatures below its melting point. Obviously, this data shows that the crystalline monomer is considerably more dense than the molten monomer. Similarly, the crystalline monomer was much more dense than the liquid polycarbonate. Thus, when the expansion in volume is calculated from the density of the crystalline monomer, the expansion was 9% at 25° C up to 17% at 130° C. Under ideal conditions the expansion might even be somewhat larger since the density of the crystalline monomer was determined by measuring the volume of a given weight of a solidified molten monomer. Under these conditions it is almost impossible to avoid the presence of some voids or the inclusion of a small amount of amorphous material. By coincidence, however, the 17% expansion is quite close to the expansion already calculated for the conversion of adamantane to polycyclopentenomer [2]. Figure 1 gives the plot of the densities of the monomer and polymer vs temperature.

It is obvious from the data that the conversion of a crystalline monomer to an amorphous polymer represents the ideal case for the large expansion in volume since in most cases the crystalline monomer would be expected to be considerably more dense than the corresponding liquid monomer. This is just the opposite of the case in which a liquid monomer is converted to a crystalline polymer. For example, when liquid ethylene monomer is converted to crystalline polyethylene at 5° C, a 66% shrinkage occurs [3]. It appears that the conversion of a liquid monomer to a crystalline polymer represents the ideal case to get the largest shrinkage during polymerization.

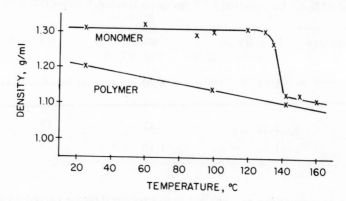

FIG. 1. Densities of the monomeric spiro ortho carbonate and related polyoxycarbonate vs temperature.

An inspection of Fig. 1 indicates that the densities of the monomeric spiro ortho carbonate and the polymer appear to cross above 200°. At that point, one would expect no change in volume during polymerization since the two materials have the same density. Above this critical temperature, one would expect to get shrinkage during the polymerization. Unfortunately, the polymerization cannot be carried out conveniently in this temperature range with the catalysts now available since carbon dioxide is liberated and the polycarbonate is not obtained in a pure form. At the lower end of the temperature scale the two lines appear to intersect again, but one would expect below the glass transition of the polymer that the density line would become more nearly horizontal and become essentially parallel to the line of the density of the monomer.

While at first it appeared difficult to find a polymerization procedure that would take full advantage of this large expansion, since the solid monomer is hard to introduce into a mold or a composite, it was found possible to make a slurry of this crystalline material in liquid epoxy monomer and copolymerize the two with controlled shrinkage. Expansion, contraction, or zero change in volume could be obtained depending on the concentration of the crystalline monomer in the slurry. The large volume increase also suggests that the polymerization may be used to replace explosives in cracking rocks in a quarry or for excavations.

A variety of analogs of this spiro ortho carbonate can be prepared and polymerized. One that is of special interest is the 3,9-dimethylene derivative which can be polymerized either cationically or by a free radical technique to produce cross-linked resins at high conversions and linear soluble polymers at low conversions [13]:

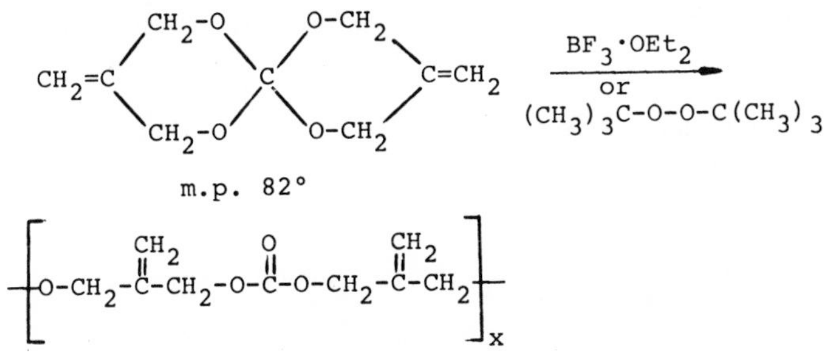

m.p. 82°

When the polymerization was run at room temperature, a 4.5% expansion in volume occurred, but when the polymerization was carried just below the melting point, an 8% expansion in volume was noted. Polymerization above the melting point of the unsaturated monomer produced a 1.4% expansion at 100° C, while at 115° C the densities of the monomer and the polymer appeared to be equal. Above 115° C, the polymerization produced a shrinkage.

A higher softening polymer in this same series could be prepared from the trispiro analog, which has a melting point of 112° C. Polymerization at room temperature produced a material with an increase in volume of 4%. Just below the melting point it gave a high melting polymer with an expansion of 7%. The spiro ortho carbonates appear to be a very versatile class of compounds for polymerization with expansion in volume.

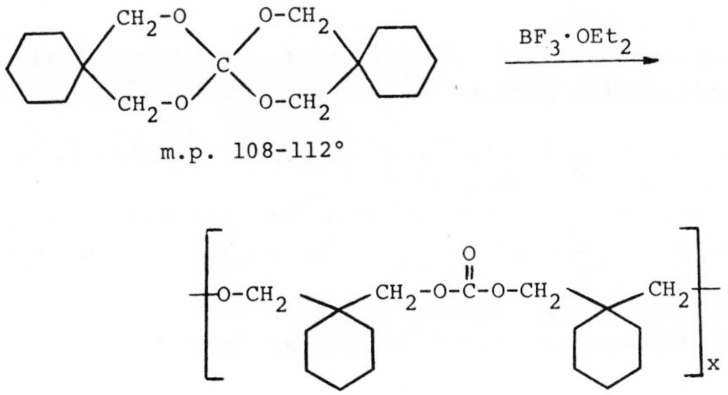

m.p. 108-112°

In order to demonstrate that expansion in volume would take place
with bicyclic materials other than spiro derivatives, a ketal lactone
was prepared by the method of Lange, Wamhoff, and Korte [14].
Polymerization of this material with either boron trifluoride or a
base produced the keto-containing polyester with essentially no
change in volume. We have demonstrated, therefore, that polymerization with no change in volume or with expansion in volume is a
general phenomena possible with a wide variety of liquid and crystalline cyclic and bicyclic monomers.

REFERENCES

[1] F. S. Nichols and R. G. Flowers, Ind. Eng. Chem., 42, 292
     (1950).
[2] W. J. Bailey and R. L. Sun, Amer. Chem. Soc., Div. Polym.
     Chem., Prepr., 13(1), 400 (1972).
[3] A. V. Tobolsky, F. Leonard, and G. P. Roeser, J. Polym.
     Sci., 3, 604 (1948).
[4] K. Bodenbenner, Justus Liebigs Ann. Chem., 625, 183 (1959).
[5] J. Kashiro, M. Kanaoka, and A. Kosakada (Toyo Rayon Co.),
     Japanese Patent 67 3495 (January 22, 1967); Chem. Abstr.,
     67, 54116v (1967).
[6] S. Inoue and T. Kataoka (Toyo Rayon Co.), Japanese Patent
     65 3708 (February 26, 1965); Chem. Abstr., 63, 11732h
     (1965).
[7] Toyo Rayon Co., French Patent 1,409,957 (September 3, 1965);
     Chem. Abstr., 65, 4052g (1966).
[8] Y. Yamase and T. Kuzuma (Yawata Chemical Industry Co.),
     Japanese Patent 69 28,111 (November 20, 1969); Chem. Abstr.,
     72, 56094n (1970).
[9] W. J. Bailey, J. Elastoplast., 5, 142 (1973).
[10] W. J. Bailey and H. Iwama, Unpublished Results.

[11] S. Sakai, Y. Kobayashi, and Y. Ishii, J. Org. Chem., 36, 1176 (1971).
[12] W. J. Bailey and H. Katsuki, Amer. Chem. Soc., Div. Polym. Chem., Prepr., 14, 1679 (1973).
[13] W. J. Bailey, H. Katsuki, and T. Endo. Ibid., 15, 445 (1974).
[14] C. Lange, H. Wamhoff, and F. Korte, Chem. Ber., 100, 2312 (1967).

Mechanism of Donor-Acceptor Alternating Copolymerization

JUNJI FURUKAWA

Department of Synthetic Chemistry
Kyoto University
Kyoto 606, Japan

ABSTRACT

The alternating copolymerization of butadiene and an acrylic compound in the presence of ethyl aluminum dichloride and vanadium oxychloride as complexing agents was studied kinetically for the comparison of two mechanisms, i.e., one involving an intermediate of a ternary complex of butadiene–acrylic monomer–EtAlCl$_2$ and the other without the complex formation. The rate of propagation was found to attain a maximum at a definite monomer composition, and this composition is not varied by changing the amount of EtAlCl$_2$ but decreased with increasing the concentration of total monomer. This fact is explained only by the mechanism of the ternary complex intermediate. In relation to the mechanism, NMR study of the ternary complex, ESR study of the growing radical, NMR study of the regularity of the copolymer, and the elementary reaction of the propagation are reviewed with discussion.

INTRODUCTION

Several kinds of alternating copolymerizations controlled by such factors as the steric hindrance of the substituent of monomers, the coordination ability of monomers, and the polar nature of the reaction sites of monomers have been found. Among these, the alternating copolymerization of electron-donating monomer and electron-accepting monomer is of special interest because it gives a copolymer with an extremely high degree of alternation. However, the mechanism is not fully established; a mechanism involving the formation of a donor-acceptor complex as an intermediate is believed to be correct, but there is a possible alternative mechanism involving a Lewis-Mayo-type copolymerization without the formation of the intermediate complex.

The author studied in detail the copolymerization of butadiene and such acrylic monomers as acrylonitrile and methyl methacrylate in the presence of ethylaluminum dichloride ($EtAlCl_2$) and vanadyl chloride ($VOCl_3$), which are a complexing agent and a promotor, respectively.

This paper includes a summary of previously published work together with the recent study.

RESULTS AND DISCUSSION

Kinetic Study [1, 2]

In the polymerization with varied concentrations of ethylaluminum dichloride [Al], vanadyl chloride [V], butadiene [BD], and acrylic monomer [A], the overall rate of polymerization R_p was found to be expressed by

$$R_p = k_p [Al]^{1.5} [V]^{0.5} [BD]^m [A]^n \tag{1}$$

where k_p is a rate constant of polymerization. The order m and n for butadiene and acrylic monomer, respectively, were both zero for the copolymerization of butadiene and methyl methacrylate (MMA) while they were 1 and 2 for the copolymerization of butadiene and acrylonitrile (AN).

The molecular weight of the resulting polymer increases linearly with an increasing extent of polymerization or polymer yield, but the increase in molecular weight is slowed down at a large degree of polymerization. This fact is compatible with the assumption that the polymerization involves rapid initiation, step-growth propagation, and some chain transfer reaction. Equation (2) describes the relationship of the weight of polymer W and its molecular weight M:

$$W/M = a + bW \tag{2}$$

where a is the number of active species and b the ratio of the transfer reaction to the propagation reaction if these reaction rates are proportional to each other. The constants a and b were measured at various concentrations of the catalyst components, and the following relationship was obtained:

$$a \propto [Al]^{0.5} [V]^{0.5} \tag{3}$$

Since the step-growth polymerization is proportional to the concentrations of the active species C* and the monomer:

$$R_p = k_p [C*][monomer] \tag{4}$$

Comparing Eqs. (1), (2), (3), and (4), it follows that

$$[C*] \propto [Al]^{0.5} [V]^{0.5} \tag{5}$$

and

$$[monomer] \propto [Al] \tag{6}$$

It is concluded that the actual monomer responsible for propagation contains 1 mole of aluminum compound as a complexing agent. A ternary complex of butadiene, acrylic monomer, and ethylaluminum dichloride was assumed to be formed and to be responsible for the alternating copolymerization. The order of monomer concentration can be explained by taking into account the formation of the ternary complex. There are several kinds of complexes among ternary components:

$$A + EtAlCl_2 \xrightleftharpoons{K_1} A \cdot EtAlCl_2$$

$$2A + EtAlCl_2 \xrightleftharpoons{K_1'} A_2 \cdot EtAlCl_2$$

$$A + BD + EtAlCl_2 \xrightleftharpoons{K_2} A \cdot BD \cdot EtAlCl_2 \text{ or } A \cdot BD \cdot al$$

The concentration of the ternary complex is expressed by a Langumuir-type equation:

$$[A \cdot BD \cdot al] = \frac{K_2 [al][A][BD]}{1 + k_1 [A]^n + K_2 [A][BD]} \tag{7}$$

Equation (7) can be simplified according to the magnitude of K_1 and K_2. The following equation hold for the case of $K_2 \gg K_1$, and 1:

$$[A \cdot BD \cdot al] = [al] \tag{8}$$

and for the case of $K_1 \gg K_2$, and 1:

$$[A \cdot BD \cdot al] = K_2 [al][A]^{1-n}[BD] \tag{9}$$

Equations (8) and (9) seem to correspond to the copolymerization of butadiene and methyl methacrylate and that of butadiene and acrylonitrile, respectively.

Formation of Binary Complex or Ternary Complex among Acrylic Monomer, Butadiene, and Aluminum Compounds [3]

The formation of binary complex between an acrylic monomer and $EtAlCl_2$ was confirmed by the cryoscopic method. The following equilibria were observed:

$$2MMA + [EtAlCl_2]_2 \xrightleftharpoons{} [MMA \cdot EtAlCl_2]_2$$

$$2MAN + [EtAlCl_2]_2 \xrightleftharpoons{} [MAN \cdot EtAlCl_2]_2$$

$$AN + [EtAlCl_2]_2 \rightleftharpoons AN \cdot [EtAlCl_2]_2$$

$$AN \cdot [EtAlCl_2]_2 + AN \rightleftharpoons 2AN \cdot EtAlCl_2$$

The formation of the ternary complex was not detected by the cryoscopic method but observed by NMR and UV spectra. In the NMR spectrum of a system of MMA and $EtAlCl_2$, an up-field shift of the ethyl group of $EtAlCl_2$ and a down-field shift of various protons of MMA were observed. Further addition of benzene as a donor agent to the binary system caused a change in the above shifts, with the MMA-proton shifts being recovered to some extent whereas the $EtAlCl_2$-protons remained unaffected. This fact was taken as evidence of the direct complexation of benzene to MMA without changing the electronegativity of $EtAlCl_2$.

In a system of butadiene, MMA, and $EtAlCl_2$, polymerization takes place at room temperature. The NMR measurement was made at $-78°C$ without polymerization. The addition of butadiene to the binary system $MMA-EtAlCl_2$ induces a shift of the MMA protons, which attains a maximum value at a 1:1 ratio of butadiene to $MMA-EtAlCl_2$. This fact suggests the formation of a ternary complex of 1:1:1 with respect to three components.

In the UV spectrum of ternary systems of BD, AN, and $EtAlCl_2$ or of BD, MMA, and $EtAlCl_2$ the absorption was observed at 360 or 340 nm, respectively. These absorptions are not observed in binary systems. The ternary complex may be associated with polymerization because its rate of polymerization is greatly enhanced by irradiation with UV light, especially at the above-mentioned wavelengths. In the experiment with light of various wavelengths, a maximum rate of initiation and propagation is shown at the above wavelengths.

Comparison of Two Mechanisms of the Lewis-Mayo Type and a Ternary Complex One by Kinetic Study [4]

In the alternating copolymerization of butadiene and MMA there exists a maximum rate of polymerization at a definite molar ratio of MMA to butadiene almost equal to 1:1. This fact can be explained by the Lewis-Mayo mechanism as well as ternary complex one, but in a different way. In the Lewis-Mayo mechanism the rate of the step-growth type of copolymerization of A and B monomers is expressed by

$$R_p = \frac{2k_{AB}k_{BA}[A][B]}{k_{AB}[B] + k_{BA}[A]}[P^*] \tag{10}$$

Equation (10) can be rewritten in terms of the fraction x of monomer A and M (total monomer):

$$\frac{[M][P*]}{R_p} = \frac{1}{k_{BA}(1-x)} + \frac{1}{k_{AB}x} \qquad (11)$$

Equation (11) gives a maximum at

$$\frac{x}{1-x} = \frac{[A]}{[B]} = \left(\frac{k_{AB}}{k_{BA}}\right)^{1/2} \qquad (12)$$

In the presence of $EtAlCl_2$ the monomer A is complexed with $EtAlCl_2$. If the complexed monomer $A \cdot EtAlCl_2$ is an actual monomer co-polymerized with B monomer (butadiene), their ratio is changed by changing not only [A] : [B] ratio but also the [Al] : [A] ratio. The concentration of the A·Al-complex is almost equal to the concentration of [Al] when [Al] is less than [A], and the equilibrium constant of complex formation is high enough to give

$$[A \cdot al] = \frac{K[Al][A]}{1 + K[A]} \qquad (13)$$

Then, instead of Eq. (12), the following equation is used:

$$\frac{x}{1-x} = \left(\frac{k_{AB}}{k_{BA}[Al]K}\right)^{1/2} \qquad (14)$$

If K is very large, and [A–Al] is almost equal to [Al], there is no maximum.

Experiments were carried out with varied amount of $EtAlCl_2$. It was found that the maximum rate exists at a MMA fraction of 0.3, and this is not affected by the amount of aluminum compound added. In the experiment in which the total monomer concentration was varied from 1.6 to 8.0 mole/liter, the MMA fraction for the maximum rate decreases with an increase in the MMA fraction. That is, Eq. (14) is not in agreement with the experimental results.

The above fact is compatible with the ternary complex mechanism. In this mechanism the rate of polymerization is proportional to the concentration of the ternary complex $\lfloor$al·A·B$\rfloor$ if the concentration of the growing species $[P^*]$ is kept constant.

This assumption is based on the fact that the amount of polymer is mainly produced by the transfer reaction, or, in other words, $bW \gg a$ in Eq. (2):

$$R_p \propto [Al·A·B][P^*] = \frac{K_2 [M]^2 x(1 - x) [Al]}{1 + K_1 [M]^n x^n + K_2 [M]^2 x(1 - x)} [P^*] \quad (15)$$

Equation (15) gives a maximum for the cases of n = 1 or 2 at

$$x = \frac{[A]}{[A] + [B]} = \frac{(K_1 [M]^n + 1)^{1/2} - 1}{K_1 [M]^n} \leq \frac{1}{2} \quad (16)$$

where $[M]$ is a total monomer concentration. In this equation, x, the mole fraction of A in the monomer giving the maximum rate, is not affected by $[Al]$ but by $[M]$. An increase of the total monomer concentration $[M]$ causes a decrease of x from a value of 0.5, and this is in complete agreement with the experimental results.

Hirooka [5] proposed a modified mechanism in which a ternary complex is formed in the vicinity of the growing polymer radical. This mechanism is useful in the case where no complex is detected or the concentration of the complex is very small, for example, where the formation of the ternary complex occurs only at the growing polymer end. In some cases a 1:1 monomer-monomer complex is formed, but there is no maximum rate at a definite monomer composition. This is also explicable for the case where the actual complex forming at the growing polymer terminal is different from the complex formed in the absence of a growing polymer terminal. However, Hirooka's mechanism leaves a problem concerning the structure of the complex at the polymer terminal. He preferred the following scheme for the copolymerization of styrene or propylene with MMA in the presence of EtAlCl$_2$:

$$\sim\sim MMA\cdot\ +\ ST \rightleftharpoons\ \sim\sim MMA\cdots ST$$

$$\vdots\qquad\qquad\qquad\vdots$$

$$Al\qquad\qquad\qquad Al$$

$$\sim\sim MMA\cdots ST\ +\ MMA \rightleftharpoons\ \sim\sim MMA\cdots ST\cdots MMA$$

$$\vdots\qquad\vdots\qquad\qquad\vdots\qquad\vdots$$

$$Al\qquad Al\qquad\qquad Al\qquad Al$$

The growing polymer terminal composed of a MMA radical is somewhat stabilized by complexation with styrene and is not able to propagate. The propagation takes place when the ternary complex is formed by the further attack of the MMA·Al monomer. Applying this mechanism to the system of butadiene (B) and an acrylic compound (A), the following ternary complexes were considered:

$$\sim\sim A\cdots B\cdots A \qquad\qquad\qquad \sim\sim A\cdots B\cdots A$$
$$\ \ \ |\qquad\ \ |\qquad\qquad\qquad\qquad\qquad |$$
$$\ \ \ Al\qquad Al\qquad\qquad\qquad\qquad\ \ Al$$
$$\qquad a \qquad\qquad\qquad\qquad\qquad\qquad\qquad b$$

The concentration for complex a is

$$[complex] = \frac{K_1{}'K_2{}'\,[B][Al]}{1 + K_1{}'[B] + K_1{}'K_2{}'[Al][B]}$$

or

$$\frac{1}{[complex]} = \frac{1}{K_1{}'K_2{}'[M][Al](1-x)} + \frac{1}{K_2{}'[Al]} + 1 \qquad (17)$$

and for complex b:

$$\frac{1}{[complex]} = \frac{1}{K_1{}'K_2{}'[M]^2\,x(1-x)} + \frac{1}{K_2{}'[M]x} + 1 \qquad (18)$$

There is no maximum for the complex formation in Eq. (17), whereas a maximum occurs in Eq. (18) at

$$\frac{x}{1 - x} = (1 \quad K_1{}'[M])^{1/2} \geqq 1 \tag{19}$$

However, Eq. (19) does not fit the experimental data. For other types of complexes occurring at the growing terminal, such as

$$\sim\!\!\!\sim B\cdots A\cdots B \qquad\qquad\qquad \sim\!\!\!\sim B\cdots A\cdots B$$
$$\qquad |$$
$$\qquad Al$$

c   d

the corresponding equations are used for complex c:

$$\frac{1}{[complex]} = \frac{1}{K_1{}'K_2{}'[M][Al](1 - x)} + \frac{1}{K_2{}'[M](1 - x)} + 1 \tag{20}$$

and for complex d:

$$\frac{1}{[complex]} = \frac{1}{K_1{}'K_2{}'[M]^2 x(1 - x)} + \frac{1}{K_2{}'[M](1 - x)} + 1 \tag{21}$$

Equation (20) does not indicate a maximum. Only Eq. (21) gives a maximum similar to Eq. (15), but unlike Eq. (15), it does not involve the function of aluminum compounds.

Consequently, the mechanism through the complex occurring at the polymer terminal is ruled out in this case.

Model Reaction of Alternating Addition

The main difference between a Lewis-Mayo-type mechanism and a ternary complex mechanism is in the propagation step. The former is composed of two kinds of addition reactions:

$$\sim\!\!\underset{\underset{\text{Al}}{|}}{A}\!\cdot\; + B \longrightarrow \sim\!\!\underset{\underset{\text{Al}}{|}}{A}\!-\!B\!\cdot \tag{a}$$

$$\sim\!\!\underset{\underset{\text{Al}}{|}}{B}\!\cdot\; + A \longrightarrow \sim\!\!B\!-\!\underset{\underset{\text{Al}}{|}}{A}\!\cdot \tag{b}$$

and the latter

$$\sim\!\!\underset{\underset{\text{Al}}{|}}{A}\!\cdot\; + B\!-\!\underset{\underset{\text{Al}}{|}}{A} \longrightarrow \sim\!\!\underset{\underset{\text{Al}}{|}}{A}\!-\!B\cdots\underset{\underset{\text{Al}}{|}}{A} \tag{c}$$

$$\sim\!\!\underset{\underset{\text{Al}}{|}}{A}\!-\!B\cdots\underset{\underset{\text{Al}}{|}}{A} \longrightarrow \sim\!\!\underset{\underset{\text{Al}}{|}}{A}\!-\!B\!-\!\underset{\underset{\text{Al}}{|}}{A}\!\cdot \tag{d}$$

Consequently, the difference exists in (a) and (c) and/or in (b) and (d). For a comparison of (a) with (c), a reaction of isobutyronitrile radical with α-methylstyrene was investigated in the absence and in the presence of methacrylonitrile (MAN).

$$
\underset{\underset{\text{CN}\cdots\text{Al}}{|}}{\overset{\overset{\text{CH}_3}{|}}{\text{CH}_3\!-\!\text{C}}}\!\cdot\;\; + \;\; \underset{\underset{\text{C}_6\text{H}_5}{|}}{\overset{\overset{\text{CH}_3}{|}}{\text{CH}_2\!=\!\text{C}}} \longrightarrow \underset{\underset{\text{CN}\cdots\text{Al}}{|}}{\overset{\overset{\text{CH}_3}{|}}{\text{CH}_3\!-\!\text{C}}}\!-\!\text{CH}_2\!-\!\underset{\underset{\text{C}_6\text{H}_5}{|}}{\overset{\overset{\text{CH}_3}{|}}{\text{C}}}\!\cdot \tag{e}
$$

$$
\underset{\underset{\text{CN}\cdots\text{Al}}{|}}{\overset{\overset{\text{CH}_3}{|}}{\text{CH}_3\!-\!\text{C}}}\!\cdot\;\; + \;\; \underset{\underset{\text{C}_6\text{H}_5}{|}}{\overset{\overset{\text{CH}_3}{|}}{\text{CH}_2\!=\!\text{C}}}\;\; + \;\; \underset{\underset{\text{CN}\cdots\text{Al}}{|}}{\overset{\overset{\text{CH}_3}{|}}{\text{CH}_3\!=\!\text{C}}} \longrightarrow
$$

$$\tag{f}$$

$$
\underset{\underset{\text{CN}\cdots\text{Al}}{|}}{\overset{\overset{\text{CH}_3}{|}}{\text{CH}_3\!-\!\text{C}}}\!-\!\text{CH}_2\!-\!\underset{\underset{\text{C}_6\text{H}_5}{|}}{\overset{\overset{\text{CH}_3}{|}}{\text{C}}}\!\cdot\;\; + \;\; \underset{\underset{\text{CN}\cdots\text{Al}}{|}}{\overset{\overset{\text{CH}_3}{|}}{\text{CH}_3\!=\!\text{C}}}
$$

In reaction (e) without MAN, it was found that the product ISI or ISSI is formed without any polymer. Consequently, the reactions involved are

$$
\underset{\substack{\mid \\ CN}}{\overset{\substack{CH_3 \\ \mid}}{CH_3-C-N}} = \underset{\substack{\mid \\ CN}}{\overset{\substack{CH_3 \\ \mid}}{N-C-CH_3}} \xrightarrow{\ k_0\ } \quad 2\underset{\substack{\mid \\ CN}}{\overset{\substack{CH_3 \\ \mid}}{CH_3-C\cdot}}
$$

I_2 $I\cdot$

$$
I\cdot \ + \ \underset{\substack{\mid \\ C_6H_5 \\ S}}{\overset{\substack{CH_3 \\ \mid}}{CH_2 = C}} \xrightarrow{\ k_1\ } \quad IS\cdot
$$

$$
2IS\cdot \xrightarrow{\ k_2\ } IS-SI
$$

$$
IS\cdot + \ I\cdot \xrightarrow{\ k_3\ } ISI
$$

At the stationary state, Eq. (22) is obtained:

$$
\frac{ISI}{ISSI} = \frac{k_3{}^2 \, [I\cdot]^2}{k_2 \{ k_0 \, [I_2] - k_1 [I\cdot][S] \}} \tag{22}
$$

which indicates that the ratio of ISI/ISSI is decreased as k_1 or k_3, viz., the reactivity of the $I\cdot$ radical, is increased. The ratio is calculated from mass analysis of the reaction products.

It was found that the amount of the IS fragment or its ratio to the amount of AIBN decomposed is not significantly affected by the addition of $EtAlCl_2$. In other words, the reactivity of the radical of methacrylonitrile is almost the same irrespective of the presence of the complexing agent. The same result is obtained by the calculation of the IS fragment from the molecular weight of the product. The ratio of the yield to the molecular weight or the number of molecule produced is not changed.

The experiment in the presence of methacrylonitrile is complicated, because a considerable amount of the polymeric product is formed. The mass analysis of product seems to give an increased amount of IS fragment as compared with that obtained in the absence of methacrylonitrile, but the true amount of IS fragment is not obtained. The number of molecule of the reaction product is almost constant irrespective of the presence of methacrylonitrile, but the calculation of IS fragment has not been made. The comparison of the reactivity in reactions (b) and (d) has not been completed. However, the fact that the reactivity of methacrylonitrile radical is not significantly affected by the aluminum compound suggests that the high alternating tendency arises from reaction (d) enhanced by the complexation with $EtAlCl_2$. The enhancement of reaction (d) is explained not merely by the activation of methacrylonitrile monomer but by the formation of such clusters as $\sim$B· $\cdots$A$-$Al. These results seem to favor the mechanism involving the ternary complex intermediate which is formed prior to or during the propagation.

Binary and Ternary Complex among Acrylic Monomer, $EtAlCl_2$, and Toluene

As previously reported, the protons of methyl, methoxyl, and olefinic protons (cis and trans with respect to the ester group) of methyl methacrylate are shifted downfield by the complexation with $EtAlCl_2$, and the ethyl protons of $EtAlCl_2$ are shifted upfield. The addition of benzene to the binary system affects the above shift of MMA but not that of $EtAlCl_2$. The product of MMA fraction, f, and the additional shift, $\Delta\tau_c$, caused by further complexation of toluene shows a maximum value at a 1:1 composition in MMA-$EtAlCl_2$ and toluene.

These facts suggest the formation of the 1:1:1 complex of $EtAlCl_2$ - acrylic monomer-toluene.

In the measurement of NMR at various temperatures, a significant temperature dependence is observed in protons of acrylic monomer but not in those of $EtAlCl_2$. The magnitude of the temperature dependence of each protons is in the order

$$\text{MAN: } H_c > H_t > \alpha\text{-CH}_3$$

and

$$\text{MMA: } OCH_3 > H_t \cong \alpha\text{-CH}_3 > H_c$$

These are almost the same as the changes in the magnitude of the shift caused by the complexation of an equimolar amount of toluene, i.e.,

MAN: $H_c > H_t > \alpha\text{-} CH_3$

MMA: $OCH_3 \cong H_t > \alpha\text{-}CH_3 > H_c$

These changes were also compared with the shift caused by the addition of toluene in various amounts. The order of the shift difference between toluene and methylene dichloride solutions, $\delta_{(toluene)} - \delta_{(methylene\ dichloride)}$, is

MAN: $H_c > H_t > \alpha\text{-}CH_3$

MMA: $OCH_3 > H_t > \alpha\text{-}CH_3 > H_c$

An almost similar tendency in shift change in these experiments suggests that these change run parallel to the effect of toluene in complexation, from which the location of toluene in the complex is assumed to be in the vicinity of olefinic protons in MAN and in that of olefinic H_t protons and methoxyl protons in MMA. However, the slope of the temperature dependence curves suggests that the energy for complexation is rather small. In addition to this, the fact that the NMR is composed of an intermediate shift between the ternary complex and the component indicates the formation of a loose complex whose rate of formation or dissociation is rapid as compared with the NMR time scale.

Other Methods for Comparison

Thus the mechanism through an intermediate complex seems to be more likely than a simple Lewis-Mayo mechanism. In an ESR study made by Zubov [6] to determine the intermediate radical during the polymerization of dimethylbutadiene and t-butyl acrylate in the presence of diethylaluminum monochloride, it was claimed there was a simple Lewis-Mayo mechanism because the two radical species were observed in an almost equimolar ratio at $-100°C$. However, his experiment was carried out at temperature too low to examine the radical species responsible for the polymerization. In recent work by Ohtsu et al. [7] using 2-methyl-2-nitrosopropane as a spin-trapping agent, only one radical species was observed in the copolymerization of styrene and acrylic monomers in the presence of zinc chloride, and it arose from the styrene unit.

Recent improvements in NMR enables the sequence regularity to be determined more precisely at times. If a detailed regularity of more than triad is obtained, the following relation may be examined among diads (F_{ij}), triads (F_{ijk}), etc. in a Lewis-Mayo mechanism of the first Markov chain statistics:

$$\frac{F_{11}}{F_{12}} = \frac{F_{111}}{F_{112}} = \frac{F_{211}}{F_{212}} = \frac{F_{ij11}}{F_{ij12}} = r_1 u \qquad (23)$$

and

$$\frac{F_{22}}{F_{21}} = \frac{F_{122}}{F_{121}} = \frac{F_{222}}{F_{221}} = \frac{F_{ij22}}{F_{ij21}} = r_2/u \qquad (24)$$

where $u = [M_1]/[M_2]$ and r_1 and r_2 are monomer reactivity ratios.

In the intermediate complex mechanism, Eqs. (23) and (24) similar to the Lewis-Mayo type are also available, if the following alternating processes are considered:

$$M_2* + M_1\cdots M_2 \xrightarrow{k_{21}} M_2-M_1*\cdots M_2$$

$$M_2-M_1*\cdots M_2 \xrightarrow{k_{12}} M_2-M_1-M_2*$$

Only special condition for this case is $r_1 = 0$ in contrast to the condition for the Lewis-Mayo mechanism that $r_1 \cong 0$. However, these two conditions are not necessarily discriminated.

There has been some reports on the tacticity of the copolymer in relation to its alternating regularity. However, the role of the complex formation may not always be the same in stereo and sequence regulations if the intermediate complex is of a rather loose structure; it may not be as effective for stereo control as for sequence control.

REFERENCES

[1] J. Furukawa, Y. Iseda, K. Haga, and N. Kataoka, J. Polym. Sci.,
     A-1, 8, 1147 (1970).
[2] J. Furukawa, E. Kobayashi, and Y. Iseda, Polym. J. (Japan), 1,
     155 (1970).
[3] J. Furukawa, E. Kobayashi, Y. Iseda, and Y. Arai, Ibid., 1,
     442 (1970).
[4] J. Furukawa, E. Kobayashi, Y. Arai, and S. Nagata, 22nd
     Polymer Symposium, Tokyo, November, 1973, preprint I, 19.
[5] M. Hirooka, 23rd IUPAC, Boston, 1971, Macromolecular
     Preprint 1, 311.
[6] V. P. Zubov, J. Polym. Sci., Polym. Chem. Ed., 11, 2463
     (1973).
[7] T. Sato, M. Abe, K. Hibino, and T. Ohtsu, 23rd Polymer
     Symposium, Tokyo, October, 1974, Preprint 573.

Ternary Molecular Complex in the Alternating Copolymerization of Methyl Methacrylate with Styrene Using Stannic Chloride

HIDEFUMI HIRAI

Department of Industrial Chemistry
Faculty of Engineering
The University of Tokyo
Hongo-7, Bunkyo-ku, Tokyo 113, Japan

ABSTRACT

The equimolar alternating copolymerization of methyl methacrylate (MMA) with styrene (St) in the presence of stannic chloride in toluene (Tl) is investigated kinetically. The concentrations of the ternary molecular complexes, $[SnCl_4 \cdot MMA \cdots St]$ and $[SnCl_4 \cdot MMA \cdots Tl]$, are calculated by use of the formation constants of the ternary molecular complexes. The rates of copolymerization under photo-irradiation and with tri-n-butyl boron—benzoyl peroxide as an initiator are proportional to the 1.5th order and 1.0th order, respectively, of the concentration of the ternary molecular complex $[SnCl_4 \cdot MMA \cdots St]$. The alternating copolymerization precedes the homopolymerization of the methyl methacrylate charged in excess. The alternating regulation of the copolymerization is ascribed to the homopolymerization of the ternary molecular complex from the kinetic results. The magnitudes of the shifts for

243

the four groups of protons in the coordinated methyl
methacrylate on the ternary molecular complex formation
are kept in a specific ratio which indicates a specific time-
averaged orientation of benzene ring to the coordinated
methyl methacrylate. A charge transfer absorption band
and the thermodynamic parameters for the formation of
the ternary molecular complex are also discussed.

INTRODUCTION

Several kinetic studies have revealed the various complicated
features of the equimolar alternating copolymerization of polar
vinyl monomer with donor monomer in the presence of metal halide.
On the copolymerization of methyl methacrylate (MMA) with styrene
(St) in toluene using ethylaluminum sesquichloride, the rate order
with respect to the concentration of the ternary molecular complex
$\lfloor AlEt_{1.5} Cl_{1.5} \cdot MMA \cdots St \rfloor$ varied between 1.0 and 2.0 with the charged
concentration of alkylmetal halide, and the reaction did not attain
steady-state conditions [1]. The rate of copolymerization of methyl
acrylate with styrene in the presence of ethylaluminum sesquichloride
was proportional to the concentration of the ethylaluminum sesqui-
chloride-methyl acrylate complex, being independent of both the
concentrations of free methyl acrylate and styrene [2]. The copolymerization of methyl methacrylate and butadiene using an $AlEtCl_2$-$VOCl_3$
catalyst proceeded by a rapid initiation followed by stepwise propagation, and the rate did not depend on the charged concentrations of
either methyl methacrylate and butadiene [3]. Whether these kinetic
features are essential for the alternating regulation or not is a significant problem for elucidation of the alternating copolymerization
mechanism.

In our previous paper [4], the equilibrium constants (K_{A1}, K_{A2})
for the complex formation between stannic chloride and methyl methacrylate were determined at several different temperatures by use of
the absorbance in the 355 nm region, e.g., $K_{A1} = 65$ and $K_{A2} = 12$
liter/mole at -50°C in toluene. The continuous variation plots using
the ^1H-chemical shifts revealed a 1:1 interaction between the methyl
methacrylate coordinated to stannic chloride and styrene or toluene
(Tl) [5-7]. Furthermore, the equilibrium constants (K_{St}, K_{T1}) for
the formation of the ternary molecular complex have been determined
in n-hexane at the temperature range -50 to +20°C by use of the
chemical shift [5-7].

In the alternating copolymerization solution of methyl methacrylate with styrene in the presence of stannic chloride in toluene, consequently, there exist four equilibria:

$$SnCl_4 + MMA \xrightleftharpoons{K_{A1}} SnCl_4(MMA) \tag{1}$$

$$SnCl_4(MMA) + MMA \xrightleftharpoons{K_{A2}} SnCl_4(MMA)_2 \tag{2}$$

$$SnCl_4 \cdot MMA + St \xrightleftharpoons{K_{St}} [SnCl_4 \cdot MMA \cdots St] \tag{3}$$

$$SnCl_4 \cdot MMA + Tl \xrightleftharpoons{K_{Tl}} [SnCl_4 \cdot MMA \cdots Tl] \tag{4}$$

where $SnCl_4 \cdot MMA$ represents the methyl methacrylate coordinated to stannic chloride, and $[SnCl_4 \cdot MMA \cdots St]$ and $[SnCl_4 \cdot MMA \cdots Tl]$ are the ternary molecular complexes of the coordinated methyl methacrylate with styrene and toluene, respectively.

In this paper the alternating copolymerizations of methyl methacrylate with styrene in the presence of stannic chloride, both under photoirradiation and with an initiator, are kinetically investigated on the basis of the concentration of the ternary molecular complex in the polymerization mixture, which can be calculated by use of the equilibrium constants. Furthermore, the ternary molecular complex formation is examined by the ^1H-chemical shift, the charge transfer absorption band, and the thermodynamic parameters.

EXPERIMENTAL

Materials

The monomers and the solvents were purified and thoroughly dehydrated in the usual manner. Stannic chloride was stored under an atomosphere of nitrogen after distillation from phosphorous pentoxide and led directly into a polymerization ampule by distillation in vacuo.

Copolymerization

   The copolymerization solution in toluene in a Pyrex glass ampule,
which cut off the light shorter than 300 nm, was irradiated at -50°C
with a 500-W high-pressure mercury lamp. The charged molar ratio
of methyl methacrylate to stannic chloride was kept at 2.4. The
alternating regularity of the resulting polymer was confirmed by the
same procedure as described in the previous paper [8]. The rate of
the alternating copolymerization (R_p) is expressed as the copolymer
yield per unit volume of the solution in 0.5 hr. All the conversions
were kept below 7.5%, and the variations of the concentrations of the
ternary molecular complexes with the progress of polymerization
were neglected.
   The copolymerization was also conducted using a radical initiator,
tri-n-butyl boron–benzoyl peroxide (2:1 molar ratio), instead of
photoirradiation. The copolymerization rate (R_p) represents the
copolymer yield per unit volume of the solution in 2 hr. All other
conditions were the same as described above.

Spectrometry

   Absorption spectra were obtained in methylene chloride at -70 ±
1°C by a Shimazu MPS-50L spectrophotometer using a cell of 1 cm
path length set in a transparent quartz Dewar vessel with a Dry
Ice-methanol coolant.
   ^{1}H-NMR spectra were run on a 100 MHz JEOL spectrometer using
tetramethylsilane as internal standard.

RESULTS

Molecular Complex Formation

   The formation constants of the ternary molecular complex between
the coordinated methyl methacrylate and styrene or toluene in n-
hexane were determined more precisely in consideration of the
equilibriums of Eqs. (1) and (2) by use of the ^{1}H-chemical shifts according to the previous reports [5-7]. For the $SnCl_4 (MMA)_2$ complex-
aromatic donor system, the charged concentrations of methyl methacrylate and stannic chloride were 0.20 and 0.04 mole/liter, respectively, in order to suppress the dissociation of the $SnCl_4 (MMA)_2$

TABLE 1. Thermodynamic Parameters for the Molecular Complex Formation in n-Hexane

System	$\Delta_{AD}{}^a$ (ppm)	K^b (liter/mole)	$-\Delta H$ (kcal/mole)	$-\Delta S$ (eu)
SnCl₄ (MMA)₂ -St	0.67	0.75	2.4 ± 0.1	11.2 ± 0.6
SnCl₄ (MMA)₂ -Tl	0.74	1.65	3.3 ± 0.2	13.8 ± 0.8
SnCl₄ (MMA)-St	0.74	0.99	2.7 ± 0.2	12.3 ± 0.8
SnCl₄ (MMA)-Tl	0.74	1.29	3.1 ± 0.2	13.4 ± 0.8
MMA-St	0.42	0.27	1.5 ± 0.1	9.4 ± 0.4
MMA-Tl	0.46	0.33	1.8 ± 0.1	10.2 ± 0.4

[a]Chemical shift change of the methoxy protons for the formation of pure molecular complex.
[b]Equilibrium constant at -50°C.

complex [7]. For the $SnCl_4$ (MMA) complex-aromatic donor system, those of methyl methacrylate and stannic chloride were 0.05 and 0.30 mole/liter, respectively, in order to prevent the formation of the $SnCl_4$ (MMA)₂ complex. The resulting equilibrium constants at -50°C and the thermodynamic parameters obtained from those at four different temperatures are listed in Table 1.

The copolymerization conditions where the charged stannic chloride exclusively forms the 1:2 complex with methyl methacrylate can be chosen in order to simplify the population of the complex species in the copolymerization solution. Under the conditions where the molar ratio of the charged methyl methacrylate to stannic chloride is 2.4 ([MMA] = 4.80, [SnCl₄] = 2.00 mole/liter) in toluene and at -50°C, the concentrations of the 1:2 stannic chloride-methyl methacrylate complex, the 1:1 stannic chloride-methyl methacrylate complex, and free methyl methacrylate in the copolymerization solution are estimated to be 1.84, 0.16, and 0.96 mole/liter respectively, by the use of K_{A1} = 65 and K_{A2} = 12 liter/mole.

Accordingly, the kinetic investigations have been conducted on the copolymerization in toluene at -50°C with a [MMA]/[SnCl₄] molar ratio of 2.4.

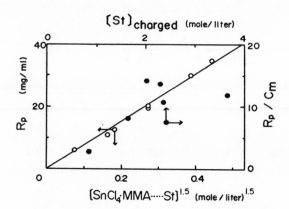

FIG. 1. Rate of alternating copolymerization vs concentration of
ternary molecular complex at -50°C in toluene under photoirradiation. R_p is expressed as the yield per unit volume in 0.5 hr.

Copolymerization under Photoirradiation

As shown in Fig. 1, the rate of the alternating copolymerization
under photoirradiation depends clearly on the 1.5th order of the
concentration of the ternary molecular complex [SnCl₄·MMA····St]
calculated by the use of K_{St} and K_{T1} in Table 1 [9]. Plotting
either the rate (R_p) itself or the rate normalized by the concentration
of the coordinated methyl methacrylate (R_p/C_m) against the charged
concentration of styrene cannot give a good correlation (see Fig. 1).

The conversion increases proportionally to the polymerization
time, while the intrinsic viscosity of the resulting copolymer remains constant throughout the polymerization period, as shown in
Fig. 2. These results indicate that the present copolymerization
process is in a steady state.

As shown in Fig. 3, the copolymerization rate is proportional
to the square root of the intensity (I) of the incident light, which
was controlled by variation of the distance between the lamp and
the ampule.

Copolymerization with Initiator

The rate of the alternating copolymerization using tri-n-butyl-
boron—benzoyl peroxide (2:1 molar ratio) as an initiator depends

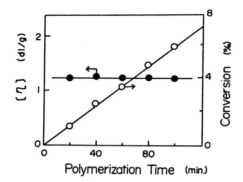

FIG. 2. Intrinsic viscosity and conversion vs polymerization time; $[\eta]$ obtained in toluene at 30°C. Copolymerization conditions: photoirradiation at -50°C in toluene; $[MMA] = 2.76$, $[St] = 1.63$, $[SnCl_4] = 1.15$ mole/liter.

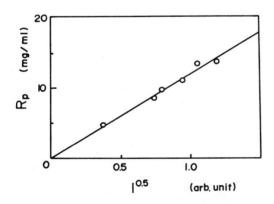

FIG. 3. Rate of alternating copolymerization vs intensity of incident light. Copolymerization conditions: photoirradiation at -50°C in toluene; $[MMA] = 2.00$, $[St] = 1.00$, $[SnCl_4] = 0.83$ mole/liter.

primarily on the first order of the concentration of the ternary molecular complex, as shown in Fig. 4. There is a poor correlation between the rate (R_p), even normalized (R_p/C_m), and the charged concentration of styrene.

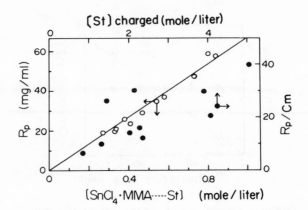

FIG. 4. Rate of alternating copolymerization vs concentration of ternary molecular complex at -50°C in toluene using B(n-Bu)₃–BPO initiator. R$_p$ is expressed as the yield per unit volume in 2 hr.

The conversion is proportional to the polymerization time, and the intrinisc viscosity is independent of the time. This is similar to the results under photoirradiation (Fig. 2).

The rate is proportional to the concentration of tri-n-butyl boron used as the initiator of B(n-Bu)$_3$·(BPO)$_{0.5}$. The copolymerization of methyl methacrylate with styrene in the absence of stannic chloride in toluene at 30°C was also carried out using the same initiator, and the rate was found to be proportional to the concentration of tri-n-butyl boron as well. In this case the result was a random copolymer similar to that obtained by the conventional radical mechanism with a biomolecular termination.

Other Features of the Copolymerization

When methyl methacrylate is used in excess ([MMA] = 3.84, [St] = 0.36 mole/liter) for the alternating copolymerization under photoirradiation at -20°C, the rate becomes very small and almost negligible, leaving the excess methyl methacrylate unreacted after the charged styrene is consumed entirely by the alternating copolymerization, as shown in Fig. 5.

All the above copolymerizations proceeded entirely homogenously throughout the reaction time. A concurrent occurrence of cationic

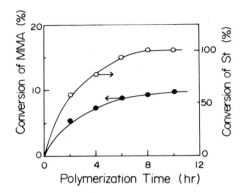

FIG. 5.   Conversion vs polymerization time using the charged
methyl methacrylate in excess.   Copolymerization conditions: photoirradiation at -20°C in toluene; [MMA] = 3.84, [St] = 0.36, [SnCl₄] =
1.60 mole/liter.

homopolymerization of styrene was not observed.   When drying of
the materials, expecially of stannic chloride, was insufficient,
however, the copolymerization behavior changed remarkably.
The appearance of white precipitates or gelation often occurred
in the course of the polymerization, resulting in a mixture of the
alternating copolymer and polystyrene.   Moreover, the intrinsic
viscosity of the alternating copolymer increased with the polymerization time.

Ternary Molecular Complex

    The absorption spectrum of the $SnCl_4 (MMA)_2$ complex-styrene
system in methylene chloride at -70°C gives a new absorption in
the 315 nm region which cannot be associated with either of the
component, as shown in Fig. 6.   The charge-transfer bands of
the solution of the complex-aromatic donor or butadiene systems
are listed in Table 2.   In the cases of the free methyl methacrylate,
the methyl isobutylate (1) complex and the 3-butenyl methyl ketone
(2) complex-aromatic donor systems, no new band could be
observed.

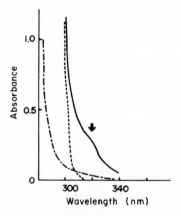

1   2

Table 3 collects the relative ratios of the shifts for each of the groups of proton to the shift for the methoxy protons of methyl methacrylate on the molecular complex formation with styrene or toluene. The absolute limiting values of the shift (Δ_{AD}) for the methoxy protons are given in Table 1. Both the 1:2 complex-styrene and toluene systems exhibit the magnitudes of the shifts for methoxy, trans-vinyl, cis-vinyl, and α-methyl protons in a ratio of approximately 1.0:0.9: 0.5:0.5. On the other hand, both the free methyl methacrylate-styrene and toluene systems exhibit the magnitudes in a ratio of

FIG. 6. Absorption spectrum of the $SnCl_4 (MMA)_2$ -styrene system in methylene chloride at -70°C. (—): [MMA] = 0.26, [St] = 0.19, [$SnCl_4$] = 0.11 mole/liter. (- ·): [MMA] = 0.26, [$SnCl_4$] = 0.11 mole/liter. (- -) [St] = 0.19 mole/liter.

TABLE 2. Absorption Bands on the Molecular Complex Formation in Methylene Chloride at $-70°C$

System[a]		New band (λ_{max}, nm)
Acceptor	Donor	
$SnCl_4 (MMA)_2$	St	+ (315)
	Tl	+
	Ms	+ (340)
	BD	+
MMA	St	-
	Ms	-
$SnCl_4 (MIB)_2$	St	-
	Ms	-
$SnCl_4 (BMK)_2$	St	-

[a]Ms, Mesitylene; BD, butadiene; MIB, methyl isobutyrate; BMK, 3-butenyl methyl ketone.

TABLE 3. Relative Ratios of the Shifts for the Protons in Methyl Methacrylate to That for the Methoxy Protons on the Molecular Complex Formation

System	trans-H	cis-H	α-CH$_3$
$SnCl_4 (MMA)_2$ -St	0.91	0.55	0.49
$SnCl_4 (MMA)_2$ -Tl	0.92	0.51	0.51
MMA-St	1.06	0.04	0.28
MMA-Tl	0.97	0.01	0.28

approximately 1.0:1.0:0.0:0.3. The difference between the effects of styrene and toluene on the ratios is very small.

DISCUSSION

Mechanism of the Alternating Copolymerization

From the kinetic results stated above, the scheme of the present alternating copolymerization can be proposed as follows:

1. Using photoirradiation,

$$[SnCl_4 \cdot MMA \cdots St] \xrightarrow{h\nu} [SnCl_4 \cdot MMA \cdots St] * \longrightarrow 2R \cdot \qquad (5)$$

$$R \cdot + [SnCl_4 \cdot MMA \cdots St] \xrightarrow{k_p} M \cdot \qquad (6)$$

$$2M \cdot \xrightarrow{k_t} polymer \qquad (7)$$

where $[SnCl_4 \cdot MMA \cdots St]*$, $R \cdot$, and $M \cdot$ represent the ternary molecular complex in the excited state, the initiation radical, and the propagating radical, respectively.

Under the steady-state conditions,

$$2fI [SnCl_4 \cdot MMA \cdots St] = 2k_t [M \cdot]^2 \qquad (8)$$

where 2f is the quantum yield.

Generally,

$$R_p = k_p [M \cdot] [SnCl_4 \cdot MMA \cdots St] \qquad (9)$$

From Eqs. (8) and (9),

$$R_p = k_p (f/k_t)^{0.5} I^{0.5} [SnCl_4 \cdot MMA \cdots St]^{1.5} \qquad (10)$$

2. Using initiator,

$$[B(n-Bu)_3 \cdot BPO_{0.5}] \xrightarrow{k_d} 2R \cdot \qquad (11)$$

$$R\cdot + [SnCl_4 \cdot MMA \cdots St] \xrightarrow{k_p} M\cdot \qquad (12)$$

$$2M\cdot \xrightarrow{k_t} polymer \qquad (13)$$

Under the steady-state conditions,

$$2fk_d [B(n-Bu)_3][BPO] = 2k_t [M\cdot]^2 \qquad (14)$$

where f is the initiator efficiency.
Under the experimental conditions,

$$[BPO] = \frac{1}{2}[B(n-Bu)_3] \qquad (15)$$

From Eqs. (9), (14), and (15),

$$R_p = k_p (f k_d / 2k_t)^{0.5} [B(n-Bu)_3][SnCl_4 \cdot MMA \cdots St] \qquad (16)$$

Equations (10) and (16) can reasonably explain all of the experimental results.

Consequently, it can be concluded that the present alternating copolymerization proceeds through radical homopolymerization of the ternary molecular complex in the steady state.

The ternary molecular complex may be excited by the light absorption using the charge-transfer band in the 310 to 340 nm range.

As shown in Fig. 1, the concentration of the $[SnCl_4 \cdot MMA \cdots St]$ complex is far less than the charged concentration of monomers, but still the ternary molecular complex governs the process of alternating copolymerization. The reactivity of the $[SnCl_4 \cdot MMA \cdots St]$ complex should be overwhelmingly larger than those of the free monomers and other molecular complexes under alternating copolymerization.

Figure 5 indicates that alternating copolymerization precedes homopolymerization of the monomer charged in excess. This fact also reflects the greater reactivity of the ternary molecular complex.

Ternary Molecular Complex

The new band which appeared in the $SnCl_4 (MMA)_2$ complex-donor system shifted to the longer wavelength with the lower ionization

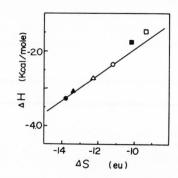

FIG. 7. Relation between enthalpy change and entropy change for
the molecular complex formation: (○) SnCl₄(MMA)₂-styrene,
(△) SnCl₄(MMA)-styrene, (□) MMA-styrene, (●) SnCl₄(MMA)₂-
toluene, (■) SnCl₄(MMA)-toluene, and (▲) MMA-toluene.

potentials of donors; absorption shoulders were observed for styrene
(I_p = 9.00 eV [10]) and mesityrene (I_p = 8.74 eV [10]) at 317 and 340
nm, respectively. These red shifts support the assignment of the
band to the charge-transfer band. The modification of the acceptor
monomer, as shown in Table 2, indicates that both the coordination
to the Lewis acid and the conjugation of the polar group with the vinyl
group are essential for the appearance of the charge-transfer band.

Figure 7 exhibits a linear relationship between enthalpy changes
(ΔH) and entropy changes (ΔS) in Table 1 for the ternary molecular
complex formation. This fact indicates that the same kind of interaction force functions in common with the ternary molecular complex.
The values of ΔH and ΔS for the free methyl methacrylate systems
seem to indicate another linear relationship. The variations of ΔH
and ΔS suggest that the ternary molecular complex formation is
governed by the enthalpy factor.

A specific time-averaged orientation of benzene ring to methyl
methacrylate was indicated by the ratio of the shifts for methoxy,
trans-vinyl, cis-vinyl, and α-methyl protons in common with the
coordinated methyl methacrylate system (1.0:0.9:0.5:0.5) and with
the free methyl methacrylate system (1.0:1.0:0.0:0.3), as shown in
Table 3.

The binding force of methyl methacrylate with the benzene ring
may be ascribed mainly to the dispersion force and to the interaction
between the dipole of the carbonyl group of methyl methacrylate and

the quadrupole of the benzene ring, and partly to the charge-transfer
interaction of methyl methacrylate and the aromatic compound.

REFERENCES

[1] N. G. Gaylord and B. Matyska, J. Macromol. Sci.—Chem., A4,
1519 (1970).
[2] M. Hirooka, Paper Presented at the 23rd IUPAC Congress,
Boston, 1971, Preprint p. 311.
[3] J. Furukawa, E. Kobayashi, Y. Iseda, and Y. Arai, Polymer J.,
1, 442 (1970).
[4] H. Hirai and M. Komiyama, J. Polym. Sci., Polym. Lett. Ed.,
10, 925 (1972).
[5] H. Hirai, M. Komiyama, and N. Toshima, J. Polym. Sci.,
B, 9, 883 (1971).
[6] H. Hirai, M. Komiyama, and N. Toshima, Ibid., 9, 789
(1971).
[7] H. Hirai and M. Komiyama, J. Polym. Sci., Polym. Chem. Ed.,
12, 2701 (1974).
[8] T. Ikegami and H. Hirai, J. Polym. Sci., A-1, 8, 463 (1970).
[9] H. Hirai and M. Komiyama, J. Polym. Sci., Polym. Lett. Ed.,
12, 673 (1974).
[10] G. F. Crable and G. L. Kearns, J. Phys. Chem., 66, 436
(1962).

Monomer-Isomerization Polymerization of Some Branched Internal Olefins with a Ziegler-Natta Catalyst

TAKAYUKI OTSU and KIYOSHI ENDO

Department of Applied Chemistry
Faculty of Engineering
Osaka City University
Sugimotocho, Sumiyoshi-ku, Osaka, Japan

ABSTRACT

New examples for the monomer-isomerization polymerizations of some branched internal olefins, 4-methyl-2-pentene and 4-phenyl-2-butene, are presented. When these olefins are polymerized with $Al(C_2H_5)_3$-$TiCl_3$ ($[TiCl_3]$ = 120 mmole/liter, Al/Ti = 3.0) catalyst at 80°C, considerable amounts of high polymers [27.5%/60 hr ($[\eta]$ = 0.68 dl/g) and 35.6%/100 hr, respectively] were obtained. From the additional fact that the isomerization from these 2-olefins to the mixture of their positional isomers including 1-olefins was observed during the polymerization, it is assumed that the polymerizations from these 2-olefins are performed with the 1-olefins which isomerized from the starting 2-olefins.

INTRODUCTION

In previous papers [1-4] it has been reported that internal olefins such as 2-butene, 2-pentene, 2-hexene, 2-heptene, 3-heptene, and 2-octene can homopolymerize in the presence of $Al(C_2H_5)_3$ - $TiCl_3$ (Al/Ti = 2 to 3) catalyst to give high polymers consisting of the corresponding 1-olefin units:

$$\underset{\substack{| \quad | \\ CH_3 \quad R}}{CH{=}CH} \overset{\text{isomerization}}{\underset{}{\rightleftharpoons}} \underset{\substack{| \\ CH_2R}}{CH_2{=}CH} \overset{\text{polymerization}}{\longrightarrow} \left(\underset{\substack{| \\ CH_2R}}{CH_2{-}CH} \right)_n$$

where $R = CH_3$, C_2H_5, C_3H_7, C_4H_9, and C_5H_{11}.

In these monomer-isomerization polymerizations, two independent steps on the two independent catalyst sites are involved. In the first step the internal olefin is isomerized to an equilibrium mixture of various olefins, including 1-olefin; in the second step, this 1-olefin is polymerized. Accordingly, the total rate of polymerization of the internal olefin is determined by its isomerization rate, its concentration in an equilibrium mixture, and the polymerizability of the 1-olefin formed by isomerization. Since the concentration of 1-olefin in an equilibrium mixture isomerized from linear 2-olefin is generally higher than that from branched 2-olefins, it is expected that the former olefins are more favorable to monomer-isomerization polymerization than the latter olefins [2, 5, 6].

We have recently found that some branched internal olefins such as 4-methyl-2-pentene and 4-phenyl-2-butene undergo monomer-isomerization polymerization in the presence of a Ziegler-Natta catalyst. The present paper deals with the results of monomer-isomerization polymerizations of these branched internal olefins.

EXPERIMENTAL

Material

4-Methyl-1-pentene, 4-methyl-2-pentene, 4-phenyl-1-butene, and 4-phenyl-2-butene were used after fractional distillation over calcium hydride of the respective commercial reagents. Purities determined by gas chromatography were 99.9, 99.9, 96.0, and 99.9%, respectively; 4-phenyl-2-butene consisted of 97.3% cis and 2.7% trans isomers.

Triethylaluminum, ethylaluminum dichloride (Ethyl Corp.), diethylaluminum chloride (Texas Alkyls), titanium trichloride (Stauffer
Chem. Co.), and transition metal compounds (commercial pure grade
reagents) were used without further purification. Solvents and pre-
cipitants were used after purification by the conventional method.

Polymerization Procedure

Titanium trichloride in a sealed small ampule, and if necessary a
transition metal compound, was placed in a glass tube provided with
a rubber stopper and with a connection to the vacuum system. The
required amount of solvent was added through a syringe after de-
gassing, and the required amount of alkylaluminum solution was
charged through a syringe in a dry nitrogen atmosphere. The $TiCl_3$
ampule was destroyed, and the mixture was aged for an hour at room
temperature. After aging, olefin monomer was charged through a
syringe. The system was then sealed under vacuum.

Polymerizations were carried out by shaking in a thermostat
maintained at a constant temperature for a given time. The tube was
opened after polymerization, and the unreacted olefin was recovered
by distillation and its isomer distribution was analyzed by gas chromatography by using a bis(2-methoxyethyl)-adipate column at 17°C
for methylpentenes and Bentone 34 and tricresylphosphate columns at
110°C for phenylbutenes. The residues of the tube were then poured
into a large amount of methanol containing hydrochloric acid in order
to precipitate the polymer formed. The polymer yield was calculated
from the weight of the dry polymer obtained.

Characterization of the Polymers

The structure of the resulting polymers was checked by the IR
spectra of their films. The isotacticity of the resulting polymer was
determined from the weight of hot n-hexane-insoluble polymer [7].
The intrinsic viscosity of the polymer was determined by viscosity
measurement of dilute Decalin solution at 130°C with an Ubbelohde
viscometer.

RESULTS AND DISCUSSION

Monomer-Isomerization Polymerization of 4-Methyl-2-pentene

Polymerization and Isomerization

The results of polymerization of 4-methyl-2-pentene by $Al(C_2H_5)_3$-$TiCl_3$ catalyst in n-heptane are shown in Table 1, in which the results of isomerization are also indicated.

From Table 1 it is observed that this olefin easily polymerizes and the yield increases with an increase of the reaction time. For comparison, the corresponding 1-olefin, 4-methyl-1-pentene, was polymerized at a very fast rate (95.8%/1 hr at 30°C) without isomerization under similar conditions. The polymer obtained from this 2-olefin shows the same IR spectrum as the polymer obtained from 4-methyl-1-pentene, as is seen from Fig. 1. The polymers are also found to contain the hexane-insoluble fractions (43.1%) which consist of a high molecular weight isotactic structure ($[\eta]$ = 0.68 dl/g).

TABLE 1. Results of Polymerization and Isomerization of 4-Methyl-2-pentene by $Al(C_2H_5)_3$-$TiCl_3$ Catalyst[a]

Time (hr)	Yield (%)	Composition of methylpentenes recovered after polymerization (%)[b]				
		4M1P	4M2P	2M2P	2M1P	2MP
5	5.2	0.5	93.1	4.6	0.3	1.5
10	10.6	0.5	88.7	8.3	0.7	1.8
20	15.9	0.4	84.5	11.5	1.1	2.5
40	16.5	0.5	83.5	13.1	0.6	2.3
60	20.4	0.8	91.2	6.8	0.6	0.6
		0.5[c]	3.7[c]	82.6[c]	13.2[c]	

[a]Polymerization conditions: $[TiCl_3]$ = 117 mmole/liter, Al/Ti = 3.0, $[4M2P]$ = 2.7 mole/liter at 80°C.

[b]Abbreviations are as follows: 4M1P, 4-methyl-1-pentene; 4M2P, 4-methyl-2-pentene; 2M2P, 2-methyl-2-pentene; 2M1P, 2-methyl-1-pentene; 2MP, 2-methylpentane.

[c]Calculated from the thermodynamic stabilities of the isomers.

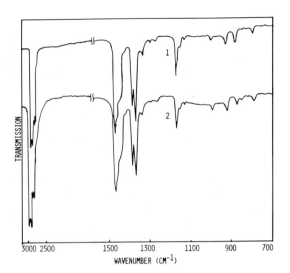

FIG. 1. IR spectra of polymers: (1) obtained from 4-methyl-1-pentene with $Al(C_2H_5)_3$-$TiCl_3$ catalyst; (2) obtained from 4-methyl-2-pentene with $Al(C_2H_5)_3$-$TiCl_3$ catalyst.

It is also found from Table 1 that the isomerization of the starting olefin to the corresponding positional isomers and the formation of a saturated alkane, 2-methylpentane, are observed during the polymerization, and that the concentration of 4-methyl-1-pentene isomer is in good agreement with that (0.5%) calculated from the thermodynamic stabilities of these isomers. From these results it is clear that this polymerization is an example of the monomer-isomerization polymerization, similar to those of linear internal olefins such as 2-butene [2].

From the change in isomer distribution of this 2-olefin with polymerization time (Table 1), the isomerization seems to proceed via a stepwise process:

$$
\underset{\substack{|\\CH_3}}{CH_3-CH=CH-CH-CH_3} \underset{\text{isomerization}}{\rightleftharpoons} \underset{\substack{|\\CH_3}}{CH_3-CH_2-CH=C-CH_3} \underset{\text{isomerization}}{\rightleftharpoons} \underset{\substack{|\\CH_3}}{CH_3-CH_2-CH_2-C=CH_2}
$$

$$\Big\updownarrow \text{isomerization}$$

$$
\underset{\substack{|\\CH_3}}{CH_2=CH-CH_2-CH-CH_3} \xrightarrow{\text{polymerization}} \left(CH_2-\underset{\substack{|\\CH_2\\|\\CH\\ \diagup\ \diagdown \\ CH_3\ \ CH_3}}{CH}- \right)_n
$$

It is also interesting that the rate of isomerization from 4-methyl-2-pentene to 4-methyl-1-pentene is faster than that to 2-methyl-2-pentene, which is the most stable of these isomers. In connection with this result, it was found that 2-methyl-2-pentene and 2-methyl-1-pentene can undergo monomer-isomerization polymerization at a very slow rate [8].

Effect of Al/Ti Molar Ratio and Transition Metal Compounds

The effect of the Al/Ti molar ratio in the catalyst system on the polymerization and the isomerization of 4-methyl-2-pentene is shown in Table 2.

From this table it is seen that the maximum rates of both polymerization and isomerization are observed at a molar ratio of Al/Ti of about 3.0 to 3.5. The result regarding the polymerization rate is in good agreement with that reported for the polymerization of 4-methyl-1-pentene by $Al(C_2 H_5)_3$ -$TiCl_3$ catalyst [9].

TABLE 2. Effects of Al/Ti Molar Ratio on the Monomer-Isomerization Polymerization of 4-Methyl-2-pentene[a]

Al/Ti molar ratio	Yield (%)	Composition of methylpentenes recovered after polymerization (%)[b]				
		4M1P	4M2P	2M2P	2M1P	2MP
0.5	0	0	99.8	0	0	0
1.0	10.6	0.2	92.3	7.0	0.3	0.2
2.0	22.4	0.4	91.3	7.0	0.5	0.3
3.0	27.5	0.1	84.8	14.1	0.6	0.4
3.5	33.1	0.2	87.7	11.4	0.2	0.5
4.0	17.6	0.8	89.7	8.0	0.3	1.2

[a]Polymerization conditions: [$TiCl_3$] = 117 mmole/liter, [4M2P] = 2.7 mole/liter, in p-xylene at 80°C for 60 hr.
[b]Abbreviations as in Table 1.

TABLE 3. Effects of Transition Metal Compounds[a]

Me(acac)$_x$	Yield (%)
None	24.6
Ni(acac)$_2$	14.4
Fe(acac)$_3$	5.9

[a]Polymerization conditions: $Al(C_2H_5)_3$-$TiCl_3$ catalyst, [$TiCl_3$] = 105 mmole/liter, [4M2P] = 3.0 mole/liter, Al/Ti = 3.0, Me/Ti = 0.5, in p-xylene at 80°C for 100 hr.

As shown in Table 3, transition metal compounds such as metal acetylacetonates [Me(acac)$_x$] are found to exhibit a retarding effect on the rate of this polymerization (Me/Ti = 0.5), contrary to the marked accelerating effect they showed on the monomer-isomerization polymerizations of 2-butene and 2-pentene [3]. To check this point further, the effect of the Ni/Ti molar ratio of the catalyst system [$Al(C_2H_5)_3$-$TiCl_3$-$Ni(acac)_2$, Al/Ti = 3.0] on the rate of polymerization and isomerization of 4-methyl-2-pentene was investigated. The results are shown in Fig. 2.

From Fig. 2 it is seen that both polymerization and isomerization are affected by the concentration of Ni(acac)$_2$, and their maximum rate is also found to appear at a molar ratio Ni/Ti of ~0.2. Accordingly, it seems that Ni(acac)$_2$ behaves as both activator and deactivator in the formation of the active sites as a function of its concentration, and when the concentration of Ni(acac)$_2$ increases to more than a molar ratio Ni/Ti of 0.5, the active sites for isomerization and polymerization are completely destroyed.

Monomer-Isomerization Polymerization of 4-Phenyl-2-butene

Polymerization with Various Catalyst System

The results of the polymerization of 4-phenyl-2-butene with various Ziegler-Natta-type catalysts are shown in Table 4, as are the results for 4-phenyl-1-butene.

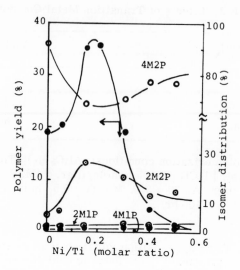

FIG. 2. Effects of Ni/Ti molar ratio on the polymerization and
the isomer distribution when 4-methyl-2-pentene is reacted with
$Al(C_2H_5)_3$ -$TiCl_3$ -Ni(acac)$_2$ as the catalyst: [$TiCl_3$] = 117 mmole/
liter Al/Ti = 2.0, [4M2P] = 2.7 mole/liter at 80°C for 60 hr.

TABLE 4. Polymerization of Phenylbutenes[a]

Monomer[b]	Catalyst system	Time (hr)	Yield (%)
4Ph1B	$(C_2H_5)_3Al$-$TiCl_3$	0.25	100
4Ph2B	$(C_2H_5)_3Al$-$TiCl_3$	100	35.6
4Ph2B	$(C_2H_5)_2AlCl$-$TiCl_3$	100	Trace
4Ph2B	$C_2H_5AlCl_2$ -$TiCl_3$	100	0.5

[a]Polymerization condition: [$TiCl_3$] = 120 mmole/liter, Al/Ti =
3.0, [Monomer] = 3.0 mole/liter, in heptane at 80°C.
[b]Abbreviations are as follows: 4Ph1B, 4-phenyl-1-butene;
4Ph2B, 4-phenyl-2-butene.

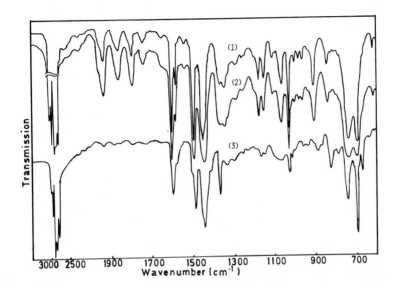

FIG. 3. IR spectra of polymers: (1) obtained from 4-phenyl-1-butene with $Al(C_2H_5)_3$-$TiCl_3$ catalyst; (2) obtained from 4-phenyl-2-butene with $Al(C_2H_5)$-$TiCl_3$ catalyst; (3) obtained from 4-phenyl-2-butene with $C_2H_5AlCl_2$-$TiCl_3$ catalyst.

As can be seen from Table 4, 4-phenyl-1-butene quite readily polymerizes with $Al(C_2H_5)_3$-$TiCl_3$ catalyst, but 4-phenyl-2-butene gives polymer at a slower rate under similar conditions. The catalyst systems consisting of $(C_2H_5)_2AlCl$ and $C_2H_5AlCl_2$ instead of $Al(C_2H_5)_3$ show very weak activity for the polymerization of 4-phenyl-2-butene, and the polymer obtained by the latter system was a viscous brownish material which was confirmed from IR spectrum (Fig. 3) to be an oligo(4-phenyl-2-butene).

However, the polymers obtained from both phenylbutenes with $Al(C_2H_5)_3$-$TiCl_3$ catalyst show an identical IR spectrum (Fig. 3), indicating that 4-phenyl-2-butene underwent monomer-isomerization polymerization with $Al(C_2H_5)$-$TiCl_3$ catalyst. This conclusion was also supported by the fact that the rate of polymerization corresponded to that of isomerization (see Table 5):

TABLE 5. Results of Polymerization and Isomerization of 4-Phenyl-2-butene with $Al(C_2H_5)_3$-$TiCl_3$ catalyst[a]

Al/Ti molar ratio	Time (hr)	Yield (%)	Composition of phenylbutenes recovered after polymerization (%)[b]					
			4Ph1B	cis-4Ph2B	trans-4Ph2B	cis-1Ph1B	trans-1Ph1B	BuBz
2.0	60	17.5	1.1	61.2	26.7	0.3	10.5	0.2
3.0	100	35.6	2.2	50.3	15.4	0.3	31.7	0.1
4.0	60	14.4	2.5	51.2	15.5	1.5	24.7	2.6
5.0	60	14.9	1.0	65.7	19.8	0.4	12.8	1.0

[a]Polymerization conditions: $[TiCl_3]$ = 120 mmole/liter, $[4Ph2B]$ = 3.0 mole/liter, in p-xylene at 80°C.
[b]Abbreviations are as follows: 1Ph1B, 1-phenyl-1-butene; BuBz, n-butylbenzene; 4Ph1B and 4Ph2B, see Table 4.

TABLE 6. Effects of Transition Metal Compounds (MeX) on the Monomer-Isomerization Polymerization of 4-Phenyl-2-butene[a]

MeX	Yield (%)	Composition of phenylbutenes recovered after polymerization (%)[b]					
		4Ph1B	cis-4Ph2B	trans-4Ph2B	cis-1Ph1B	trans-1Ph1B	BuBz
None	12.5	1.9	52.7	15.8	0.4	25.3	3.5
Ni(acac)$_2$	12.7	1.9	36.3	10.4	0.7	47.5	3.2
Ni(DMG)$_2$	15.9	1.0	17.9	4.0	1.9	69.9	5.3
Co(acac)$_3$	20.6	1.9	52.1	15.6	0.4	27.6	2.4
Fe(acac)$_3$	13.5	2.3	49.2	15.5	0.4	27.2	5.5

[a]Polymerization conditions: $[TiCl_3]$ = 120 mmole/liter, Al/Ti = 3.0, $[4Ph2B]$ = 3.0 mole/liter, MeX/Ti = 0.1, in n-heptane at 80°C for 60 hr.
[b]Abbreviations as in Tables 4 and 5.

$$
\begin{array}{cc}
\underset{\underset{C_2H_5}{|}}{CH}\!\!=\!\!\underset{\underset{C_6H_5}{|}}{CH} & \xrightleftharpoons{\text{isomerization}}
\end{array}
\quad
\begin{array}{cc}
\underset{\underset{CH_3}{|}}{CH}\!\!=\!\!\underset{\underset{CH_2}{|}}{CH} \\
\underset{\underset{C_6H_5}{|}}{}
\end{array}
\quad \xrightleftharpoons{\text{isomerization}} \quad
\begin{array}{c}
CH_2\!\!=\!\!\underset{\underset{\underset{\underset{C_6H_5}{|}}{(CH_2)_2}}{|}}{CH}
\end{array}
\quad \xrightarrow{\text{polymerization}}
$$

$$
\left(
\begin{array}{c}
CH_2\!\!-\!\!\underset{\underset{\underset{\underset{C_6H_5}{|}}{(CH_2)_2}}{|}}{CH}
\end{array}
\right)_n
$$

Effects of Al/Ti Molar Ratio and Transition Metal Compounds

The effects of the Al/Ti molar ratio in the $Al(C_2H_5)_3$-$TiCl_3$ catalyst on the polymerization and the isomerization of 4-phenyl-2-butene are shown in Table 5.

From Table 5 it is also observed that the rate of polymerization goes through a maximum at a molar ratio of Al/Ti of 2 to 3.

The effect of transition metal compounds on this monomer-isomerization polymerization is shown in Table 6. It is found that $Ni(DMG)_2$ and $Co(acac)_3$ serve as accelerators for this polymerization when the Me/Ti molar ratio is 0.1, but the other compounds are not effective. However, the effect of the concentration of these metal compounds on polymerization is uncertain.

CONCLUSION

Although from a consideration of the thermodynamic stabilities of isomers in 4-methyl- and 4-phenyl-branched 2-olefins the isomerization of the 2-olefin to the most unstable but polymerizable 1-olefin seemed to occur very slightly as compared with that to the most stable 3-olefin, 4-methyl-2-pentene and 4-phenyl-2-butene were found to undergo monomer-isomerization polymerization easily to give high molecular weight homopolymers of 4-methyl-1-pentene and 4-phenyl-1-butene, respectively. From gas chromatographic analysis of olefin isomers in the reaction mixture during the polymerization, it was confirmed that the rate of isomerization from 2-olefin to 1-olefin was larger than that to 3-olefin, and hence the concentration of the 1-olefin necessary to homopolymerize was always supplied through the fast isomerization. The mechanism of isomerization and polymerization in these monomer-isomerization polymerizations seemed to be similar to that reported for 2-butene.

REFERENCES

[1] T. Otsu, A. Shimizu, and M. Imoto, J. Polym. Sci., B, 3, 449, 1031 (1965).
[2] T. Otsu, A. Shimizu, and M. Imoto, J. Polym. Sci., A-1, 4, 1579 (1966).
[3] T. Otsu, A. Shimizu, and M. Imoto, Ibid., 7, 3111, 3119 (1969).
[4] T. Otsu, H. Nagahama, and K. Endo, J. Polym. Sci., B, 10, 601 (1972).
[5] J. P. Kennedy and T. Otsu, Adv. Polym. Sci., 7, 369 (1970).
[6] T. Otsu and A. Shimizu, Newer Polymer Synthesis (Kagaku-zokan No. 53) (T. Otsu and K. Takemoto, eds.), Kagakudojin, 1972, p. 147.
[7] S. Tani, S. Hamada, and A. Nakajima, Polym. J., 5, 86 (1973).
[8] T. Otsu and K. Endo, Unpublished Results.
[9] Y. Atarashi, Kobunshi Kagaku, 21, 264 (1964).

Selectivity Aspects in Cross Metathesis Reactions

WALTER J. KELLY and NISSIM CALDERON

The Goodyear Tire and Rubber Co.*
Research Division
Akron, Ohio 44316

ABSTRACT

When certain catalysts that display a low apparent metathesis activity on terminal olefins are employed on mixtures of terminal and internal olefins, they lead to a selective formation of cross metathesis products. Critical experimentation using deuterated 1-pentene reveals that terminal olefins prefer to scramble "head-to-tail." A study of the macrocyclics distribution produced at various conversions during 1,5-cyclooctadiene polymerization suggests that these are being formed exclusively via an intramolecular transalkylidenation. The significance of the two sets of results is discussed in terms of two basic mechanistic schemes.

INTRODUCTION

Molecular weight regulation in ring-opening polymerization of cyclo-olefins is accomplished by the introduction of controlled amounts of

---

*Contribution #535.

acyclic olefins [1-3]. The cross metathesis of the acyclic olefin with the high molecular weight chain results in a net scission of the poly-alkenamer chain:

$$\tag{1}$$

Thus the alkylidene moieties of the acyclic olefins [$R_1 CH=$] and [$R_2 CH=$] end up as chain ends. It can be easily shown that at ex-tremely high acyclic/cyclic olefin ratios the main reaction products will be homologous series of low molecular weight polyenes.

Gunther et al. [2] compared the chain scission efficiency of ter-minal vs internal olefins in cyclopentene polymerization under iden-tical reaction conditions. They concluded that 1-butene is consider-ably more efficient than either cis- or trans-2-butene in lowering the molecular weight of polypentenamer; suggesting a higher affinity of terminal than internal olefins to cross metathesis with vinylene double bonds. Herisson and Chauvin [4] studied the distribution of polyene reaction products produced by the cross metathesis of 2-pentene and cyclopentene by employing $WOCl_4/Sn(C_4H_9)_4$ or $WOCl_4/(C_2H_5)_2 AlCl$ catalytic systems. The cross metathesis of cyclopentene with 2-pentene provides three homologous series (symmetric/unsymmetric/symmetric) of polyenes:

$$\tag{2}$$

If the controlling factor of the process is primarily entropy and the reaction is carried to equilibrium, one expects a molar distribution of 1/2/1 between the respective symmetric/unsymmetric/symmetric series. Indeed, Herisson and Chauvin observed approximately a 1:2:1 distribution for the respective homologous series.

Several three-component metathesis catalysts, prepared by com-bining WCl_6 with an alcohol and further reacting with an organoaluminum halide, are known to be sluggish toward the self-metathesis of terminal olefins [5]. For example, the catalyst combination $C_2H_5AlCl_2/WCl_6/C_2H_5OH$ (Al/W/O molar ratio 4/1/1), when employed at normal levels of olefin/W ratio of 5,000 to 10,000, exhibit very little metathesis of 1-pentene; about 1.0 mole % of the respective ethylene and 4-octene is produced. The lack of apparent metathesis activity of terminal olefins could not be rationalized on grounds of steric hindrance that prevents accommodation of the terminal olefin within the coordination sphere of the metal. It is a well-established fact that the steric requirements of terminal olefins are less than those of internal olefins. Further, in view of Gunther's results, one could not invoke electronic considerations to rationalize the lack of self-metathesis of terminal olefins. Gunther's results obtained with $R_2AlCl/WCl_6/ClC_2H_4OH$ catalyst point to the fact that terminal olefins are more reactive than internal olefins in cross metathesis reactions with a $-CH=CH-$ type of unsaturation.

To resolve this apparent dilemma, an extensive study of the cross metathesis reactions of 1-pentene with 2-pentene and 1-pentene with cyclopentene was conducted. The results demonstrate that certain metathesis catalysts exhibit a considerable degree of selectivity in rendering a products mixture which is not dictated by a random scrambling of alkylidene moieties. The observed selectivity in the metathesis of terminal with internal olefins is accountable if one assumes a higher affinity of terminal olefins toward the catalyst and a specific geometry of the olefinic substrates on the catalyst site.

The formation of low molecular weight oligomers during cycloolefin polymerization has been reported [6]. Three basic features have been established:

1. The oligomers are macrocyclic.
2. The population of macrocyclics is controlled primarily by the double bond frequency along the polymeric chain.
3. The oligomeric macrocyclics and their corresponding high molecular weight polyalkenamer are interconvertible.

Hence 1,5-cyclooctadiene has been shown to yield two series of macrocyclics; one which contains "whole" multiples of the monomer $(C_8H_{12})_n$ and a second consists of the "sesqui"-oligomers $(C_8H_{12})_n-C_4H_6$, where $n \geq 2$.

By determining quantitatively the relative occurrence of "sesqui"- and "whole" oligomers at various conversions, it was possible to elucidate a pathway by which the macrocyclics are being formed.

The results reported in the present paper are accountable by either mechanism proposed heretofore. Two basic schemes for the trans-

alkylidenation step have been advanced. One requires the initial formation of a bisolefin-metal entity, bearing two olefinic ligands in a
cis configuration about the metal:

$$
\begin{array}{c}
M \\
R-CH \diagup \diagdown CH-R \\
\| \diagup \quad \diagdown \| \\
R-CH \qquad CH-R
\end{array}
\tag{3}
$$

Several views regarding the nature of the electronic transformation
occurring within the coordination sphere of the metal, which provide
the actual alkylidene scrambling, have been proposed, e.g., quasicyclobutane [7, 8], tetramethylene-metal complex [9], and metallocycle
[10]. Recently, a carbene-type mechanism was proposed [4, 11, 12]
that does not necessarily require the initial formation of a bisolefin-
metal complex (Eq. 3).

EXPERIMENTAL

Materials

Benzene solvent (reagent grade) was dried by passing over silica
gel and alumina prior to use. Cyclohexane (reagent grade ACS, MCB)
was distilled over CaH_2 and kept under a nitrogen atmosphere. Tungsten hexachloride (Shieldalloy Inc.) was used without prior purification
as were the catalytic modifiers: absolute ethanol (Gold Shield), 2,2,2-
trichloroethanol, and 2-chloroethanol (Aldrich Chemical Co.). Cyclopentene (Arapahoe), 1-pentene (Phillips Pure Grade), and cis-2-pentene
(Phillips Technical Grade) were purified by distillation and dried by
passing over activated silica gel and alumina. 1-Pentene-d_{10} (Merck
Sharp & Dohme of Canada Ltd., 99% isotopic purity) was used without
prior purification. 1,5-Cyclooctadiene (Columbian Carbon) was purified
by distillation over sodium.

Analytical Procedures

Analyses by gas-liquid chromatography were performed on an F&M
810 Model Gas Chromatograph, employing either a 20-ft silicone gum
Hi Pak column (Hewlett Packard Co), programmed from 50 to 300°C at

10°C/min, or a 10-ft carborane-siloxane column (Analabs Inc.), programmed from 70 to 350°C at 10°C/min. Additional gas chromatography was performed on a Varian Aerograph Model 1200 employing a 200-ft squalene-coated capillary column at 27° C. Mass spectrometric analyses were conducted on a low resolution CEC 21-103 mass spectrometer.

Preparation of Catalyst Solutions

The 0.05 $\underline{M}$ WCl_6 solutions were prepared by dissolving 1.0 g WCl_6 in 50 ml of the appropriate solvent. The specified amounts of alcoholic modifiers were added to the tungsten solutions and allowed to react at room temperature for a minimum period of 30 min prior to use. (WCl_6/ ROH molar ratios of 1/1 and 1/2 were employed.) Ethylaluminum dichloride (EADC) and diethylaluminum chloride (DEAC) (Texas Alkyls) were diluted in the appropriate solvent to form 0.2 $\underline{M}$ solutions.

Metathesis of 1-Pentene with cis-2-Pentene

A mixture consisting of 10 ml of each olefin (91.3 mmoles 1-pentene and 94.1 mmoles cis-2-pentene) was prepared in a 2-oz bottle, equipped with a self-sealing gasket and Teflon liner, in the absence of diluent. Catalyst was introduced by syringe, 0.02 mmole of $WCl_6/2$ $ClCH_2CH_2OH$ in benzene followed by 0.04 mmole of DEAC in benzene, and the progress of the reaction was monitored by gas-liquid chromatography.

Metathesis of 1-Pentene with Cyclopentene

In a typical experiment the reactants were passed under a nitrogen atmosphere over a silica gel/alumina column and collected in a 2-oz bottle equipped with a self-sealing gasket and Teflon liner. The bottles were sparged with nitrogen prior to the addition of catalyst. The tungsten component (presparged) followed by the aluminum cocatalyst were injected into the solution with hypodermic syringes. The reactions were shortstopped at the appropriate time by the addition of 0.1 ml methanol.
In the first series, experiments were conducted at 0 and 25°C using an acyclic/cyclic olefin ratio of 2.4/1 at neat conditions (no solvent added except catalyst carrier) and employing an olefin/W ratio of 3800/1. Catalyst: EADC/WCl_6/Cl_3CCH_2OH (Al/W/O = 2/1/2). The reactions were terminated after various times. In a second series, experiments were conducted at 0 and 25°C employing a similar catalyst

combination as above, using an acyclic/cyclic olefin ratio of 0.77/1 in the presence of benzene diluent (50%) and an olefin/W ratio of 2100/1. In the third series, conducted at 0 and 25°C, a EADC/WCl_6/$ClCH_2CH_2OH$ catalyst system was employed (Al/W/O = 2/1/2) in the presence of benzene diluent (50%) at an acyclic/cyclic olefin ratio of 4.7/1, and an olefin/W ratio of 3000/1. In the fourth series, the effect of reaction time on conversions at varied temperature was studied employing DEAC/WCl_6/$ClCH_2CH_2OH$ catalyst (Al/W/O = 2/1/2) with other variables maintained as in the third series.

Metathesis of 1-Pentene-d_{10} with 1-Pentene

A sample vial containing 1.0 g of 1-pentene-d_{10} was cooled in liquid nitrogen, cracked open, and 5.0 ml of freshly distilled 1-pentene was syringed into the vial. The contents were then transferred to a flask containing 45 ml of freshly distilled cyclohexane. The vial was rinsed a second time with 5.0 ml of 1-pentene which was also transferred to the cyclohexane solution.

Purification of the premix solution was achieved by passing through an alumina/silica gel column, and finally diluted to a total of 125 ml. The column-passed solution was stored in the cold under nitrogen.

Two 15-ml samples were transferred to 2-oz vials and used for this study. Into one of the vials catalyst solutions, 0.0075 mmoles of WCl_6/CH_3CH_2OH(1/1) in cyclohexane followed by 0.02 mmoles of EADC in cyclohexane were added, and after 10 min it was quenched with 0.1 ml of isopropyl alcohol. The second vial containing unreacted material was used as a standard solution.

The samples were analyzed for composition on a capillary gas-liquid chromatographic column and the mass spectrometric samples were obtained via preparative gas chromatography using a 10-ft, 0.5 in. diameter SGR column on an F&M 810 instrument.

Metathesis of 1,5-Cyclooctadiene to Form Macrocyclics

A mixture consisting of 1,5-cyclooctadiene (11.4%), n-eicosane (n-$C_{20}H_{42}$, 0.13%), antioxidant (2,6-di-t-butyl-p-cresol), and benzene was column passed under nitrogen over silica gel/alumina/silica gel. Samples were collected as 50 ml aliquots in 2-oz bottles, sealed and capped as described earlier, and were sparged with nitrogen prior to catalyst addition. First the tungsten component, 0.02 mmole of

WCl_6/CH_3CH_2OH in benzene, was introduced followed by the aluminum cocatalyst, 0.04 mmole of EADC in benzene ($Al/W/O = 2/1/1$). At appropriate time intervals the polymerization was shortstopped by the addition of 0.15 ml of CH_3OH. Each cement was then quantitatively analyzed by gas-liquid chromatography for oligomer composition.

RESULTS AND DISCUSSION

Metathesis of 1-Pentene with cis-2-Pentene

Table 1 presents the relative distribution of the various cross metathesis products after the reaction proceeded to a point where ~25% of the original 2-pentene has been consumed (50% from theoretical equilibrium).

The self-metathesis of either olefin leads to symmetric products; ethylene and 4-octene from 1-pentene, and 2-butene and 3-hexene from 2-pentene. The products from the cross metathesis are unsymmetric. Depending on the alignment of the olefin substrates on the catalyst site, one can obtain either propylene and 3-heptene, or 1-butene and 2-hexene. The significance of the results of Table 1 is in the fact that, although 1-pentene undergoes little self-metathesis, it readily reacts with 2-pentene to produce the cross reaction unsymmetric products. Furthermore, the data suggest that the affinity of 2-pentene to cross metathesize with 1-pentene is eight times greater than its affinity to self-metathesize.

TABLE 1. Cross Metathesis of 1-Pentene and cis-2-Pentene

$$C_3H_7CH=CH_2 + C_2H_5CH=CHCH_3$$

Relative concentrations[a]			
Symmetric		Unsymmetric	
$CH_2=CH_2$	1	$CH_3CH=CH_2$	4
$C_3H_7CH=CHC_3H_7$	1	$C_2H_5CH=CHC_3H_7$	4
$CH_3CH=CHCH_3$	2	$C_2H_5CH=CH_2$	12
$C_2H_5CH=CHC_2H_5$	2	$CH_3CH=CHC_3H_7$	12

[a]Values at 50% from theoretical equilibrium.

Metathesis of 1-Pentene with Cyclopentene

Three series of polyenes can be produced by the cross metathesis of 1-pentene and cyclopentene:

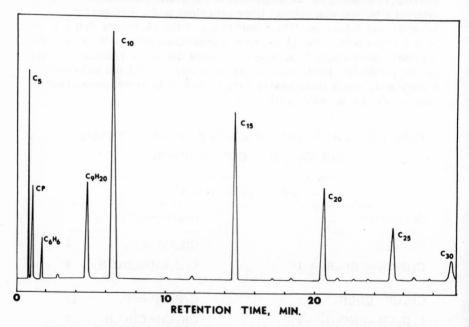

The unsymmetric series have carbon numbers C_{10}, C_{15}, C_{20}, C_{25}, etc. while the two symmetric series have the carbon numbers C_7, C_{12}, C_{17}, C_{22}, etc. and C_{13}, C_{18}, C_{23}, C_{27}, etc., respectively. Figure 1 illustrates a typical chromatogram of the 1-pentene + cyclopentene reaction products. The main peaks are attributed to the unsymmetric homologous series $CH_2 \!=\!\!= CH(CH_2)_3 CH \!=\!\!=_x CHC_3H_7$. Integration of the total minor

FIG. 1. Gas chromatogram of polyenes obtained in the [1-pentene + cyclopentene] cross metathesis.

peaks vs the major C_{10}, C_{15}, C_{20} components indicates the symmetric polyenes constitute less than 5% of the total product.

Table 2 summarizes the results obtained in the cross metathesis of 1-pentene with cyclopentene under various reaction conditions. The data were treated in a manner to account for the conversion of cyclopentene into the various C_{10}, C_{15}, C_{20}, and C_{25} unsymmetric polyenes vs its conversion to other products. In all cases the reaction was not allowed to proceed to maximum conversion in order to minimize loss of reactants to side products. The conversion vs time curves, illustrated in Figs. 2 and 3, indicate a leveling-off trend in the yield of the desired polyenes after 60 mins, except in the case where the reaction was carried out in benzene diluent using $EADC/WCl_6/CCl_3CH_2OH$ catalyst at 0°C (see Table 2, series 2). The best overall conversion and selectivity to unsymmetric polyenes was experienced in the experiment represented in Table 2, series 2 (0°).

Three reaction pathways for the formation of the polyenes are to be considered. Schematically, these are:

$$(5)$$

Path 1 assumes the selective cross metathesis of 1-pentene with cyclopentene followed by further incorporation of cyclopentene units. If this process is dominating unsymmetric polyenes will be formed preferentially. Path 2 assumed a selective self-metathesis of 1-pentene forming ethylene and 4-octene. Each of these symmetric olefins can incorporate cyclopentene units, thus resulting in the two symmetric polyene series. Path 3 assumes the initial self-metathesis of cyclopentene to a high molecular weight polypentenamer, followed by

TABLE 2. Conversion of Cyclopentene into Polyenes (%)

Series (°C)	Unreacted cyclopentene	Cyclopentene in unsymmetric polyenes: C_{10}, C_{15}, C_{20}, C_{25}	Cyclopentene in other products	Total % selectivity
1[a] (0°)	67	26.6	6.4	80.6
1 (25°)	72	21.2	6.8	75.7
2[b] (0°)	21	67.5	11.5	85.4
2 (25°)	67	18.0	15.0	54.5
3[c] (0°)	70	11.1	18.9	36.6
3 (25°)	58	22.9	19.1	54.7
4[d] (0°)	64	12.3	23.7	34.1
4 (25°)	54	27.7	18.3	60.5

[a]Series 1: neat; 1-pentene/cyclopentene = 2.4/1; EADC/WCl_6/CCl_3CH_2OH; reaction 30 min.
[b]Series 2: 50% benzene; 1-pentene/cyclopentene = 0.77/1; EADC/WCl_6/CCl_3CH_2OH; reaction 30 min.
[c]Series 3: 50% benzene; 1-pentene/cyclopentene = 4.7/1; EADC/WCl_6/$ClCH_2CH_2OH$; reaction 60 min.
[d]Series 4: 50% benzene; 1-pentene/cyclopentene = 4.7/1; DEAC/WCl_6/$ClCH_2CH_2CH$; reaction 60 min.

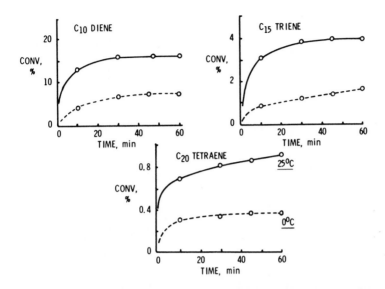

FIG. 2. Molar conversion of polyenes vs time in the [1-pentene +
cyclopentene] metathesis. Data related to Table 2, series 4.

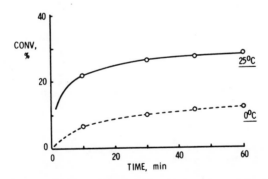

FIG. 3. Conversion of cyclopentene to unsymmetric polyenes: C_{10},
C_{15}, C_{20}, and C_{25} vs time. Data related to Table 2, series 4.

random cross metathesis scissions with 1-pentene. In the latter case
one should observe a 1/2/1 molar distribution of the symmetric/unsym-
metric/symmetric polyenes. The preferential formation of the unsymmetric polyene series indicates that, under the prevailing conditions
employed herein, Path 1 is dominating.

The observations described heretofore are summarized as follows:

A catalyst that displays a high metathesis activity for internal olefins and an apparent poor activity for terminal olefins, when employed on a mixture of internal plus terminal olefinic reactants, leads to a selective formation of the cross metathesis products.

A possible hypothesis for this behavior may be that terminal olefins, in fact, are more prone to metathesis than internal olefins, and the reason why the products of their reaction are not being observed is that they prefer to metathesize "head-to-tail," yielding back the starting material:

$$
\begin{array}{c}
C_3H_7CH = CH_2 \\
+ \\
CH_2 = CHC_3H_7
\end{array}
\ \rightleftarrows \
\begin{array}{c}
C_3H_7CH \\
\parallel \\
CH_2
\end{array}
\ + \
\begin{array}{c}
CH_2 \\
\parallel \\
CHC_3H_7
\end{array}
\tag{6}
$$

When terminal and internal olefins are mixed, the terminal olefins "flood" the catalyst sites, inhibiting the self-metathesis of the internal olefins. Hence terminal olefins keep "spinning their wheels" until an internal olefin manages to enter the complex, which after metathesis will yield an unsymmetric cross product. This hypothesis was confirmed experimentally.

Metathesis of 1-Pentene-d_{10} and 1-Pentene

A 90/10 mixture of 1-pentene (m/e = 70) and 1-pentene-d_{10} (m/e = 80) was exposed to the EADC/WCl$_6$/C$_2$H$_5$OH catalyst and analyzed by gas chromatography/mass spectrometry. The results of this experiment are presented in Table 3.

TABLE 3. Gas Chromatography/Mass Spectroscopy of 1-Pentene and 1-Pentene-d_{10} Metathesis[a]

m/e	T_0 (min)	T_{10} (min)	Equilibrium (calc)
80	0.1074	0.0395	0.0115
78	-	0.0686	0.0959
72	-	0.0771	0.0959
70	0.8926	0.8148	0.7967

[a]Traces of C_8H_{16}, $C_8H_8D_8$, and C_8D_{16} only.

Assuming that 1-pentene and 1-pentene-d_{10} experience a "head-to-tail" scrambling exclusively, forming $C_3D_7 CD{=}CH_2$ (m/e = 78) and $C_3H_7 CH{=}CD_2$ (m/e = 72), it is possible to calculate the theoretical equilibrium concentrations of the four respective 1-pentenes. The data in Table 3 indicate that after a 10-min reaction time the original mixture underwent about 70% of the theoretical scrambling with only trace amounts of the various possible 4-octenes (C_8H_{16}, $C_8H_8D_8$, and C_8D_{16}). (Monitoring of ethylene in this experiment was not attempted due to the interference posed by N_2, m/e = 28.)

Macrocyclics Formation from 1,5-Cyclooctadiene

As stated earlier, the population of macrocyclics in a given cyclo-olefin polymerization system depends on the frequency of occurrence of double bonds along the polymeric chain.

The relative concentration of macrocyclics obtained during the polymerization of 1,5-cyclooctadiene is presented in Table 4. The "whole" oligomers are multiples of the starting monomer—C_{16}, C_{24}, C_{32}, . . ., etc; the "sesqui"-oligomers are members of the C_{12}, C_{20}, C_{28}, . . ., etc. homologous series. The results of Table 4 indicate that the C_{12}/C_{16}, C_{20}/C_{16}, C_{24}/C_{16}, C_{28}/C_{16}, and C_{32}/C_{16} ratios are essentially constant throughout the polymerization. (The only variant is the C_{12}/C_{16} value at 98.5% conversion, which can be rationalized by taking into account changes in the thermodynamic equilibrium value of the C_{12} component, 1,5,9-cyclododecatriene, due to cis/trans isomerization of its double bonds.) These results are quite significant. The

TABLE 4. Mole Ratio of Cyclic Oligomers and Sesquioligomers to Cyclohexadecadiene (C_{16}/C_{16} = 1.0)

Conversion (%)	C_{12}/C_{16}	C_{20}/C_{16}	C_{24}/C_{16}	C_{28}/C_{16}	C_{32}/C_{16}
1.3	.47	.65	-	-	-
3.2	.51	.63	.32	.18	-
7.9	.55	.63	.27	.24	-
9.5	.48	.62	.28	.25	.10
19.5	.52	.59	.31	.27	.10
36.3	.52	.56	.32	.18	.13
51.1	.51	.58	.33	.26	.13
61.2	.53	.54	.30	.19	.10
98.5	.93	.65	.27	.23	.14

constant ratios of "sesqui"- vs "whole" oligomers throughout the polymerization strongly suggest that these are being formed exclusively via an intramolecular metathesis occurring within a high molecular weight macromolecule, and not via a condensation of two smaller rings. This finding is highly compatible with the recently proposed views regarding the mechanism of olefin metathesis involving carbene-metal intermediates [4, 11, 12], but it does not exclude the "traditional" mechanisms that require the initial formation of bisolefin-metal complex.

MECHANISTIC IMPLICATIONS

Two basic views regarding the mechanism of olefin metathesis have been advanced. The quasicyclobutane proposal, first suggested by Bradshaw [7] and later subscribed to by Calderon and co-workers [8, 13], as well as Pettit's [9] tetramethylene-metal complex and Grubbs [10] five-membered ring metallocycle, are all transition states that attempt to describe the rearrangement:

$$\begin{array}{ccc}
\underset{\substack{C \\ \| \\ C}}{C} \overset{M}{\diagup\diagdown} \underset{\substack{C \\ \| \\ C}}{C} & \rightleftharpoons & \underset{\substack{C \\ = \\ C}}{C} \overset{M}{=} \underset{\substack{C \\ = \\ C}}{C}
\end{array} \qquad (7)$$

They all assume a priori the existence of a bisolefin-metal complex. Lately, Herisson and Chauvin [4], Dolgoplosk and co-workers [12], and Casey [11] have advanced the carbene-to-metallocycle scheme which does not necessarily require the initial formation of a bisolefin-metal complex. The mechanism is illustrated by:

$$\begin{array}{c}
\underset{W\cdot}{\overset{R_1CH}{\|}} + \underset{HCR_2}{\overset{HCR_1}{\|}} \rightleftharpoons \underset{W\cdot \leftarrow \|}{\overset{R_1CH}{\|}} \overset{HCR_1}{\underset{HCR_2}{}}
\\[2em]
\Big\downarrow\!\!\nwarrow
\\[1em]
\left[\begin{array}{c} R_1CH - HCR_1 \\ | \quad\quad | \\ W\cdot - HCR_2 \end{array}\right]
\\[2em]
\nwarrow\!\!\Big\downarrow
\\[1em]
R_1CH \underset{W\cdot = HCR_2}{\overset{=}{\Big\downarrow}} CHR_1
\end{array} \qquad (8)$$

$$\begin{array}{c}
R_1CH = CHR_1 \\ + \\ W\cdot = HCR_2 \end{array} \rightleftharpoons \begin{array}{c} R_1CH \overset{=}{\underset{\Big\downarrow}{}} CHR_1 \\ W\cdot = HCR_2 \end{array}$$

An incoming olefin undergoes π-coordination with the active metal site (W*) which possesses a bound carbene moiety. A carbene inter-change between the metal and the olefin is accomplished by rearrange-ment of the complex, forming a four-membered ring metallocycle along the reaction coordinate. Finally, the newly formed olefin mole-cule dissociates away from the active site.

When applied to cycloolefins, the carbene-to-metallocycle mech-anism does not postulate a polymerization via a macrocyclization pathway as is the case for the concerted mechanism (Eq. 7), rather it is consistent with a ring-opening polymerization by chain-end growth:

$$(9)$$

The formation of macrocyclic oligomers is accomplished by the intramolecular "back-biting" transalkylidenation of the growing car-bene-W chain end with any internal double bond on the same chain:

$$(10)$$

whereas in the concerted mechanism macrocyclic oligomers can be formed either by a "pinching-off" or "condensation" processes as illus-trated in Eqs. (11) and (12), respectively:

$$(11)$$

$$
\begin{array}{c}
(CH_2)_n - CH \\
| \quad\quad || \\
M_x - CH
\end{array}
\;+\;
\begin{array}{c}
CH - M_y \\
|| \quad\quad | \\
CH - (CH_2)_n
\end{array}
\;\underset{\longleftarrow}{\longrightarrow}\;
\begin{array}{c}
(CH_2)_n - CH = CH - M_y \\
| \\
M_x - CH = CH - (CH_2)_n
\end{array}
\qquad (12)
$$

The results of Table 4 are inconsistent with Eq. (12). If a given oligomer is being formed primarily by a bimolecular condensation of two smaller rings, one expects to find little or no "sesqui"-oligomers at the early stages of polymerization of 1,5-cyclooctadiene. The constant ratios of oligomers found throughout the polymerization suggest that they are being formed either via Eq. (10) or Eq. (11), depending on what mechanism one elects to apply. Both schemes require the formation of a high molecular weight polymer at the outset. Dolgoplosk [12] argued that the formation of high molecular weight polymer at the early stages of reaction serves as an indication that cycloolefin polymerization proceeds via chain-end growth. In a previous publication [13] it has been shown that a rapid increase in molecular weight is also accountable by a macrocyclization scheme, where growth may occur at any double bond along the chain.

To accommodate the observations related to the selectivity in cross metathesis reactions of terminal with internal olefins by the carbene-to-metallocycle reaction scheme, one must invoke preference to certain transition states over others. For example, to account for 1-pentene's tendency to generate itself upon metathesis, either I $\gg$ II or III $\gg$ IV:

$$
\begin{array}{c}
CH_2 - CHR_1 \\
| \quad\quad | \\
W - CH_2
\end{array}
\qquad
\begin{array}{c}
CH_2 - CH_2 \\
| \quad\quad | \\
W - CHR_1
\end{array}
\qquad
\begin{array}{c}
R_1CH - CH_2 \\
| \quad\quad | \\
W - CHR_1
\end{array}
\qquad
\begin{array}{c}
R_1CH - CHR_1 \\
| \quad\quad | \\
W - CH_2
\end{array}
$$

$$
\quad\;\; I \quad\quad\quad\quad\quad II \quad\quad\quad\quad\quad III \quad\quad\quad\quad\quad IV
$$

If steric hindrance around the metal is the controlling factor, the most favored metallocycle transition state will be I, and III the least favored. Thus, in analogy with the "head-to-tail" configuration proposed earlier for 1-pentene metathesis by the concerted mechanism, one may suggest that metallocycle I is preferred over II for the carbene mechanism. This type of speculation can be further extended to account for the selectivity in cross products observed in the 1-pentene/2-pentene and 1-pentene/cyclopentene metathesis.

The carbene in a metal-carbene complex is viewed as a singlet carbene; that is, having a pair of electrons occupying one orbital of an

sp^2 set. The bonding nature in a carbene-metal complex is somewhat analogous to CO or olefin bonded to a metal. Two bonding elements are involved: 1) a forward donation of an electron pair from the occupied sp^2 of the $C_{carbene}$ to the metal, and 2) a back donation element from the metal into an empty p_z orbital of the $C_{carbene}$. As illustrated in Fig. 4, the metal, $C_{carbene}$, and the two substituents (R_1 and H) all lie on the same plane. On the other hand, the complexed olefin molecule (its cis configuration is arbitrary) is on a plane that bisects the carbene-metal plane on an adjacent coordination site of the metal. (The olefinic carbons are situated above and below the carbene-metal complexation plane.) In order to achieve a transformation to a four-membered ring metallocycle, the $C_{carbene}$ expands from an sp^2 to an sp^3 hybridization. In the process the carbene-to-metal bond transforms into a conventional σ-C—M bond, and the carbon assumes a tetrahedral configuration. Simultaneously, the π-bonded olefin undergoes a process analogous to Ziegler's cis-ligand insertion, resulting in one of the C=C carbons σ-bonded to the metal and the other to the original $C_{carbene}$. The intimate electronic transformations described above are accompanied by a set of steric manipulations (Fig. 4). The sp^2 to sp^3 expansion involves trigonal to tetrahedral change of configuration,

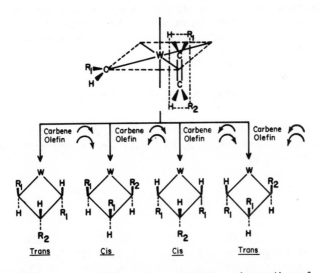

FIG. 4. Carbene-to-metallocycle mechanism; formation of various metathesis products by proper rotations of carbene and olefin substrates.

which requires either clockwise ($\curvearrowright$) or counterclockwise rotation ($\curvearrowleft$) of the $C_{carbene}$ in relation to the C-W axis. Furthermore, to form the metallocycle, the complexed olefin must "slide" and rotate (clockwise or counterclockwise) to align the olefinic carbons with their respective bonding counteratoms.

To account for the ease of formation of all possible cis and trans $R_1CH=CHR_1$, $R_1CH=CHR_2$, and $R_2CH=CHR_2$ olefins from either cis or trans starting olefins, one ought to assume low stearic barriers for clockwise and counterclockwise rotations of the carbene and olefin substrates. For the specific configuration illustrated in Fig. 4, applying various clockwise and counterclockwise rotations of the respective substrates, metallocycle transition states leading to trans-$R_1CH=CHR_2$, cis-$R_1CH=CHR_1$, cis-$R_1CH=CHR_2$, and trans-$R_1CH=CHR_1$ are obtainable.

CONCLUSION

Selectivity aspects of cross metathesis reactions between terminal and internal olefins, as well as the reaction pathway for the formation of macrocyclics in cycloolefin metathesis, are compatible with concerted and nonconcerted mechanistic schemes. The nonconcerted carbene-to-metallocycle mechanism, recently proposed for olefin metathesis, involves a basic concept which is common to most transition metal catalyzed reactions; namely, the cis ligand insertion concept.

ACKNOWLEDGMENTS

The authors wish to acknowledge Dr. E. A. Ofstead for stimulating discussions and Mr. N. J. Faircloth, Jr. for conducting the mass spectrometric analysis.

REFERENCES

[1] K. W. Scott, N. Calderon, E. A. Ofstead, W. A. Judy, and J. P. Ward, Rubber Chem. Technol., 44, 1341 (1971).

[2] P. Gunther, F. Hess, G. Merwede, K. Nutzel, W. Oberkirch, G. Pampus, N. Schon, and J. Witte, Angew. Makromol. Chem., 14, 87 (1970).

[3] N. Calderon, J. Macromol. Sci.—Revs. Macromol. Chem., C7(1), 105 (1972).
[4] J. L. Herisson and Y. Chauvin, Makromol. Chem., 141, 161 (1971).
[5] N. Calderon and W. J. Kelly, Unpublished Results.
[6] K. W. Scott, N. Calderon, E. A. Ofstead, W. A. Judy, and J. P. Ward, Adv. Chem. Ser., 91, 399 (1969).
[7] C. P. C. Bradshaw, E. J. Howman, and L. Turner, J. Catal., 7, 269 (1967).
[8] N. Calderon, E. A. Ofstead, J. P. Ward, W. A. Judy, and K. W. Scott, J. Amer. Chem. Soc., 90, 4133 (1968).
[9] G. S. Lewandos and R. Pettit, Tetrahedron Lett., 1971, 789.
[10] R. H. Grubbs and T. K. Brunck, J. Amer. Chem. Soc., 94, 2538 (1972).
[11] C. P. Casey and T. J. Burkhardt, Ibid., 96, 7808 (1974).
[12] B. A. Dolgoplosk, K. L. Makovetsky, T. G. Golenko, Yu. V. Korshak, and E. I. Tinyakova, Eur. Polym. J., 10, 901 (1974).
[13] N. Calderon, Acc. Chem. Res., 5, 127 (1972).

Vinyl Polymerization with
Transition Metal Alkyls and Hydrides

AKIO YAMAMOTO and SAKUJI IKEDA

Research Laboratory of Resources Utilization
Tokyo Institute of Technology
Ookayama, Meguro, 152, Tokyo, Japan

ABSTRACT

Transition metal alkyls and hydrides isolated from
Ziegler-type catalyst mixtures serve as appropriate models
for coordination polymerization. These transition metal complexes initiate the polymerization of some vinyl compounds
and aldehydes. In some cases the monomer-coordinated
complexes may be isolated and the interaction of the monomer
with the transition metal complexes may be studied. Comparison
of the polymerization kinetics of the vinyl compounds with the
decay kinetics of the initial transition metal complexes on interaction with the monomer provided important information with
respect to the mechanism of coordination polymerization by
these complexes. The effect of organoaluminum compounds on
the reactivity of the transition metal complexes has been also
studied. These transition metal alkyls initiate the polymerization of acetaldehyde to give a polyether-type polymer at -78°C
and a polymer with OH groups at room temperature.

Ziegler-type catalysts represent one of the most versatile and excellent catalyst systems and have attracted, over two decades, the interest of very many polymer chemists. Nevertheless, their complexity and heterogeneity have hindered a clear understanding of the reaction mechanism, and some unsolved problems still remain concerning the details of the polymerization mechanism. It has been generally accepted, however, that an organotransition metal complex constitutes the active center, and the polymerization proceeds by coordination of monomers and their ensuing insertion into the transition metal-carbon or metal-hydrogen bond.

An approach to understanding the mechanism of the coordination polymerization is to isolate aluminum-free transition metal complexes with the metal-carbon or metal-hydrogen bond and study their behavior toward monomers [1, 2]. We have found that the transition metal alkyls and hydrides can be isolated from Ziegler-type reaction mixtures containing transition metal compounds and alkylaluminum compounds under appropriate conditions. Some of the isolated complexes show polymerization activities, and others form coordination compounds with some olefins. Table 1 summarizes the transition metal alkyls and hydrides which have been isolated and characterized, and also which of their polymerization activities have been examined in our group [2]. Tertiary phosphines and 2,2'-bipyridyl (bipy) proved to be particularly useful in stabilizing alkyl and hydrido complexes of later transition metals, whereas their stabilization effect was less pronounced in early transition metal complexes. Alkyls or hydrides of Ti [3], Cr [4], and Mn [5] have been prepared by the reactions of these transition metal halides with alkylaluminum compounds whereas the rest of the complexes shown in Table 1 have been prepared from transition metal acetylacetonates. Hydride complexes are considered to be formed by β-elimination reaction of unstable intermediate alkyl transition metal complexes.

Most of the alkyls and hydrides of Group VIII transition metals and copper reacted with various olefins and converted many vinyl monomers to high polymers, whereas some of the alkyls of early transition metals initiated cationic polymerization. Detailed studies have been made on the reactions of $R_2Ni(bipy)$ [6-9], $R_2Fe(bipy)_2$ [10-16], $HCo(N_2)(PPH_3)_3$ [17-20], and $H_2Ru(PPh_3)_4$ [21-24] with various vinyl compounds, and we describe here an outline of the results.

$R_2Ni(bipy)$ (I) behaves somewhat differently from the other three complexes mentioned. It reacts with a variety of olefins but no polymerization is initiated in the complete absence of oxygen, and as such it serves as a good model compound for studying the interaction of a transition metal complex with olefins. The reaction

TABLE 1. Transition Metal Alkyls and Hydride Isolated from
Ziegler-Type Catalyst Systems

Group IV	$MeTiCl_3$, $MeTiCl_3 \cdot L$ (L = Py, THF, bipy, PPh_3)
VI	$RCrCl_2 \cdot L_3$ (R = Me, Et, Pr, i-Bu; L = THF, Py, α, β, γ-picolines, $PrNH_2$, $BuNH_2$) $MoH(acac)(dpe)_2$ dpe = 1,2-bis(diphenylphosphino)ethane
VII	$MnHI(THF)_{1,5}$
VIIIa	$R_2Fe(bipy)_2$ (R = Me, Et, Pr), $Me_2Fe(dpe)_2$, $H_2Fe(dpe)_2$, $Me_2Fe(PPh_3)_3$, $EtFe(acac)(PPh_3)_3$, H_2RuL_4 (L = PPh_3, Ph_2Me, PPh_2H)
VIIIb	$RCo(bipy)_2$ (R = Me, Et), $MeCo(dpe)_2$, $HCo(N_2)(PPh_3)_3$, $MeCo(PPh_3)_3$, $HRh(PPh_3)_4$
VIIIc	$R_2Ni(bipy)$ (R = Me, Et, Pr, i-Bu), $R_2Ni(dpe)$ (R = Me, Et), $MeNi(acac)(PPh_3)_2$, $EtNi(acac)(PPh_3)$, R_2NiL_2 (L = PEt_3, PBu_3)
IB	$MeCu$, $MeCu(bipy)$, $MeCuL_n$ (R = Me, Et, Pr, i-Bu; L = PPh_3, PCy_3, PPh_2Me, PPh_2Et, $PPhMe_2$, PBu_3, PEt_3), $(RCu)_2(dpe)_3$ (R = Me, Et, Pr, i-Bu)

---

proceeds through intermediate formation of an olefin-coordinated
dialykyl(bipyridyl)nickel complex (II) [6, 8, 9]:

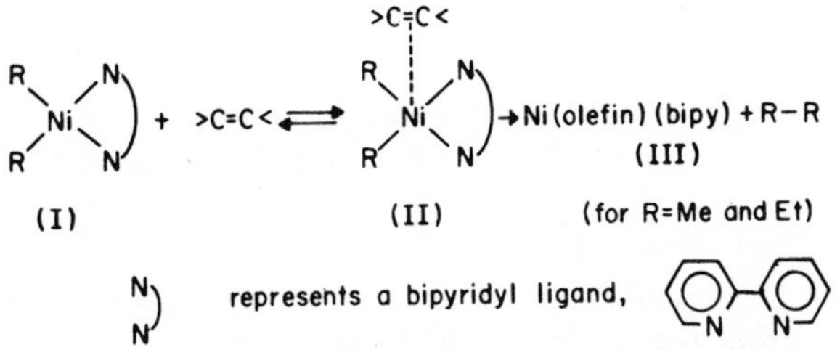

$$\text{(1)}$$

The intermediate olefin-coordinated alkylnickel complex (II) can be
isolated in the cases of acrylonitrile and acrolein at low temperatures.
The coordination of olefin to the alkylnickel complex leads to the
activation of the metal-carbon bonds, and the splitting of the M—C

bonds yields a zero-valent nickel complex coordinated with the olefin. The strength of interaction between the olefin and the zero-valent bipyridyl-nickel complex has been determined spectrophotometrically by examining the dissociation equilibrium of the zero-valent olefin-coordinated complexes (III) in solution. The study revealed that the olefin which interacts with nickel the more strongly activates the M–C bond in $R_2Ni(bipy)$ to the greater extent [9]. This finding has formed the basis for further study of the mechanism of the coordination polymerization of the vinyl monomers with iron, cobalt, and ruthenium complexes.

Examination of the reactions of $R_2Fe(bipy)_2$, $HCo(N_2)(PPh_3)_3$, and $H_2Ru(PPh_3)_4$ with various olefins and comparison of the kinetics of the decay of these complexes in the presence of olefins with those of the polymerization of vinyl compounds by these complexes revealed some common features in the reactions of the alkyl and hydrido transition metal complexes with olefins.

The behavior of $HCo(N_2)(PPh_3)_3$ (IV) is illustrative in demonstrating the characteristic reactions and will be described here [18-20].

BEHAVIOR OF $HCo(N_2)(PPh_3)_3$ TOWARD OLEFINS

Complex IV reacts with a very electronegative olefin such as tetracyanoethylene (TCNE) to yield an olefin-coordinated complex $CoH(TCNE)_2(PPh_3)_2$. Olefins of medium Alfrey-Price's e-values such as acrylonitrile, methacrylonitrile, acrolein, and methyl methacrylate are polymerized by IV. Olefins of lower e-values such as styrene form olefin-coordinated complexes and liberate 1 mole equivalent of the hydrogenated olefin, i.e., ethylbenzene from styrene:

$$HCo(N_2)(PPh_3)_3 + 2CH_2=CH \longrightarrow CH_2=CH-Ph+N_2+PhEt$$

Ph $(Ph_2P)_2$Co–PPh_2

(V)

(2)

Ethylbenzene is considered to be formed by insertion of styrene into the Co–H bond followed by abstraction of hydrogen at the ortho-position of one of the coordinated triphenylphosphines. On the other hand, a less electronegative olefin, isobutyl vinyl ether (IBVE), displaced all the triphenylphosphine ligands from IV to give $CoH(IBVE)_2$. Vinyl acetate showed an unique reactivity to give cobalt acetate with evolution of ethylene [24].

The corresponding methylcobalt complex showed similar reactivities toward various olefins [18].

The course of the reactions of $HCo(N_2)(PPh_3)_3$ with various olefins can be conveniently followed by observing the spectral change of the complex in solution. The electronic spectrum of complex IV with a peak at 390 nm decays in the presence of olefins following the first-order kinetics with respect to the concentration of IV:

$$- \frac{d}{dt} [HCo(N_2)(PPh_3)_3] = k[HCo(N_2)(PPh_3)_3] \tag{3}$$

The pseudo-first-order rate constant k increases with an increase in olefin concentration and approaches a constant value k_1 of 1.6×10^{-3} sec^{-1} at 25°C which is independent of the kind of olefin. Analysis of the dependence of the pseudo-first-order rate constant on the olefin concentration and the examination of the inhibition effects of the addition of triphenylphosphine and of nitrogen pressure revealed that k can be expressed by

$$\frac{1}{k} = \frac{1}{k_1} + A \{B + C [N_2]\} \frac{[PPh_3]}{[M]} \tag{4}$$

where [M] is the olefin concentration and A, B, and C are constants. The kinetic results expressed by Eqs. (3) and (4) are compatible with the mechanism

$$HCo(N_2)(PPh_3)_3 \underset{k_{-1}}{\overset{k_1}{\rightleftharpoons}} HCo(N_2)(PPh_3)_2 + PPh_3$$
$$[C_1] \qquad\qquad\qquad [C_2]$$

$$[C_2] + \text{olefin} \underset{k_{-2}}{\overset{k_2}{\rightleftharpoons}} HCo(\text{olefin})(N_2)(PPh_3)_2$$
$$[C_3]$$

$$[C_3] \underset{k_{-3}}{\overset{k_3, -N_2}{\rightleftharpoons}} HCo(\text{olefin})(PPh_3)_2 \overset{k_4}{\longrightarrow} \text{irreversible reaction} \tag{5}$$
$$[C_4] \qquad\qquad\qquad\qquad \text{(polymerization)}$$

$CoH(N_2)(PPh_3)_3$ is a coordinatively saturated species and it slowly liberates one of the triphenylphosphine ligands in solution to accommodate a coordination site for the incoming olefin. The coordination of olefin leads to the activation of the Co–H bond and of the coordinated N_2, and the olefin may be inserted into the Co–H bond, thus liberating N_2. This process is considered as the rate-determining step. The insertion of the olefin may be followed by ensuing insertions of monomers into the Co–C bond to yield a polymer attached to cobalt which is spontaneously and irreversibly terminated. Assumptions of the above mechanism and of steady-state concentrations for $[C_2]$, $[C_3]$, and $[C_4]$ lead to the first-order kinetics of Eq. (3), and the pseudo-first-order rate constant k expressed as

$$\frac{1}{k} = \frac{1}{k_1} + \frac{k_{-1}}{k_1 k_2 k_3} \frac{[PPh_3]}{[M]} \left\{ (k_{-2} + k_{-3}) + \frac{k_{-2} k_{-3}}{k_4} [N_2] \right\} \tag{6}$$

Equation (6) is in agreement with the experimental results expressed by Eq. (4).

Examination of the ^{31}P NMR spectrum of $CoH(N_2)(PPh_3)_3$ in toluene supports the slow dissociation of triphenylphosphine. The proton-decoupled ^{31}P NMR spectrum of IV in the absence of added triphenylphosphine shows no peak of free triphenylphosphine. In the presence of an equivalent amount of added triphenylphosphine, the ^{31}P NMR spectrum shows very sharp peaks of the coordinated triphenylphosphine and free triphenylphosphine in a peak ratio of 3:1 at -80°C, and these peaks approach each other accompanied by broadening with an increase of the temperature. These results show that the dissociation of triphenylphosphine from IV is negligible at room temperature but the added triphenylphosphine does exchange with the coordinated triphenylphosphine, suggesting that a slow S_N1 type dissociation of triphenylphosphine from $CoH(N_2)(PPh_3)_3$ is taking place.

A quite similar kinetic behavior is observed with $CoH(CO)(PPh_3)_3$ which can be prepared by displacement of N_2 with CO from IV [19]. The kinetic results are in agreement with the mechanism

$$CoH(CO)(PPh_3)_3 \underset{k_{-1}}{\overset{k_1{'}}{\rightleftarrows}} CoH(CO(PPh_3)_2 + PPh_3$$

$$k_{-2}{'} \Big\Updownarrow k_2{'}$$

$$+ \text{ olefin}$$

$$CoH(\text{olefin})(PPh_3)_2$$

$$\downarrow$$

$$\text{irreversible reaction} \tag{7}$$
$$\text{(polymerization)}$$

In the polymerization of vinyl monomers with IV, the polymerization starts after a short induction period of about 2 min and, with time, the polymer yield (Y_t) approaches a certain maximum value (Y_∞) which depends on the concentration of IV. The molecular weight of the polymer was independent of the polymer yield, polymerization time, and the initiator concentration, and increased linearly with the monomer concentration. Plotting of $\log(Y_\infty - Y_t)$ against polymerization time gave straight lines indicating the relationship.

$$\log(Y_\infty - Y_t) = -\bar{k}t + \text{const} \tag{8}$$

The $\bar{k}$ value increases with the monomer concentration, approaching a limiting value of 1.6×10^{-3} sec^{-1} at $25°C$.

On the basis of the spectroscopic study of the decay kinetics of $CoH(N_2)(PPh_3)_3$ in the presence of the monomer and on the assumption that a polymer molecule is produced from the complex with an initiator efficiency f, followed by further rapid insertion of monomers into the Co–C bond (propagation) and spontaneous termination after the polymer reached a certain degree of polymerization $\overline{DP}$, the rate of polymerization R_p can be expressed by

$$R_p = \frac{d[Y]}{dt} = f\overline{DP} \left\{ -\frac{d[CoH(N_2)(PPh_3)_3]}{dt} \right\} \tag{9}$$

Integration of Eq. (9) utilizing the relationship of Eq. (3) leads to

$$\log(Y_\infty - Y_t) = -kt + \log\{f\overline{DP}\,[CoH(N_2)(PPh_3)_3]_0\} \tag{10}$$

where $[CoH(N_2)(PPh_3)_3]_0$ represents the initial concentration of IV.

The form of Eq. (10) is identical with Eq. (8). The comparison of $\bar{k}$, obtained from polymerization kinetics, with k, observed by the spectroscopic method following the decay kinetics of IV, indicated a reasonable agreement, supporting the validity of our assumptions.

A linearity was observed between the molecular weight of the polymer and the monomer concentration. Addition of triphenylphosphine to the polymerization system caused a decrease in polymer yield as well as in the molecular weight. Employment of solvents of higher coordinating abilities also caused a decrease in the polymer yield.

These results suggest that coordination of monomer to the complex

with a growing polymer chain leads to insertion of the monomer between the cobalt-carbon bond whereas the coordination of triphenylphosphine or the solvent molecule leads to spontaneous termination
with certain probabilities. These assumptions lead to the following
equation which is in agreement with the experimental results:

$$\overline{DP} = (k_p/k_t)\,[M] \tag{11}$$

The linearity between $\overline{DP}$ and $[M]$ passing through the origin indicates
the absence of a chain transfer with the monomer.

In the polymerization of methyl methacrylate with $Et_2Fe(bipy)_2$,
we have confirmed the formation of ethane-d_1 when cis-
$CHD{=}C(CH_3)CO_2CH_3$ was used [15]. This fact was taken as evidence
indicating the participation of a termination mechanism involving the
hydrogen abstraction from the β-position of the polymer chain bonded
to iron. In the case of $HCo(N_2)(PPh_3)_3$ such a termination process is
unlikely since it would regenerate the active cobalt hydride species
to initiate the polymerization again. A possible termination mechanism is the one which involves the participation of the ortho-hydrogen
of the triphenylphosphine ligand:

$$\tag{12}$$

This scheme is in line with the formation of the previously described
ortho-metallated species, Eq. (2).

STEREOREGULARITIES OF POLYMERS PRODUCED
WITH TRANSITION METAL ALKYLS AND HYDRIDES

Methyl methacrylate can be polymerized with $HCo(N_2)(PPh_3)_3$ and
$R_2Fe(bipy)_2$, and the stereoregularity of the polymers was examined
by NMR spectroscopy. The polymers produced with $HCo(N_2)(PPh_3)_3$

were rich in syndiotactic fraction, and the polymerization process appears to obey the Bernoulli statistics with the probability of meso-placement p_m ranging from 0.15 to 0.27 depending on the solvent employed. In contrast, $R_2 Fe(bipy)_2$ gave two types of poly(methyl methacrylate): one was rich in syndiotactic fraction and the other was a stereoblock-type polymer. The latter was prepared by polymerization of methyl methacrylate by $Et_2 Fe(bipy)_2$ in solvents of weak coordinating abilities, and the polymerization process that gave this type of polymer obeyed the first-order Markovian statistics with a penultimate effect, whereas the former was obtained in strongly coordinating solvents and the propagation process obeyed the Bernoulli statistics. Based on these results, a possible mechanism of stereoregulation involving the control of monomer coordination toward the complex with the remaining alkyl group, bipyridyl, and the growing polymer chain attached to iron has been discussed [14].

NMR examination of the mode of double bond opening in the polymerization of cis-$CHD=C(CH_3)CO_2 CH_3$ with $Et_2 Fe(bipy)_2$ revealed trans-opening of the double bond, and mechanisms compatible with the result have been discussed [16].

COPOLYMERIZATION

Some sets of vinyl monomers may be copolymerized with $HCo(N_2)(PPh_3)_3$, $R_2 Fe(bipy)_2$, and $H_2 Ru(PPh_3)_4$. Examination of the compositions of these copolymers indicated a trend of electronegative monomers with stronger coordinating abilities to transition metal being introduced into the copolymer in preference to less electronegative monomers with weaker coordinating abilities (12-14, 20, 23].

CONCLUSIONS OF THE STUDY ON THE MECHANISM OF VINYL POLYMERIZATION WITH TRANSITION METAL ALKYLS AND HYDRIDES

1. Predissociation of a part of the ligands from the coordinatively saturated transition metal complex is required to accommodate the incoming monomer.

2. The coordination of the monomer to the transition metal complex leads to the activation of the M–C or M–H bond, and the rate-determining initiation step probably involves the insertion of the monomer into the activated M–C or M–H bond. The slow initiation step is followed by a rapid propagation step involving the successive

insertion of the monomers into the M–C bonds and the spontaneous termination process.

3. Coordination of the monomer with the transition metal complex constitutes the most important factor in controlling copolymerization and stereospecific polymerization.

EFFECT OF ORGANOALUMINUM COMPOUNDS

While the main characteristics in the polymerization of vinyl monomers with transition metal alkyls and hydrides isolated from Ziegler-type mixed catalyst systems may be summarized by the above conclusions, the actual Ziegler catalysts prepared in situ from transition metal halides and alkylaluminum compounds are more complicated and the effect of the alkylaluminum component cannot be neglected. In order to study the role of the alkylaluminum component in Ziegler catalysts, the behavior of transition metal alkyls in the presence of alkylaluminum compounds were examined [25]. The aluminum-free alkyltransition metal complexes CH_3TiCl_3, $R_2Ni(bipy)$, $C_2H_5CrCl_2 \cdot Py_3$, $R_2Fe(bipy)_2$, and $CH_3Cu(PPh_3)_3 \cdot$(toluene) [26] are stable compounds but they are decomposed by the addition of alkylaluminum compounds, whereas $HMnI(THF)_{1.5}$ was somewhat stabilized [5]. Kinetic study of the decomposition of M–C bond in CH_3TiCl_3 and $R_2Ni(bipy)$ revealed that the decomposition reactions were first order in the concentrations of the transition metal alkyls. Comparison of the effect of different aluminum compounds showed that the destabilizing effect of the aluminum compound increased in the order of Lewis acidity, $AlR_2(OEt) < AlR_3 < AlCl_3$.

In explaining the role of the aluminum component in the actual Ziegler catalyst, some binary models, such as the one shown here, have been proposed [29-31]:

In this model the alkylaluminum components are bridged with titanium through chlorine or alkyl bridges and indirectly modify the activity of the active center–in this case, alkyltitanium complex. We consider that there are two main effects of the alkylaluminum component: 1) as a modifier of the stability of the transition metal-alkyl bond, and hence the activity toward insertion of the monomer into the M–R bond, and 2) as a modifier of the reactivity of the transition metal complex toward olefin. The striking destabilization of the transition metal alkyls by the addition of alkyl aluminum compounds suggests that the above binary model may be useful in accounting for the role of alkyl-aluminum compounds. The alkylaluminum compound complexed with an alkyltransition metal may be considered as acting as a Lewis acid by withdrawing electrons from the alkyl transition metal. Thus the situation may resemble that of the complexation of electronegative olefins, which may be regarded as π-acids, with $R_2 Ni(bipy)$ leading to the destabilization of R–Ni bonds.

POLYMERIZATION OF OTHER MONOMERS

In addition to vinyl monomers, other types of monomers may be polymerized with alkyltransition metal complexes [2]. Butadiene can be converted to its oligomers; cyclooctadiene, vinyl cyclohexene can be obtained with $R_2 Fe(bipy)_2$ [10]; methyl heptatriene with $RCo(bipy)_2$ [27]; and cyclododecatriene with $R_2 Ni(bipy)$ [7]. Propylene can be rapidly converted to its dimers with $R_2 Ni(bipy)$ in the presence of $AlEt_2 Cl$ [28].

In this case the aluminum component appears to modify the nature of the catalytically active nickel species by complexation as discussed in the preceding section.

Polymerization of acetaldehyde gives two types of polymers. The polymerization with alkyl transition metals at -78°C gives a poly-ether-type polymer whereas polymerization above room temperature leads to viscous to powdery polymers of the "polyvinyl alcohol-type."

$$CH_3CHO \xrightarrow{-78°} (-\underset{\underset{H}{|}}{\overset{\overset{CH_3}{|}}{C}}—O)_n$$

$$CH_3CHO \xrightarrow{\text{room temp}} \text{polymer with pendant OH groups}$$

The mechanism of polymerization for the latter polymer is not completely understood at the moment, but the polymerization seems to proceed through the repetitive occurrence of aldol condensation initiated by the alkyltransition metal complexes.

REFERENCES

[1] For example, D. G. H. Ballard, Polym. Preprint, 15, 364 (1974); Adv. Catal., 23, 263 (1973); Pure Appl. Chem., Special Lectures of IUPAC Meeting, 6, 219 (1971).

[2] A. Yamamoto and S. Ikeda, Progr. Polym. Sci. Japan, 3, 49 (1972), and references cited therein.

[3] A. Yamamoto, K. Isaka, and S. Ikeda, Unpublished.

[4] K. Nishimura, H. Kuribayashi, A. Yamamoto, and S. Ikeda, J. Organometal. Chem., 37, 317 (1972); A. Yamamoto, T. Yamamoto, and Y. Kano, J. Organometal. Chem., In Press.

[5] A. Yamamoto, K. Kato, and S. Ikeda, Ibid., 60, 139 (1973).

[6] A. Yamamoto and S. Ikeda, J. Amer. Chem. Soc., 89, 5989 (1967).

[7] T. Saito. Y. Uchida, A. Misono, A. Yamamoto, K. Morifuji, and S. Ikeda, Ibid., 88, 5198 (1966).

[8] T. Yamamoto. A. Yamamoto, and S. Ikeda, Ibid., 93, 3350 (1971).

[9] T. Yamamoto, A. Yamamoto, and S. Ikeda, Ibid., 93, 3360 (1971).

[10] A. Yamamoto, K. Morifuji, S. Ikeda, T. Saito, Y. Uchida, and A. Misono, Ibid., 90, 1878 (1968).

[11] T. Yamamoto. A. Yamamoto, and S. Ikeda, Bull. Chem. Soc. Japan, 45, 1104 (1972).

[12] T. Yamamoto, A. Yamamoto, and S. Ikeda, Bull. Chem. Soc. Japan, 45, 1111 (1972).

[13] A. Yamamoto, T. Shimizu, and S. Ikeda, Makromol. Chem., 136, 297 (1970).

[14] A. Yamamoto, T. Shimizu, and S. Ikeda, Polym. J., 1, 171 (1970).

[15] T. Yamamoto, A. Yamamoto, and S. Ikeda, J. Polym. Sci., B, 9, 281 (1971).

[16] T. Yamamoto, A. Yamamoto, and S. Ikeda, J. Polym. Sci., Polym. Lett. Ed., 10, 835 (1972).

[17] A. Yamamoto, S. Kitazume, L. S. Pu, and S. Ikeda, J. Amer. Chem. Soc., 93, 371 (1971).

[18] Y. Kubo, A. Yamamoto, and S. Ikeda, J. Organometal. Chem., 59, 353 (1973).

[19] Y. Kubo, A. Yamamoto, and S. Ikeda, Ibid., 60, 165 (1973).

[20] Y. Kubo, A. Yamamoto, and S. Ikeda, Bull. Chem. Soc. Japan, 47, 393 (1974).

[21] T. Ito, S. Kitazume, A. Yamamoto, and S. Ikeda, J. Amer. Chem. Soc., 92, 3011 (1970).

[22] S. Komiya, A. Yamamoto, and S. Ikeda, J. Organometal. Chem., 42, C65 (1972).

[23] S. Komiya, A. Yamamoto, and S. Ikeda, Bull. Chem. Soc. Japan, 48, 101 (1975).

[24] S. Komiya and A. Yamamoto, J. Chem. Soc., Chem. Commun., 1974, 523; J. Organometal. Chem., 87, 333 (1975); Chem. Lett., 1975, 475.

[25] T. Yamamoto and A. Yamamoto, J. Organometal. Chem., 57, 127 (1973).

[26] A. Yamamoto, A. Miyashita, T. Yamamoto, and S. Ikeda, Bull. Chem. Soc. Japan, 45, 1583 (1973).

[27] T. Saito. Y. Uchida, A. Misono, A. Yamamoto, K. Morifuji, and S. Ikeda, J. Organometal. Chem., 6, 572 (1966).

[28] M. Uchino, A. Yamamoto, and S. Ikeda, Ibid., 24, C64, (1970); M. Uchino, K. Asagi, A. Yamamoto, and S. Ikeda, Ibid., 84, 93 (1975).

[29] G. H. Olivé and S. Olive, Kolloid-Z. Z. Polym., 228, 43 (1968); J. Organometal. Chem., 16, 339 (1969); Angew. Chem., 83, 121 (1971); 83, 782 (1971); Adv. Polym. Sci., 6, 421 (1969).

[30] A. Zambelli, I. Pasquon, R. Signorini, and G. Natta, Makromol. Chem., 112, 160 (1968); I. Pasquon, Ibid., 112, 160 (1968); 3, 465 (1967).

[31] L. A. Rodriguez and H. M. van Looy, J. Polym. Sci., A-1, 4, 1971 (1966).

Ionic Polymerization of α,α-Disubstituted Vinyl Monomers

ROBERT W. LENZ

Polymer Science and Engineering Program
Chemical Engineering Department
University of Massachusetts
Amherst, Massachusetts 01002

ABSTRACT

Alkyl α-chloroacrylates and p-substituted α-methylstyrenes were investigated for the effect of polymerization conditions on tacticity, molecular weight, and distribution, and for the relationship between tacticity, glass temperature, and crystalline properties.

INTRODUCTION

The ionic polymerization reactions of substituted α-methylstyrenes and of α-chloroacrylate esters are under investigation in this laboratory for the effect of polymerization initiators and reaction conditions on tacticity and subsequently on physical properties. Unusually high stereoregularities have been obtained in some of these systems, and the control of molecular weight distribution has been a problem in many cases. The unsolved problems connected with these observations for the cationic polymerization of the α-methylstyrene monomers

305

and the anionic polymerization of the α-chloroacrylate monomers, as well as for the ionic polymerization reactions of similar monomers, are discussed.

POLY(ALKYL α-CHLOROACRYLATES)

For the determination of the effect of tacticity on physical properties, it is necessary to be able to prepare polymers of varied stereochemical structures ranging from highly syndiotactic to highly isotactic in character [1]. In our earlier work on ethyl α-chloroacrylate, we attempted to prepare polymers of high syndiotacticity by homogeneous anionic and free-radical polymerization reactions, but it was observed that only relatively low molecular weight polymers were obtained in the former. The low molecular weights of these anionic polymers were always accompanied by a lower than theoretical chlorine analysis, and careful attempts have been made in the present investigations to eliminate the possible termination or transfer reactions which might be responsible for these two effects. Nevertheless, in spite of considerable effort and detailed investigations of a variety of homogeneous anionic systems, a few of which are listed in Table 1 for the three different ester monomers studied, it was not found possible to improve upon the molecular weights previously obtained, so that in almost all cases, number-average molecular weight values of less than 10,000 were obtained. On the other hand, free radical polymerization reactions were readily applicable to the preparation of syndiotactic polymers of high molecular weights for all three ester monomers as shown in Table 1.

Our attempts to prepare highly isotactic polymers in the previous investigations of the polymerization of ethyl α-chloroacrylate were concentrated on the use of Grignard-reagent initiators, which have been found to be very effective for this purpose in the polymerization of methyl methacrylate. Again in those investigations, only low molecular weight materials of atactic structures were obtained with these initiators. However, we were fortunate since that work because of the publication of a new catalyst system by Breslow and Kutner which was found by them and others to be very effective for the synthesis of isotactic polymers from both methyl methacrylate and α-chloroacrylate ester monomers [2, 3]. Their catalyst was obtained from the 1,4-addition reaction of a Grignard reagent to an unsaturated ketone. The solid product from this reaction was reported to be a very effective catalyst for the preparation of highly isotactic, high molecular weight crystalline polymers when used in a variety of

TABLE 1. Anionic and Free-Radical Polymerization of α-Chloroacrylate Esters

Monomer	Initiator[a]	Temp (°C)	Yield (%)	$\overline{M}_n$
Methyl α-chloroacrylate	n-BuLi	0	55	3,600
	t-BuLi	-78	45	4,300
	C_6H_5Li	-78	15	4,200
	C_6H_5Na	-78	1	4,800
	Bz_2O_2	70	90	270,000
	Benzoin-UV	-50	60	109,000
Ethyl α-chloroacrylate	n-BuLi	-78	16	6,000
	n-BuLi	0	25	4,500
	FLi	-78	9	5,000
	Bz_2O_2	70	90	362,000
	Benzoin-UV	-50	12	181,000
Isopropyl α-chloroacrylate	n-BuLi	-78	51	8,000
	Bz_2O_2	60	54	77,300
	Benzoin-UV	-65	5	2,100

[a]Anionic polymerization reactions were run in toluene; Bz_2O_2 is benzoyl peroxide, FLi is fluorenyl lithium.

solvents such as ethers, aliphatic and cyclic hydrocarbons, and aromatic hydrocarbons over a wide range of temperatures, from -60 to 100°C. These heterogeneous initiators were also applied to the polymerization of methyl, ethyl, and isopropyl α-chloroacrylate with the results shown in Table 2 for some of the products obtained.

The products of all polymerization reactions, in which molecular weights above 10,000 were obtained, were characterized for tacticity as fully as possible by IR and NMR spectroscopy and for glass transition temperature and crystalline properties by DSC. For complete tacticity analysis, both 100 and 300 MHz proton magnetic resonance spectroscopy were used for the determination of triad and tetrad structures of each of the three poly(alkyl α-chloroacrylates), and it was found possible to make assignments and carry out quantitative analyses for all of the tetrad peaks in the backbone methylene resonance of methyl, ethyl, and isopropyl ester polymers of syndiotactic atactic and isotactic structures. Representative tetrad spectra for the ethyl ester polymers are shown in Fig. 1.

In addition, the methyl singlet peaks of poly(methyl α-chloroacrylate), the methyl triplet peaks of poly(ethyl α-chloroacrylate), which had not been amenable previously to resolution at 220 MHz, and the methyl doublet peaks in poly(isopropyl α-chloroacrylate) were resolved. The latter are shown in Fig. 2 for isopropyl ester polymers of different tacticities. The triad values which were obtained from these pendant ester group resonances compared favorably with triad values

TABLE 2. Polymerization of α-Chloroacrylate Esters with Heterogeneous Grignard-Based Catalyst

Monomer	Monomer to catalyst ratio	Yield (%)[a]	$\overline{M}_n$
Methyl α-chloroacrylate	55	11	13,000
	150	25	32,000
	160	10	42,000
	300	5	10,600
Ethyl α-chloroacrylate	150	12	309,000
Isopropyl chloroacrylate	100	5	50,000

[a]At 30°C for 18 to 20 hr reaction time.

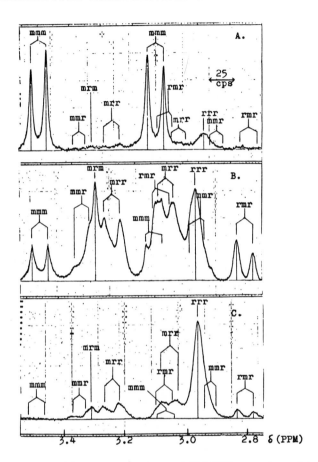

FIG. 1. Methylene proton region of 300 MHz spectra of poly(ethyl α-chloroacrylate). (A) Isotactic, (B) heterotactic, and (C) syndiotactic polymers.

calculated from experimental tetrad values as shown in Table 3 for representative samples of the isotactic ethyl ester polymers.

For mechanism interpretations, and also as an additional demonstration of the internal agreement of the triad and tetrad peak assignments, statistical calculations of the expected triad and tetrad structures of all atactic and syndiotactic polymers, which were prepared by either homogeneous anionic or free-radical polymerization, were made on the basis of random selection or Bernoullian statistics with

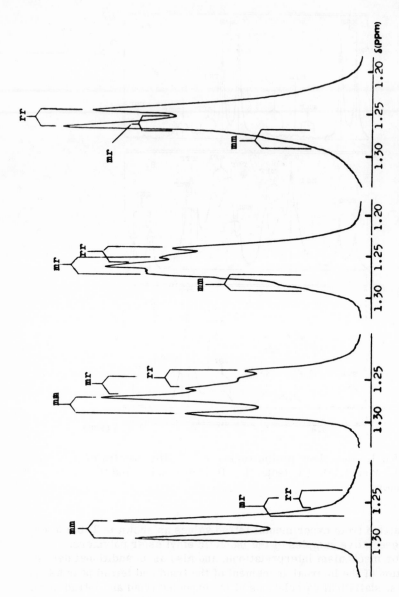

FIG. 2. Isopropoxy methyl region of 300 MHz spectra of poly(isopropyl α-chloroacrylate). Left to right: isotactic, moderately isotactic, heterotactic, and syndiotactic polymers.

TABLE 3. Observed and Calculated Triad Tacticities of Isotactic
Poly(ethyl α-Chloroacrylate)

Polymer sample	Triad	Triad amounts	
		Observed from ethoxy-methyl resonance	Calculated from methylene tetrads
1	mm	0.814	0.798
	mr	0.083	0.082
	rr	0.103	0.120
2	mm	0.650	0.678
	mr	0.125	0.112
	rr	0.225	0.210

the results collected in Table 4. Many, but not all, of these polymers
were found to adhere very closely to this type of propagation statistics,
and those that did are so indicated in the results collected in Table 4.
Surprisingly, the free-radical polymerization of the methyl and isopropyl
monomers did not yield Bernoullian polymers. For reasons not yet
obvious, these polymers had tetrad distributions more closely described
by first-order Markov statistics.

In contrast, the tetrad distributions of the moderately to highly isotactic polymers prepared with the Breslow-Kutner catalyst were found
to be consistent with a nonrandom propagation process as revealed by
their agreement with values calculated on the basis of first-order
Markov statistics as shown in Table 5. From the general tacticity results and the implications of the first-order Markov statistics, coupled
with other tacticity information, especially the large amounts of mrr
and rrr triads present, it seems likely that the isotactic stereoregularity obtained with the Breslow-Kutner catalyst is attributable primarily
to a competition between "template" and end-group control of the insertion reaction. That is, the isotactic stereoregulation is caused by
the asymmetry of the active site of the catalyst, which must counteract
a strong tendency toward syndiotactic placement by the end groups.
According to this interpretation, the mrr tetrad results from an attempt
by the catalyst to restore the proper absolute configuration to the chain
end after a propagation mistake is made, and the mrm tetrads which
would occur under conditions of isotactic "steric" control by the end

TABLE 4. Triad and Tetrad Information for Anionic and Free-Radical Polymers Assumed to Follow Bernoullian Growth Statistics

Polymer type and initiation	P_m[a]	Bernoullian statistics	Triads			Tetrads					
			mm	mr	rr	mmm	mmr	rmr	mrm	mrr	rrr
Methyl ester,[b] anionic	0.40	Obs	0.17	0.49	0.34	0.07	0.19	0.16	0.11	0.27	0.20
		Calc	0.16	0.49	0.35	0.07	0.19	0.14	0.10	0.29	0.21
Methyl ester,[c] free-radical	0.30	Obs	0.12	0.35	0.53	0.05	0.14	0.11	0.05	0.23	0.42
		Calc	0.09	0.41	0.50	0.03	0.12	0.15	0.06	0.29	0.35
Ethyl ester,[b] anionic	0.46	Obs	0.21	0.49	0.30	0.13	0.16	0.15	0.11	0.29	0.16
		Calc	0.21	0.49	0.30	0.09	0.23	0.14	0.11	0.27	0.16
Ethyl ester,[b] free radical	0.20	Obs	0.05	0.29	0.66	0.02	0.06	0.11	0.03	0.23	0.55
		Calc	0.04	0.31	0.65	0.01	0.06	0.13	0.03	0.25	0.52
Isopropyl ester,[c] free radical	0.26	Obs	0.14	0.44	0.42	0.04	0.20	0.13	0.10	0.23	0.30
		Calc	0.07	0.38	0.55	0.02	0.10	0.14	0.05	0.29	0.40

[a] Calculated from observed tetrad tacticity.
[b] Good agreement between observed and calculated tacticities.
[c] Poor agreement between observed and calculated tacticities.

TABLE 5. Observed and Calculated First-Order Markov Tetrad
Amounts for Isotactic Poly(alkyl α-Chloroacrylates)

Polymer type	Statistics	Tetrads[a]					
		mmm	mmr	rmr	mrm	mrr	rrr
Methyl ester	Obs	0.56	0.09	0.07	0.04	0.12	0.12
	Calc	0.51	0.17	0.02	0.04	0.14	0.12
Ethyl ester	Obs	0.79	0.02	0.03	0.01	0.07	0.08
	Calc	0.76	0.08	0.00	0.01	0.06	0.09
Isopropyl ester	Obs	0.92	0.01	0.02	0.01	0.04	0.01
	Calc	0.92	0.05	0.00	0.01	0.02	0.01

groups are not present to any extent. Therefore the principal propagation sequence involving an error and its subsequent correction
can be represented as shown by Scheme 1.

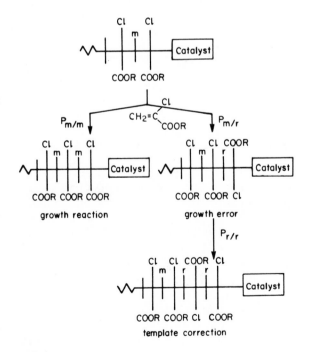

SCHEME 1.

POLY-α-METHYLSTYRENES

Cationic polymerization reactions of α-methylstyrene are known to give predominantly syndiotactic polymers [4]. We were interested to learn how this stereoregularity depends upon monomer structure and polymerization conditions, specifically counterion and solvent, and how it affects crystallinity of the polymers. For this purpose we polymerized a series of p-substituted-α-methylstyrene monomers at -78°C using various Friedel-Crafts initiators and solvents. In the initial investigations, unsubstituted, p-methyl- and p-chloro-α-methylstyrene were polymerized and polymer tacticities were determined by NMR spectroscopy. The α-methyl peaks were used for quantitative tacticity analyses in all cases, but for the p-chloro polymers the phenyl proton peaks could also be used. Typical spectra for p-chloro-α-methylstyrene polymers of high and intermediate syndiotacticities are shown in Fig. 3.

For all three monomers, solvent polarity was found to have a strong influence on tacticity, and polar solvents favored the formation of highly syndiotactic polymers (up to 96%) as shown in Table 6. The influence of initiator was of less importance in polar solvents, but it was quite appreciable in hexane (for example, 67% syndiotactic content for boron trifluoride etherate compared to 90% for boron trifluoride, see Table 6). As expected, samples of the same polymer but with different tacticity showed reasonably good correlation between stereoregularity and crystallinity as shown in Table 6. Thus the two polymers of highest content of syndiotactic triads in the p-methyl series had sufficiently high degrees of crystallinity for detection by x ray. The p-chloro polymers showed crystalline melting points in all cases except when the stereoregularity was very low. The melting points varied with tacticity roughly as expected, but the glass transition temperatures showed the expected variations only for the p-methyl polymers as shown in Table 6. In sharp contrast to the crystallinities of the p-methyl and p-chloro polymers, the unsubstituted polymer lacked crystallinity even at 92% syndiotactic content. This strong effect of p-substituent on crystallinity may depend on either a thermodynamic or kinetic ability of certain substituents to facilitate the crystallization process of such polymers. As the T_g/T_m ratios within this series of poly-α-methylstyrenes is unusually high (0.8 to 0.9), it is perhaps not surprising that their ability to crystallize is quite sensitive to structural variations.

As to the tacticity results of Table 6, polymerization conditions and monomer structure undoubtedly affect the nature of the growing ion-pair and thereby the steric course of the propagation reaction in

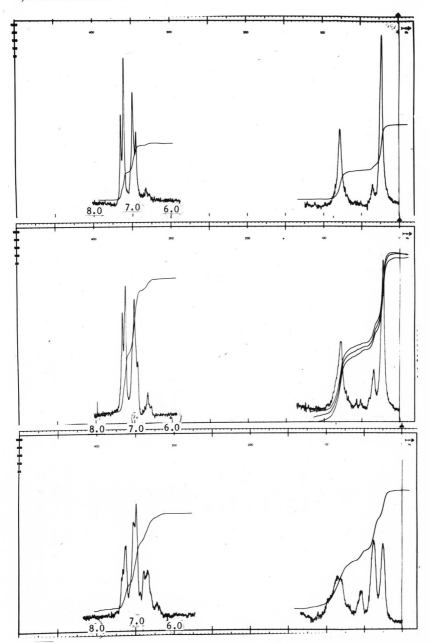

FIG. 3. 100 MHz NMR spectra of poly-p-chloro-α-methylstyrene.
Top to bottom: 90, 79, and 40% syndiotactic triad contents.

TABLE 6. Effect of Polymerization Conditions on Structure and Properties of α-Methylstyrene Polymers[a]

Monomer: p-substituent	Initiator	Solvent[b]	Triad tacticity (%)			Properties	
			S	H	I	T_g (°C)	T_m (°C)
H	TiCl$_4$	M	92	6	2	180	none
	BF$_3$·OEt$_2$	H/C[d]	81	19	0	-	-
	BF$_3$	H/C[d]	84	16	0	-	-
	BF$_3$	T	88	12	0	-	-
	BF$_3$	H	57	32	11	-	-
CH$_3$	BF$_3$·OEt$_2$	H/C[d]	94	6	0	175	225
	BF$_3$·OEt$_2$	T	96	4	0	176	217
	BF$_3$·OEt$_2$	H	67	24	9	165	204
	BF$_3$	H	90	10	0	-	-
Cl	TiCl$_4$	M	83	13	4	143	214
	TiCl$_4$	M/T[d]	79	19	2	151	209
	TiCl$_4$	M/H[d]	76	21	3	-	-
	TiCl$_4$	T	71	25	4	154	209
	TiCl$_4$	H	42	48	10	153	-
	AlCl$_3$	M	90	10	0	133	213
	BF$_3$	M	79	19	2	-	-
	SnCl$_4$	M	85	13	2	-	-

[a]All polymerization reactions were run at −78°C.
[b]M, methylene chloride; H, hexane; C, chloroform; T, toluene.
[c]S, syndiotactic triads; H, heterotactic triads; I, isotactic triads.
[d]Solvent mixture.

these polymerizations. The more polar the solvent, presumably the more separated is the counterion from the growing cation, and the syndiotactic product, which has the least steric crowding, should be formed [5]. In a nonpolar solvent a tighter ion-pair, and even a coordinate covalent structure, is possible in the propagation step, which could be more like a substitution reaction so that the sterically-strained heterotactic or isotactic sequence becomes possible [5]. Hence the products of these reactions are less stereoregular, and the lower melting points and glass transition temperatures are a result of lower syndiotactic contents. For these reasons the steric course of the propagation reaction was found to be more sensitive to change of initiator, and thus counterion structure, in nonpolar solvents than in polar ones.

Electron-donating p-substituents operating either through resonance interaction or on demand in the transition state of the addition reaction were expected to favor the formation of syndiotactic polymer by loosening the ion-pair structure, and such was the case as shown in Table 7 for a wide variety of substituted α-methylstyrene monomers. However, because all of these polymerization reactions were conducted in a highly polar solvent (methylene chloride) and ion-pair separation was apparently already large even in the unsubstituted case, electron-

TABLE 7. Tacticity and Molecular Weight as a Function of Para-Substituent in α-Methylstyrene Polymers[a]

Polymer: p-substituent	Triad tacticity (%)[b]			$\overline{M}_n \times 10^{-3}$	$\overline{M}_w \times 10^{-3}$
	S	H	I		
H	92	6	2	41	145
CH$_3$	95	5	0	-	-
CH(CH$_3$)$_2$	85	10	5	22	51
C(CH$_3$)$_3$	85	14	1	15	50
OCH$_3$	91	7	2	32	85
Cl	83	13	4	171	405
F	83	17	0	247	984
CF$_3$	45	45	10	43	102

[a]All polymers were prepared in methylene chloride at -78°C with TiCl$_4$ catalyst.
[b]See Table 6 for abbreviations.

donating groups brought about only relative small increases in syndiotactic content.

Polar solvents favored the formation of higher molecular weight
polymers, and among the initiators evaluated, stannic chloride gave
the highest molecular weight of all. When sufficiently small amounts
of titanium tetrachloride or aluminum chloride initiators were used,
0.2 mole %, the polymer obtained showed a single peak on GPC. Larger
amounts of initiator, 1%, resulted in polymers with a binodal molecular
weight distribution, and the stereoregularity of the low molecular
weight fraction was found to be somewhat inferior to that of the high
molecular weight fraction (72 compared to 83%, respectively, of syndiotactic triads). Electron-withdrawing substituents like chloro and
fluoro groups also seemed to favor the formation of higher molecular
weight polymers.

The solvent effect on molecular weight can again be rationalized
on the basis of the termination reaction mechanism as a function of
end-group ion-pair structure. Tighter ion-pair end-groups in nonpolar solvent could conceivably show enhanced counterion chain transfer, in competition with normal propagation:

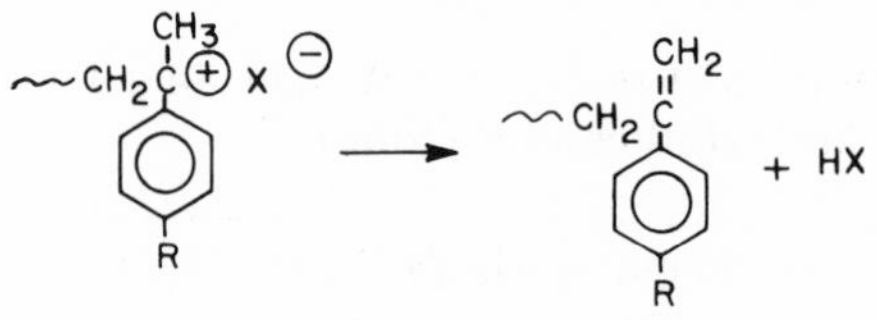

On the other hand, the substituent effect would be consistent with
an intramolecular alkylation, chain transfer reaction of the type.

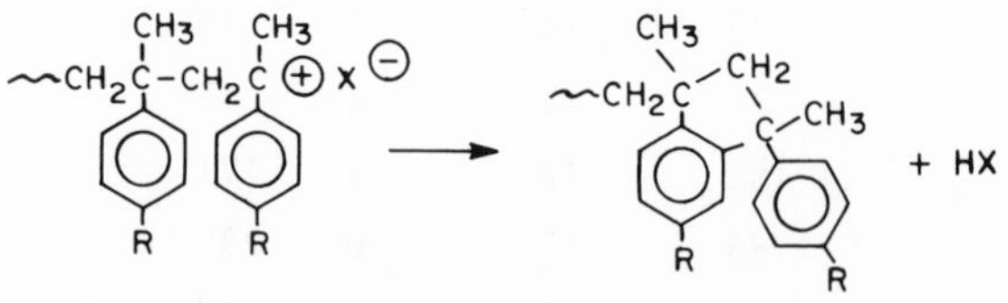

That is, electron-withdrawing substituents would be expected to
deactivate the aromatic ring to this type of alkylation reaction,
thereby enhancing the relative role of propagation and polymer
chain length. The relative contributions of these two molecular
termination reactions, and the existence of a binodal molecular

weight distribution and higher initiator amounts, remain unsolved problems in this and other closely related cationic polymerization reactions [6].

ACKNOWLEDGEMENTS

The work described in this review involved the contributions of many of the author's associates at the University of Massachusetts and elsewhere. The α-chloroacrylate investigations were carried out by Dr. Gerald Dever [7] (present address: Research Laboratories, Xerox Corporation, Rochester, New York) under the co-direction of Professors William MacKnight and Frank Karasz of our Polymer Science and Engineering Program and Materials Research Laboratory. The α-methylstyrene investigations were carried out by Dr. Judith Sutherland [8] (present address: Research Laboratories, Eastman Kodak Co., Rochester, New York) and Dr. Lars Westfelt (present address: Chemistry Department, Swedish Forest Products Research Laboratory, Stockholm, Sweden). In addition, NMR investigations on p-fluoro-α-methylstyrene were made in collaboration with Dr. W. Regel of the Institute of Macromolecular Chemistry at the University of Freiburg in Germany [9]. The financial support of the National Science Foundation which permitted the presentation of this paper is gratefully acknowledged.

REFERENCES

[1] B. Wesslen and R. W. Lenz, Macromolecules, 4, 20 (1971); B. Wesslen, R. W. Lenz, and F. A. Bovey, Ibid., 4, 709 (1971).

[2] D. S. Breslow and A. Kutner, J. Polym. Sci., B, 9, 129 (1971); A. Kutner, U.S. Patent 3,151,102 (1964).

[3] T. Uryu, K.-I. Ohaku, and K. Matsuzaki, J. Polym. Sci., Polym. Chem. Ed., 12, 1723 (1974).

[4] S. Brownstein, S. Bywater, and D. J. Worsfold, Makromol. Chem., 48, 127 (1961); Y. Ohsumi, T. Higashimura, and S. Okamura, J. Polym. Sci., A-1, 4, 923 (1966).

[5] T. Kunitake and C. Aso, J. Polym. Sci., A-1, 8, 665 (1970).

[6] D. C. Pepper, IUPAC International Symposium on Macromolecules, Madrid, 1974.

[7] G. R. Dever, Ph.D. Thesis, Polymer Science and Engineering Program, University of Massachusetts, 1974.

[8] J. E. Sutherland, Ph.D. Thesis, Polymer Science and Engineering Program, University of Massachusetts, 1972.

[9] R. W. Lenz, W. Regel, and L. Westfelt, Makromol. Chem., 176, 781 (1975).

Studies on Polymerization Activities by Soluble Catalysts Based on Compounds of Organotransition Metals and Aluminum Alkyls

NAOYUKI KOIDE, KAZUYOSHI IIMURA, and MASATAMI TAKEDA

Science University of Tokyo
Kagurazaka, Shinjuku-ku, Tokyo, Japan

ABSTRACT

Polymerization activities of the soluble Ziegler-type of catalyst systems, $Ti(OR)_4$-$AlEt_3$, $Ti(NEt_2)_4$-$AlMe_3$, and $V(NEt_2)_4$-$AlEt_3$, were investigated. In the catalyst system of $Ti(OR)_4$-$AlEt_3$, formation of two types of Ti(III) compounds, i.e., $Ti(OR)_2Et$ and its bridged complex with aluminum alkyl, was confirmed by IR and ESR measurements. With the addition of donor molecule to the system, it was found that the polymer yield decreased remarkably and that the bridged complex dissociated into a single or uncomplex Ti(III) paramagnetic species. It has been concluded that the bridged structure of Ti(III) species was responsible for the polymerization activity of styrene. Two reaction products of $Ti(NEt_2)_3Me$ and $Al(NEt_2)Me_2$ were found by NMR spectroscopic observation with the $Ti(NEt_2)_4$-$AlMe_3$ catalyst system. From the kinetic study of polymerization of styrene, it was found that $Ti(NEt_2)_3Me$ is an active species. An anionic mechanism was proposed for the styrene polymerization by $Ti(NEt_2)_3Me$. In the polymerization of MMA with the $V(NEt_2)_4$-$AlEt_3$ system, a difference in the tacticity of polymer was found to depend on the polymerization conditions, e.g., Al/V ratio and temperature. From an analysis of the tacticity of the polymer, the presence of two active sites in the propagation process is suggested.

INTRODUCTION

For several years we have been studying the polymerization activities of the soluble Ziegler-type of catalysts. In general, catalysts which are composed of a mixture of a transition metal halide and aluminum alkyl are often heterogeneous systems containing a precipitate. The heterogeneity leads to a difficulty in determining the reaction products in the catalyst system by means of physicochemical methods. This paper summerizes our works on the polymerization activities of the soluble catalysts by using some spectroscopic techniques. The catalyst systems discussed here are $Ti(OR)_4$-$AlEt_3$ [1-3], $Ti(NEt_2)_4$-$AlMe_3$ [4], and $V(NEt_2)_4$-$AlEt_3$ [5, 6].

EXPERIMENTAL

Reagents

$M(NEt_2)_4$ (M = Ti and V) was synthesized by reacting $LiNEt_2$ with MCl_4 by the method employed by Bradley et al. [7].

$Ti(NEt_2)_3Me$ was prepared according to the method of Bürger et al. [8].

$Ti(O-n-Bu)_4$ (Nihon Soda Co. Ltd.,) was purified by distillation under vacuum (160°C/2 Torr). $Ti(O-tert-Bu)_4$, $Ti(O-sec-Bu)_4$, and $Ti(O-n-Bu)_3$ were synthesized with the methods reported by Yoshino et al. [9] and Nesmeyanov et al. [10].

$AlEt_3$, $AlMe_3$, $Al(O-n-Bu)_3$, and VCl_4 were the commercial pure grade reagents and used without further purification.

$Al(O-n-Bu)Et_2$ was synthesized by the reaction of $AlEt_3$ with $Al(O-n-Bu)_3$ and distilled under reduced pressure (124 to 129°C/1 Torr).

$Al(NEt_2)R_2$ (R = Me, Et) was synthesized according to the method of Davidson et al. [11].

The purity of the reagents employed in this investigation were ascertained by measurements of their NMR and IR spectra.

Polymerization

All polymerization procedures were carried out under nitrogen atmosphere.

Analyses

The IR spectra were recorded by a Hitachi EPI-2 double-beam spectrophotometer. The spectra were observed by using a 0.203-mm thick cell with KBr windows. Absorption peaks arising from the solvent were compensated for by using a variable thickness cell in the reference beam.

NMR spectra were recorded by a Varian spectrometer (Model A-60) and a JEOL (Model JNM-4H-100) spectrometer.

ESR measurements were carried out with JEOL (Model No. 111, X-band) and Hitachi (Model MES-4002A) instruments.

A Gel Permeation Chromatographer of Waters and Associates (Model GPC 200) was used to analyze the molecular weight distribution of polymethyl methacrylate.

RESULTS AND DISCUSSION

Studies on the Active Complex in the $Ti(OR)_4$-$AlEt_3$ Systems

$Ti(O-n-Bu)_4$—$AlEt_3$ System

Polymerization of Styrene. Polymerization reactions were carried out at two different conditions in order to obtain an optimum Al/Ti ratio for polymerization activity. In the first case, the concentration of $AlEt_3$ was varied and that of $Ti(O-n-Bu)_4$ was kept constant, and in the second case, $Ti(O-n-Bu)_4$ was varied and $AlEt_3$ was kept constant. The polymerization was carried out after 20 min of aging of the catalyst system.

In both catalyst systems the polymer yield was found to be increased with an increase in Al/Ti ratio up to 1.4, where the polymer yield reached a maximum, thereafter it decreased

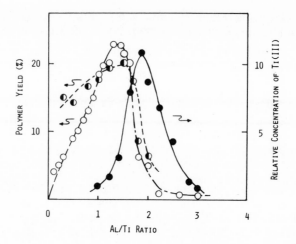

FIG. 1. Variations of polymer yield and relative concentration of Ti(III) species (g = 1.960 and g = 1.962) with Al/Ti molar ratio in the catalyst solution. (○): Concentration of Ti(O-n-Bu)$_4$ was kept constant at 1.5×10^{-3} mole and that of AlEt$_3$ was varied. (◐): Concentration of AlEt$_3$ was kept constant at 1.5×10^{-3} mole and that of Ti(O-n-Bu)$_4$ was varied. (●): Relative concentration of Ti(III) species determined from the ESR measurement. Polymerization conditions: n-hexane solvent = 10 ml, styrene = 5 ml, temperature = 60°C, time = 20 hr.

rapidly and became negligibly small when the Al/Ti ratio was over 3 (Fig. 1). These results demonstrate that the polymerization activity of the catalyst system depends strongly on the Al/Ti ratios. It was also noted that no polymerization occurred at temperatures below 0°C.

IR and ESR Spectra. With an increase in the Al/Ti ratio, the catalyst solution of n-hexane developed colors from green to reddish-brown. To determine the reaction products between the catalytic components, IR and ESR spectra were observed at the different Al/Ti ratios. When AlEt$_3$ was added to the solution of Ti(O-n-Bu)$_4$ in n-hexane, rapid decreases in the band intensities at 1085 and 1120 cm^{-1} (attributable to the Ti—O—C stretching vibration [12]) were observed, and the new bands at 1045 and 1065 cm^{-1} (attributable to the Al—O—C stretching vibration [13]) appeared. The latter new bands were caused from the formation of Al(O-n-Bu)Et$_2$. Reliable evidence for this conclusion was

provided by the experimental fact that the IR spectrum of
Al(O-n-Bu)Et$_2$, synthesized independently, was identical with that
of the reaction product at a Al/Ti ratio of 3.

The ESR spectra of the catalyst system at different Al/Ti ratios
are shown in Fig. 2. A singlet signal with a g-value of 1.960 was
observed in the ESR spectra of the green catalyst solution in which
Al/Ti ratios were less than 1.2. The singlet signal can be ascribed
to a trivalent titanium species [14]. With an increase in the Al/Ti
ratios (Al/Ti = 1.4 to 2.4), where the catalyst solution developed a
reddish-brown color, the singlet signal changed to a g-value of
1.962, having a hyperfine structure with eleven splittings (a = 2.2
gauss). The hyperfine structure can be caused by a complex (I),

I

as pointed out by Djabiev et al. [15]. The signal having a hyperfine
structure with eleven splittings arises from an interaction between
the trivalent titanium nucleus and the two aluminum nuclei. No
paramagnetic signals were observed at Al/Ti ratios higher than 5.
This could be attributed to a further reduction of the valency of
titanium to less than 3.

Changes in the relative concentration of the paramagnetic species
with a variation in the Al/Ti ratio are shown in Fig. 1. A correlation
between the polymer yield versus the Al/Ti ratios and the relative
concentration of paramagnetic species versus the Al/Ti ratios
was found.

IR and ESR Studies with Ti(O-n-Bu)$_3$—AlEt$_3$
System. Ti(O-n-Bu)$_3$ was only slightly soluble in n-hexane.
However, when AlEt$_3$ was added successively to Ti(O-n-Bu)$_3$ in
n-hexane, the color of the mixture changed from green to reddish-
brown, similar to that observed in the Ti(O-n-Bu)$_4$—AlEt$_3$ system.
In the supernatant part of the reddish-brown solution, the formation
of Al(O-n-Bu)Et$_2$ was confirmed by the observation of the IR
spectrum, and also the paramagnetic signal with eleven hyperfine
splittings (g = 1.962, a = 2.2 gauss) was observed by the ESR
measurement. These findings were the same as those observed
for the Ti(O-n-Bu)$_4$—AlEt$_3$ system.

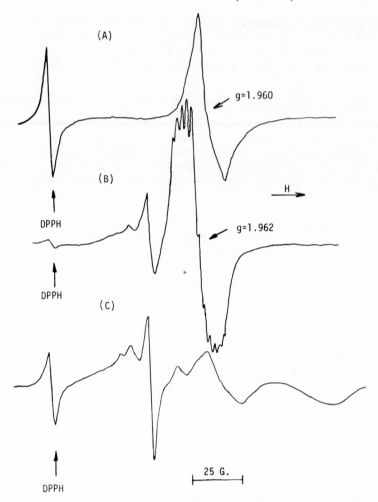

FIG. 2. ESR spectra of Ti(O-n-Bu)$_4$—AlEt$_3$ system in n-hexane. [Ti(O-nBu)$_4$] = 0.05 mole/liter. (A) Al/Ti = 0.5, (B) Al/Ti = 1.5, and (C) Al/Ti = 3.0.

The polymerization of styrene with the Ti(O-n-Bu)$_3$—AlEt$_3$ system was also carried out. The single component, Ti(O-n-Bu)$_3$ or AlEt$_3$, was not active for the polymerization of styrene, However, the Ti(O-n-Bu)$_3$—AlEt$_3$ catalyst solution yielding a reddish-brown color gave a maximum activity for the polymerization of styrene in 10% polymer yield.

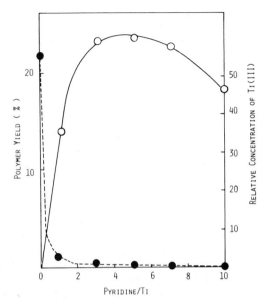

FIG. 3. Variations of polymer yield and relative concentration of
Ti(III) species (singlet signal at g = 1.960) against pyridine/Ti molar
ratio. (●): Polymer yield. (○): Relative concentration of Ti(III)
species determined from the ESR measurement. Polymerization
conditions: n-hexane solvent = 10 ml, styrene = 5 ml, temperature =
60°C, time = 20 hr, Al/Ti = 1.5, $[Ti(O-n-Bu)_4]$ = 2×10^{-4} mole.

The above experimental findings lead us to conclude that the
complex of a trivalent titanium species with 2 moles of $Al(O-n-Bu)Et_2$
having the eleven hyperfine splittings may be responsible for the
polymerization of styrene.

Effects of Electron Donor on Paramagnetic
Species and Polymerization Activities. In order to
determine the structure of active species in the $Ti(O-n-Bu)_4$—$AlEt_3$
system, changes in the polymerization activity and ESR spectra of
the catalyst solution in the presence of a electron donor, pyridine,
were examined.

In the case of the $Ti(O-n-Bu)_4$—$AlEt_3$—pyridine system (Al/Ti =
1.5), as shown in Fig. 3, the polymer yield decreased remarkably
with an increase in the pyridine/Ti ratio over 1.0. The bridged
structure of the paramagnetic trivalent titanium species with two
aluminum nuclei (g = 1.962) dissociated into a single paramagnetic
trivalent titanium species (g = 1.960) at a pyridine/Ti ratio of 1.0.

This was confirmed by the disappearance of the ESR signal having the eleven hyperfine splittings with the addition of pyridine to the catalyst solution. The relative intensity of the signal arose from the single or uncomplexed trivalent titanium species (g = 1.960), increased with an increase in the pyridine/Ti ratio, and reached to a maximum at a pyridine/Ti ratio of 3.0.

Ti(O-tert-Bu)$_4$—AlEt$_3$ System

Polymerization. Polymer yield was dependent on the Al/Ti ratios, with optimum activity at an Al/Ti ratio of ~1.0. When the Al/Ti ratio was higher than 3, no polymerization occurred. Polymerization activity of the Ti(O-tert-Bu)$_4$—AlEt$_3$ catalyst system for styrene was found to be twice that of the Ti(O-n-Bu)$_4$—AlEt$_3$ catalyst system.

ESR Spectra. ESR spectra of the Ti(O-tert-Bu)$_4$—AlEt$_3$ catalyst system are shown in Fig. 4. At an Al/Ti ratio of 0.5, a singlet signal was observed at g = 1.960. In the case of an Al/Ti ratio of 1.5, a hyperfine structure with six splittings was found, with a coupling constant of ~3.5 gauss. The hyperfine structure may arise from an interaction between a trivalent titanium nucleus and an aluminum nucleus (I = 5/2). The signal with the eleven hyperfine splittings could not be detected at any Al/Ti ratio employed in the catalyst system.

Changes in the relative intensity arising from the paramagnetic species with a variation of the Al/Ti ratio were found to be similar to those of the Ti(O-n-Bu)$_4$—AlEt$_3$ system (see Fig. 1).

Effect of Electron Donor. With the addition of pyridine to the catalyst solution (Al/Ti = 1.5), a rapid decrease in the polymer yield was noted, and also the signal with six hyperfine splittings disappeared and changed into a signal with a g-value of 1.960, which was caused by the single or uncomplexed trivalent titanium species.

Conclusion. The active species for the polymerization of styrene in these catalyst system is considered to be the bridged complexes of a trivalent titanium species with 1 or 2 moles of Al(OR)Et$_2$. This can be concluded from the following experimental facts: 1) In the catalyst system where a maximum activity for the polymerization of styrene is noted, the ESR signals attributable to these complexes are observed; 2) with the addition of an electron donor, pyridine, a remarkable decrease in polymerization activity is found accompanying the dissociation of bridged complexes into a single or uncomplexed trivalent titanium species. The reaction scheme producing the active species may be considered to be as in Scheme 1.

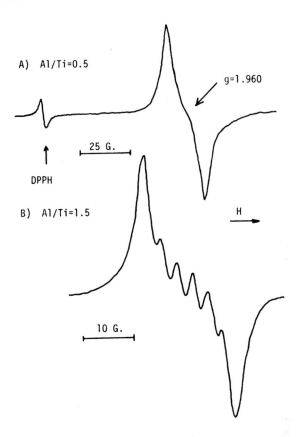

FIG. 4. ESR spectra of Ti(O-tert-Bu)₄—AlEt₃ system in n-hexane. [Ti(O-tert-Bu)₄] = 0.05 mole/liter.

NMR Studies on Ti(NEt₂)₄-AlMe₃ System and Its Polymerization Activity

In this study, an active species in the Ti(NEt₂)₄-AlMe₃ system for the polymerization of styrene was examined by NMR measurement.

Polymerization and NMR Measurement of Catalyst Solution

As shown in Fig. 6, the polymerization activity for styrene was found to depend on the Al/Ti ratio with an optimum activity at a

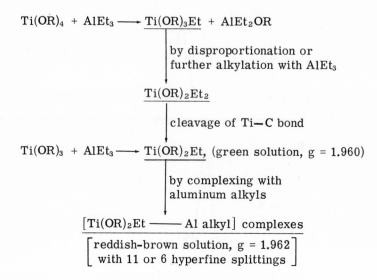

SCHEME 1. Sequence of reactions

ratio of 1.5. When the Al/Ti ratio was beyond 4, no polymer was obtained. Even with an addition of DPPH, the polymerization of styrene was not inhibited. This evidence suggests that no radical initiation occurred in the catalyst system.

NMR measurements at the various Al/Ti ratios were carried out to determine a structure of the active species. From the NMR observations the two reaction products in the catalyst system were confirmed to be $Ti(NEt_2)_3Me$ and $Al(NEt_2)Me_2$ (Fig. 5). The relative concentration of $Ti(NEt_2)_3Me$ came to a maximum at an Al/Ti ratio of about 1.5, as shown in Fig. 6. This concentration dependence on the Al/Ti ratios is well correlated with the variation of the polymerization activity against the Al/Ti ratio. The peak at $\tau = 9.52$ ppm attributed to the methyl protons in the Ti—Me linkage of $Ti(NEt_2)_3Me$ disappeared as the polymerization progressed.

It may be concluded that $Ti(NEt_2)_3Me$ plays a major role for the polymerization of styrene. The polymerization of styrene was, then, carried out by using $Ti(NEt_2)_3Me$ only.

Polymerization of Styrene by $Ti(NEt_2)_3Me$ [16]

A kinetic study of the polymerization of styrene with $Ti(NEt_2)_3Me$ itself was carried out. Variations of the rate of polymerization against the concentrations of catalyst and styrene were examined.

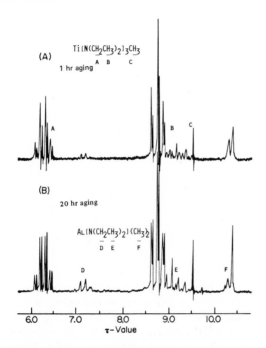

FIG. 5. NMR spectra of Ti(NEt₂)₄-AlMe₃ system at Al/Ti molar ratio of 1.0. [Ti(NEt₂)₄] = 6 × 10⁻² mole/liter benzene.

It was found that the polymer yield was proportional to the degree of polymerization. This suggests that the polymerization reaction of the catalyst is not caused by a radical mechanism (Fig. 7).

The rate of polymerization was found to be proportional to the first-order of Ti(NEt₂)₃Me concentration and to the second-order of styrene concentration (Figs. 8 and 9).

The activation energy determined from the relationship between the logarithum of the rate of polymerization and the reciprocal of absolute temperature was found to be 9.56 kcal/mole. An ionic mechanism is proposed by the above kinetical investigations. In this case, an insertion reaction of monomer to the Ti—Me bond of Ti(NEt₂)₃Me may be anticipated, because the peak at τ = 9.52 ppm arising from the Ti—Me protons disappears with the addition of monomer.

With the addition of benzoquinone (an inhibitor for radical polymerization) to the catalyst, no inhibitive effect was found

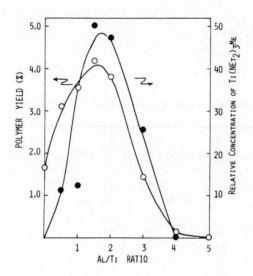

FIG. 6. Variations of polymer yield and relative concentration of $Ti(NEt_2)_3Me$ with Al/Ti molar ratio in the catalyst solution. (○): Polymer yield. (●): Relative concentration of $Ti(NEt_2)_3Me$ determined from the NMR measurement. Polymerization conditions: benzene = 5 ml, styrene = 5 ml, temperature = 60°C, time = 20 hr, $[Ti(NEt_2)_4]$ = 1×10^{-3} mole.

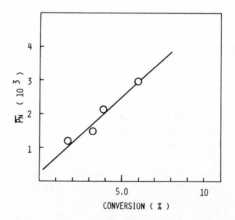

FIG. 7. Degree of polymerization against polymer yield. Polymerization conditions: toluene = 9 ml, styrene = 10 ml, temperature = 60°C, time = 2, 4, 8, and 24 hr, $[Ti(NEt_2)_3ME]$ = 5×10^{-3} mole.

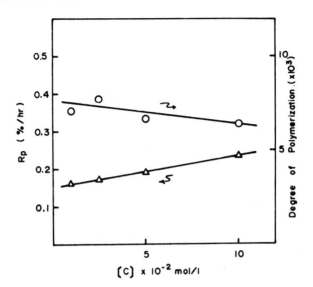

FIG. 8. Rate of polymerization and degree of polymerization against the catalyst concentration. Polymerization conditions: toluene = 10 ml, styrene = 4.4 mole, time = 6 hr, temperature = 60°C.

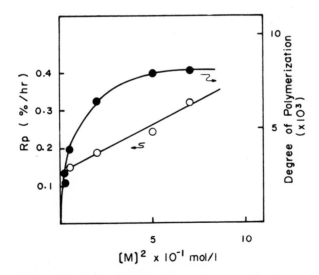

FIG. 9. Rate of polymerization and degree of polymerization against the monomer concentration. Polymerization conditions: $[Ti(NEt_2)_3Me] = 1 \times 10^{-3}$ mole, time = 6 hr, temperature 60°C, toluene solvent, ($\circ$) R_p, ($\bullet$) D_p.

for the polymerization of styrene. Moreover, in the polymerization of isobutyl vinyl ether by the catalyst, no polymer was obtained. These experimental facts suggest that $Ti(NEt_2)_3Me$ initiates the polymerization of styrene by an anionic mechanism.

By assuming a complex formation between styrene and the catalyst in the initiation step, it is possible to deduce the rate law [17]:

$$C \quad + M \xrightarrow{K} (CM) \qquad \text{(rapid equilibrium)}$$

$$(CM) + M \xrightarrow{k} (CM) + polymer \qquad \text{(slow)}$$

where C is the catalyst, M is a styrene molecule, and (CM) is the styrene-catalyst complex. Then the rate of polymerization can be written as

$$-dM/kt = k[(CM)][M] \tag{1}$$

and

$$K = [(CM)]/\{[C] - [(CM)]\}[M] \tag{2}$$

Solving Eq. (2),

$$[(CM)] = \frac{K[C][M]}{1 + K[M]} \tag{3}$$

Substituting Eq. (3) for the rate Eq. (1) yields

$$-dM/dt = kK[M]^2[C]/\{1 + K[M]\} \simeq kK[M]^2[C] \qquad \text{(when } K[M] \text{ is negligibly small)} \tag{4}$$

This rate equation agrees with the experimental facts. Therefore, the assumption of complex formation prior to polymerization may be supported.

Conclusion

The active species in the $Ti(NEt_2)_4$-$AlMe_3$ system for the polymerization of styrene is confirmed as $Ti(NEt_2)_3Me$. Kinetic investigation of polymerization by using $Ti(NEt_2)_3Me$ shows that the rate of polymerization is proportional to the first order of $Ti(NEt_2)_3Me$ and to the second order of styrene concentration.

Polymerization of MMA with $V(NEt_2)_4$-$AlEt_3$ Catalyst System and Tacticity of Polymers

Several studies have been reported on olefin polymerization with the $V(NEt_2)_4$-$AlEt_3$ catalyst system [18-21]. However, no work has been carried out on the polymerization of polar vinyl monomers using this catalyst system. In our studies on the polymerization of methyl methacrylate (MMA), it was found that the tacticities of polymer change depend upon the polymerization conditions, i.e., the polymerization temperature and the Al/V ratios in the catalyst system. Although the catalytic mechanism giving the different tacticities for PMMA is not understood, in this section we present our experimental findings on the polymerization of MMA.

Polymerization

Figure 10 shows the effects of polymerization temperatures and Al/V ratios in the catalyst system for polymer yield. It is clear that the $V(NEt_2)_4$-$AlEt_3$ catalyst system is active for the polymerization of MMA at the lower temperature of -78°C and in the region of higher Al/V ratios beyond 3.

From the observation of the NMR spectra of the catalyst solution, the formation of $Al(NEt_2)Et_2$ was confirmed at higher Al/V ratios. Since it has been reported [22] that $Al(NPh_2)Et_2$ initiates an anionic polymerization of MMA, we have tried the polymerization of MMA with $Al(NEt_2)Et_2$ at -78°C for 20 hr, but no polymer was obtained. It is concluded that $Al(NEt_2)Et_2$ shows no activity for polymerization in the catalyst system $V(NEt_2)_4$-$AlEt_3$.

The effect of aging of the catalyst for polymerization activity was also examined. After the aging of the catalyst solution (Al/V = 5.0) at 70°C for the prescribed time, the catalyst solution was cooled to -78°C, monomer was added slowly, and then polymerization was carried out for 20 hr at this temperature. The catalyst solution aged for 0.5 hr gave the polymer in 94% yield, and that aged for 24 hr gave the polymer in 92% yield. It is noted that the catalyst

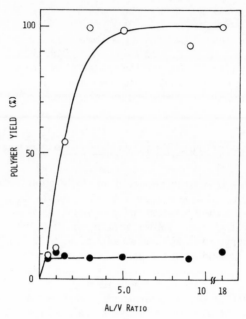

FIG. 10. Variation of polymer yield against Al/V molar ratio
in the catalyst solution at the different temperatures. ($\circ$): 20 hr
at -78°C. ($\bullet$): 20 hr at 60°C. Polymerization conditions:
toluene = 10 ml, MMA = 5 ml, $[V(NEt_2)_4]$ = 5 × 10^{-4} mole.

system holds its polymerization activity even after a long aging
time. As judged by the above experimental findings, a radical
mechanism may be excluded from consideration.

Tacticity of Polymer

The effects of the polymerization temperature and the Al/V
ratio for the triad tacticity of polymers are shown in Fig. 11. In
the region of Al/V ratios beyond 3, the tacticity of the polymer is
constant and does not depend on the Al/V ratios. In this region,
however, the tacticity of the polymer depends on the polymerization
temperature, and the syndiotactic polymer is obtained at -78°C.
Evidently, according to Fig. 11, the tacticity of polymer prepared
at 60°C does not depend strongly on the Al/V ratio. In the polymerization at -78°C, however, the tacticity of the polymer prepared
at a Al/V ratio of 0.5 is remarkably different from that prepared
at higher Al/V ratios, i.e., an isotactic-rich polymer is obtained.

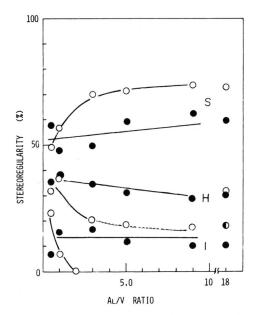

FIG. 11. Variation of triad tacticity of PMMA obtained at the different temperatures against Al/V molar ratio in the catalyst solution. Polymerization conditions are same as for Fig. 10. (○): At -78°C. (●): At 60°C.

As Fig. 12 shows, the propagation steps which yield polymer at -78°C and at Al/V ratios greater than 1, follow Bernoullian statistics, whereas the polymer prepared at a Al/V ratio of 0.5 deviates from Bernoullian statistics (denoted by the arrowheads in the figure). The polymers prepared at a Al/V ratio of 0.5 is supposed to be a stereoblock type of tacticity because the normalized probabilities of isotactic and heterotactic triad are larger in the former and smaller in the latter than those expected from Bernoullian statistics. Regardless of the Al/V ratio, the propagation steps yielding polymer at 60°C follow Bernoullian statistics.

By taking account of the molecular weight distribution described in the next section, the different propagation steps which yield polymers can be suggested to depend on the polymerization conditions.

Molecular Weight Distribution of Polymer

The molecular weight distribution (MWD) of polymers was measured by gel permeation chromatography. As Fig. 13 shows,

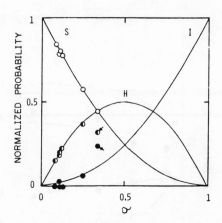

FIG. 12. Stereoregularity of PMMA prepared at -78°C and at different Al/V molar ratios. The triads are syndiotactic (○), heterotactic (◐), and isotactic (●).

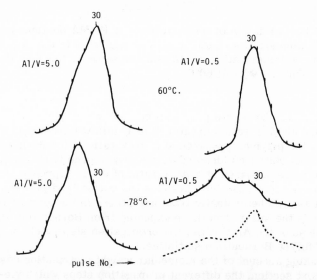

FIG. 13. Molecular weight distributions of PMMA prepared with $V(NEt_2)_4$-AlEt$_3$ system. (- -) $V(NEt_2)_4$-AlEt$_3$-PPh$_3$ system, Al/V = 0.5, PPh$_3$/V = 2.0.

there was no remarkable difference in the MWD's among the polymers prepared at 60°C and Al/V = 0.5, at 60°C and Al/V = 5.0, and at -78°C and Al/V = 5.0. In this case the MWD curves had a single peak.

However, the polymer prepared at -78°C and Al/V = 0.5 had at least two peaks in the MWD curve. This suggests that a polymerization mechanism giving polymer at the lower temperature and the lower Al/V ratio may not be simple one.

Effects of PPh₃-Addition to the Catalyst System

In Table 1 are shown the effects of PPh₃ addition to the catalyst system on the tacticities of polymers.

In spite of the addition of PPh₃, no effect was observed for the triad tacticities of polymers prepared under the polymerization conditions (at -78°C, Al/V = 5.0) following Bernoullian statistics. Other polymerization conditions (at -78°C, Al/V = 0.5) deviated from Bernoullian statistics, and the isotactic content was greatly diminished by the addition of PPh₃. As shown in Fig. 14, the value of triad tacticities of polymer prepared at -78°C and Al/V = 0.5 in the presence of PPh₃ fit Bernoullian statistics. The addition of PPh₃ also affected the MWD of polymer. As shown by the dotted

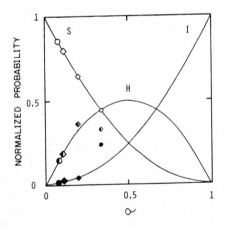

FIG. 14. Effect of the addition of PPh₃ on the triad tacticities of PMMA. (o), (O), (◇), and (◊) are sample numbers 1, 2, 3, and 4, respectively, as listed in Table 1.

TABLE 1. Effects of Addition of PPh₃ on the Triad Tacticities of Polymers[a]

No.	Al/V	PPh₃/V	Polymerization temperature (°C)	Conversion (%)	Triad tacticity		
					I	H	S
1	0.5	0	-78	10.7	0.24	0.32	0.44
2	0.5	2.0	-78	7.6	0.01	0.14	0.85
3	5.0	0	-78	98.5	0.03	0.18	0.79
4	5.0	2.0	-78	62.0	0.02	0.23	0.75

[a]Polymerization condition: $[V(NEt_2)_4] = 5 \times 10^{-4}$ mole, MMA = 5 ml, polymerization time = 20 hr, toluene = 20 ml.

curve in Fig. 13, the addition of PPh₃ induced a remarkable decrease in the peak intensity on the higher side of the molecular weight for the polymer prepared at -78°C and Al/V = 0.5

Conclusion

Two types of the propagation steps can be assumed for the polymerization of MMA with the $V(NEt_2)_4$-$AlEt_3$ system. One is a process which follows Bernoullian statistics. The other process proceeds at a lower temperature and a lower Al/V ratio, and it deviates from Bernoullian statistics. The active site causing the latter process is poisoned by the addition of PPh₃, and the propagation step proceeds in the former process.

REFERENCES

[1]   M. Takeda, K. Iimura, Y. Nozawa, M. Hisatome, and N. Koide, J. Polym. Sci., C, 23, 741 (1968).
[2]   M. Takeda, K. Iimura, and N. Koide, Kogyo Kagaku Zasshi, 71, 563 (1968).
[3]   N. Koide, K. Iimura, and M. Takeda, Ibid., 73, 1038 (1970).
[4]   N. Koide, K. Iimura, and M. Takeda, J. Polym. Sci., Polym. Chem. Ed., 11, 3161 (1973).
[5]   N. Koide, T. Akita, K. Iimura, and M. Takeda, Paper Presented at the 22nd Annual Meeting of the Society of Polymer Science, Japan, May 1973.
[6]   N. Koide, T. Akita, K. Iimura, and M. Takeda, Paper Presented at the 23rd Annual Meeting of the Society of Polymer Science, Japan, June 1974.
[7]   D. C. Bradley and I. M. Thomas, J. Chem. Soc., 1960, 3857.
[8]   H. Bürger and H. J. Neese, J. Organometal Chem., 20, 129 (1969).
[9]   T. Yoshino, I. Kijima, Y. Masuda, and I. Sugiyama, Kogyo Kagaku Zasshi, 62, 77 (1959).
[10] A. N. Nesmeyanov, O. V. Nogina, and R. Kh. Freidlina, Chem. Abstr., 48, 9254 (1954).
[11] N. Davidson and H. C. Brown, J. Amer. Chem. Soc., 64, 316 (1942).
[12] T. Takaya, T. Yoshimoto, and Y. Mashiko, Kogyo Kagaku Zasshi, 60, 1382 (1957).
[13] T. Ishino and S. Minami, Kagaku no Ryoiki, 11, 656 (1957).

[14] P. E. M. Allen, J. K. Brown, and R. M. S. Obaid, Trans.
      Faraday Soc., 85, 1808 (1963).
[15] T. S. Djabiev, R. D. Sabirova, and A. E. Shilov, Kinet. Katal.,
      5, 441 (1964).
[16] N. Koide, T. Asahi, K. Iimura, and M. Takeda, Paper Presented
      at the 23rd Annual Meeting of the Society of Polymer Science,
      Japan, June 1974.
[17] N. G. Gaylord, T. K. Kwei, and H. F. Mark, J. Polym. Sci., 42,
      417 (1960).
[18] A. Mazzei, S. Cucinella, and W. Marioni, Chim. Ind. (Milan),
      51, 374 (1969).
[19] S. Cucinella and A. Mazzei, Ibid., 53, 653 (1971).
[20] S. Cucinella and A. Mazzei, Ibid., 53, 749 (1971).
[21] C. Busetto and N. Palladion, Ibid., 53, 934 (1971).
[22] S. Murahashi, H. Yuki, K. Hatada, T. Niki, and T. Obokata,
      Kobunshi Kagaku, 24, 309 (1967).

Anionic Polymerizations and Copolymerizations of Methacrylates — Reactivity of Monomer and Tacticity of Polymer

HEIMEI YUKI, KOICHI HATADA, KOJI OHTA,
and YOSHIO OKAMOTO

Department of Chemistry
Faculty of Engineering Science
Osaka University
Toyonaka, Osaka, Japan

ABSTRACT

The anionic polymerizations and copolymerizations of methacrylates were investigated. The studies were focused on the stereoregularity of the polymers and the relative reactivity of the monomers in relation to the stereospecificity of polymerization.

INTRODUCTION

Methyl methacrylate (MMA) forms an isotactic polymer in a nonpolar solvent and a syndiotactic one in THF with n-BuLi. In these polymerizations, α-methyl and ester groups play a sterically important role in controlling the tacticity of the polymer. It is, therefore, interesting to study how the bulkiness of the ester group in a methacrylate affects the stereoregularity of the polymer as well as the reactivity of the monomer. In this paper the authors will state the outline of the anionic polymerizations and copolymerizations of various methacrylates which have been investigated in their laboratory. The studies were mainly focused on the stereoregularity of the polymers and the relative reactivity of the monomers in relation to the stereospecificity of polymerization.

The monomers employed were methacrylic esters (RMA) of benzyl (R = Bz) alcohol [1] and its derivatives, p-phenylbenzyl (PhBz) [2], α-methylbenzyl (MB) [3], diphenylmethyl (DPM) [2], α,α-dimethylbenzyl (DMB) [4], 1,1-diphenylethyl (DPE) [5], and trityl (Tr) [1, 6] alcohols (Scheme 1).

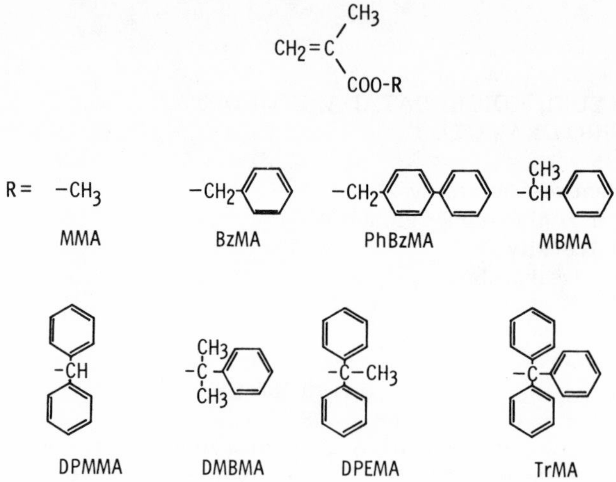

SCHEME 1.

The polymers of these methacrylates can be converted to poly-MMA via poly(methacrylic acid) which is easily formed from the original polymers with HBr in toluene and sometimes with HCl in methanol. Then the tacticity of the polymethacrylates can be determined from the NMR spectra of poly-MMA's thus obtained.

The polymerizations of methyl esters of α-ethyl, α-n-propyl, and α-phenyl acrylic acids (MEA [7], MPA [7], and MPhA [8], respectively) (Scheme 2) were also investigated in order to learn the effect of the α-substituent.

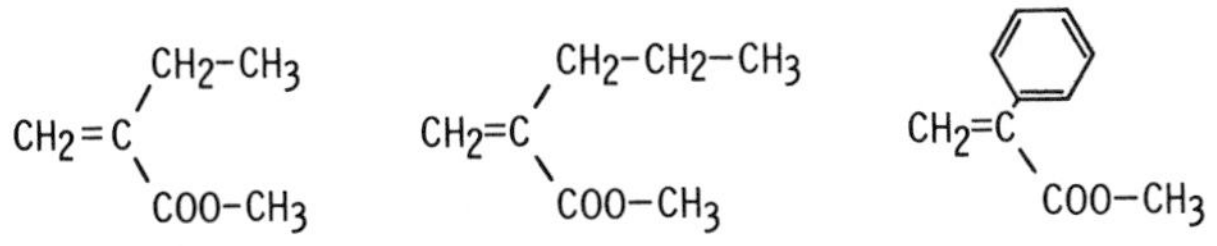

$$CH_2=C \begin{smallmatrix} CH_2-CH_3 \\ \\ COO-CH_3 \end{smallmatrix} \qquad CH_2=C \begin{smallmatrix} CH_2-CH_2-CH_3 \\ \\ COO-CH_3 \end{smallmatrix} \qquad CH_2=C \begin{smallmatrix} C_6H_5 \\ \\ COO-CH_3 \end{smallmatrix}$$

SCHEME 2.

TACTICITY OF POLYMETHACRYLATE

Table 1 shows the triad tacticity of polymethacrylates which were prepared with radical initiators. Similarly to MMA, most methacrylates formed rather syndiotactic polymers. However, atactic rather than syndiotactic polymers were obtained from DMBMA and DPEMA having tertiary ester groups. Almost complete atacticity, namely I:H:S = 1:2:1, was observed in the polymer of MPhA, which has the largest steric effect among the monomers. On the other hand, TrMA which has the largest ester group, gave a rather isotactic polymer even by a radical initiator.

The results of the polymerization by n-BuLi in toluene are listed in Table 2. Isotactic polymers were produced from most

TABLE 1. Stereoregularity of Polymethacrylates Obtained by Radical Initiator

Monomer	Initiator	Solvent	Temperature (°C)	I	H	S
MMA	AIBN	Toluene	60	6	36	58
BzMA[a]	BPO	Toluene	60	7	37	56
PhBzMA	AIBN	Toluene	80	5	38	57
DL-MBMA[a]	AIBN	Toluene	60	9	33	58
DPMMA	AIBN	Toluene	60	2	41	57
DMBMA	AIBN	Toluene	85	13	47	40
DPEMA	AIBN	Toluene	60	19	49	32
TrMA	AIBN	Toluene	60	64	22	14
MPhA	AIBN	None	60	21	50	29

[a]K. Matsuzaki et al., Makromol. Chem., 174, 215 (1973).

TABLE 2. Stereoregularity of Polymethacrylates Obtained by n–BuLi in Toluene

Monomer	Initiator	Solvent	Temperature (°C)	I	H	S
MMA	n–BuLi	Toluene	-78	72	17	11
MMA	n–BuLi	Toluene	0	81	14	5
BzMA	n–BuLi	Toluene	-78	81	15	4
BzMA	n–BuLi	Toluene	0	73	22	5
PhBzMA	n–BuLi	Toluene	-78	87	8	5
PhBzMA	n–BuLi	Toluene	0	76	20	4
D–MBMA	n–BuLi	Toluene	-78	78	17	5
DL–MBMA	n–BuLi	Toluene	-78	56	35	9
DL–MBMA	n–BuLi	Toluene	0	70	24	6
DPMMA	n–BuLi	Toluene	-78	99	1	0
DPMMA	n–BuLi	Toluene	0	93	4	3
DMBMA	n–BuLi	Toluene	-78	68	19	13

Monomer	Initiator	Solvent	Temp			
DMBMA	n-BuLi	Toluene	0	60	27	13
DMBMA	n-BuLi	$\dfrac{\text{Toluene}}{\text{THF}} = \dfrac{99}{1}$	-78	38	22	40
DPEMA	n-BuLi	Toluene	-78	23	28	49
DPEMA	n-BuLi	Toluene	0	52	37	11
TrMA	n-BuLi	Toluene	-78	96	2	1
TrMA	n-BuLi	Toluene	0	93	4	3
MEA	n-BuLi	Toluene	-78	70	11	19
MEA	n-BuLi	Toluene	0	97	2	2
MPA	n-BuLi	Toluene	-78	62	36	2
MPA	n-BuLi	Toluene	0	95	4	1
MPhA	n-BuLi	Toluene	-78	56	25	19
MPhA	n-BuLi	Toluene	-30	22	52	26
MPhA	n-BuLi	Toluene	0	-	-	-

monomers as well as MMA. Among these, poly-DPMMA and poly-
TrMA showed the highest isotacticity. On the other hand, at -78°C
DPEMA gave a syndiotactic polymer in this nonpolar solvent, although
the polymer became isotactic at 0°C. This is surprising because
DPEMA has an ester group whose bulkiness is similar to those of
DPMMA and TrMA. Most of the monomers gave polymers of lower
isotacticity at elevated temperatures, but some monomers, particularly MEA and MPA, formed highly isotactic polymers at 0°C.
MPhA gave a rather isotactic polymer at -78°C, but an atactic one
with a I:H:S = 1:2:1 ratio at -30°C. It is noticeable that the isotacticity
of the polymer of optically active D-MBMA was higher than that of the
racemic monomer. By the addition of a small amount of THF to the
solvent, DMBMA formed a so-called stereoblock or stereocomplex-
type polymer, which has been found in the polymerization of MMA.
The same phenomenon may be expected for the polymerizations of
other monomers.

The results of the polymerization by n-BuLi in THF are shown
in Table 3. At -78°C, syndiotactic polymers were mainly obtained.
DPMMA gave a polymer of especially high syndiotacticity. On the
other hand, TrMA formed a highly isotactic polymer not only in
toluene but also in this polar solvent. The large trityl ester group
may prevent the syndiotactic placement of the incoming monomer
to the growing chain end. In general, the polymers became less
stereoregular at 0°C, while DPEMA and MPhA gave atactic polymers
even at -78°C.

Table 4 shows the stereoregularity of the polymers obtained by
phenylmagnesium bromide and by dialkylaluminum diphenylamide
in toluene. These results are also very similar to those in the case
of MMA. Most methacrylates formed isotactic polymers by PhMgBr
and syndiotactic ones by R_2AlNPh_2. However, there is one exception,
i.e., by PhMgBr only an atactic polymer was obtained from DPMMA,
which produced highly isotactic and syndiotactic polymers by n-BuLi
in toluene and in THF, respectively. Contrary to the behavior of
DPMMA, DPEMA gave a polymer of extremely high isotacticity in
spite of the very low stereoregularity of the polymer obtained by
n-BuLi.

The general view of the above results, shown in Tables 1-4,
indicates that most of the methacrylates behave very similarly to
MMA in their stereospecific polymerizations. However, pronounced
steric effects come out in the polymerizations of DPMMA, DMBMA,
DPEMA, and TrMA. The steric control in their polymer formations
does not seem to depend simply on the bulkiness of the ester groups,
since the individual monomer exhibits quite different behavior in the

stereospecific polymerization. The stereoregulation may be deli-
cately concerned with the geometrical size of the ester group, as well
as the counterions produced from initiators. It is obvious that the
steric influence of α-substituent is more striking than that of the
ester group.

MONOMER REACTIVITY RATIOS AND RELATIVE REACTIVITY IN COPOLYMERIZATION [9, 10]

Although monomer reactivity ratios are important as fundamental
information, only a few data [11-13] have been reported on the
reactivity ratios in anionic copolymerizations, except for styrene
derivatives [14].

The monomer reactivity ratios in the copolymerizations of MMA
(M_1) with other methacrylates (M_2) by n-BuLi at -78°C were determined by analyzing the compositions of the initial copolymers.
Ethyl (Et), isopropyl (i-Pr), and tert-butyl (t-Bu) methacrylates
were also employed as the comonomer M_2.

It is known that in the anionic polymerization of MMA the greater
part of initiators are consumed in side reactions and the formation
of oligomers. Table 5 shows the results of the copolymerization of
MMA with BzMA. No difference was observed between the compositions of the oligomers and the corresponding polymers, regardless
of the monomer composition in the feed and the solvent used. Similar
results were obtained in the copolymerization of MMA and EtMA.
However, as shown in Table 6, the compositions of methanol-soluble
and -insoluble parts differed in the MMA—i-PrMA and MMA—t-BuMA
systems. The differences were slight, but they had a large effect on
the values of the monomer reactivity ratios. Since poly-t-BuMA is
slightly soluble in methanol, the copolymer rich in t-BuMA may be
soluble in methanol. However, it is not clear at present whether
the differences in compositions indicate different mechanisms in
the formations of the oligomer and the polymer or not.

All the copolymer composition data fitted well on the theoretical
curves with given r_1 and r_2 values as shown in Fig. 1.

The monomer reactivity ratios at -78°C are summarized in
Table 7. Only BzMA was more reactive than MMA both in toluene
and THF. In most cases the relative reactivity of the monomer
was not greatly affected by the solvents, but DMBMA and TrMA
showed extremely lower reactivity in toluene than in THF. t-BuMA
also showed low reactivity, not only in toluene but also in THF.

TABLE 3. Stereoregularity of Polymethacrylates Obtained by n–BuLi in THF

Monomer	Initiator	Solvent	Temperature (°C)	I	H	S
MMA	n–BuLi	THF	−78	6	38	56
MMA	n–BuLi	THF	0	31	32	37
BzMA	n–BuLi	THF	−78	6	31	63
BzMA	n–BuLi	THF	0	18	33	49
PhBzMA	n–BuLi	THF	−78	9	30	61
D–MBMA	n–BuLi	THF	−78	12	28	60
DL–MBMA	n–BuLi	THF	−78	8	31	60
DL–MBMA	n–BuLi	THF	0	16	39	45
DPMMA	n–BuLi	THF	−78	2	11	87
DPMMA	n–BuLi	THF	0	2	31	67
DMBMA	n–BuLi	THF	−78	10	29	61
DMBMA	n–BuLi	THF	0	11	37	52

DPEMA	n-BuLi	THF	-78	21	46	33
DPEMA	n-BuLi	$\frac{\text{Toluene}}{\text{THF}} = \frac{9}{1}$	-78	11	33	56
DPEMA	n-BuLi	THF	0	17	48	35
TrMA	n-BuLi	THF	-78	94	4	2
TrMA	n-BuLi	THF	0	81	13	6
MEA	n-BuLi	THF	-78	12	13	75
MPhA	n-BuLi	THF	-78	26	50	24

TABLE 4. Stereoregularity of Polymethacrylates Obtained by PhMgBr or R_2AlNPh_2 in Toluene

Monomer	Initiator	Solvent	Temperature (°C)	I	H	S
MMA	PhMgBr	Toluene	30	99	1	0
BzMA	PhMgBr	Toluene	30	85	11	4
PhBzMA	PhMgBr	Toluene	30	82	12	6
DL-MBMA[a]	PhMgBr	Toluene	20	75	21	4
DPMMA	PhMgBr	Toluene	20	28	36	36
DMBMA	PhMgBr	Toluene	30	74	16	10
DPEMA	PhMgBr	Toluene	0	97	3	0
TrMA	PhMgBr	Toluene	30	-	-	-
MMA	Et_2AlNPh_2	Toluene	-78	1	14	85
PhBzMA	Bu_2AlNPh_2	Toluene	-78	2	22	76
D-MBMA	Et_2AlNPh_2	Toluene	-78	4	17	79
DL-MBMA	Et_2AlNPh_2	Toluene	-78	4	20	76
DPMMA	Et_2AlNPh_2	Toluene	-40	8	27	65
DMBMA	Et_2AlNPh_2	Toluene	-78	1	17	82
DPEMA	Et_2AlNPh_2	Toluene	-40	12	34	54
MEA	Bu_2AlNPh_2	Toluene	-40	14	12	74
MPhA	Bu_2AlNPh_2	Toluene	-78	20	32	48

[a] K. Matsuzaki et al., Makromol. Chem., 174, 215 (1973).

TABLE 5. Anionic Copolymerization of MMA (M_1) and BzMA (M_2) by n-BuLi at -78°C[a]

| | | CH$_3$OH insoluble | | CH$_3$OH soluble | |
| | | Yield | | Yield | |
Solvent	f_1	(wt%)	F_1	(wt%)	F_1
Toluene	0.90	3.1	0.85	3.9	0.84
Toluene	0.70	3.0	0.55	2.4	0.58
Toluene	0.40	3.1	0.31	2.5	0.30
Toluene	0.30	2.9	0.25	2.1	0.22
Toluene	0.15	2.6	0.10	2.0	0.11
THF	0.90	2.5	0.86	4.5	0.86
THF	0.70	3.0	0.62	3.6	0.58
THF	0.50	4.7	0.42	3.7	0.37
THF	0.30	4.2	0.20	3.3	0.20
THF	0.15	3.9	0.10	2.9	0.12

[a]$[M_1]_0 + [M_2]_0$, 15 mmole; solvent, 15 ml; [n-BuLi], 0.30 mmole; time, 5 min in toluene; 1 to 2 min in THF; f_1 and F_1, mole fractions of M_1 in feed and polymer, respectively.

TABLE 6. Anionic Copolymerization of MMA (M_1) and Methacrylates (M_2) in THF by n-BuLi at -78°C[a]

| | | CH$_3$OH insoluble | | CH$_3$OH soluble | |
| | | Yield | | Yield | |
M_2	f_1	(wt%)	F_1	(wt%)	F_1
i-PrMA	0.35	2.0	0.57	7.5	0.53
i-PrMA	0.70	5.5	0.83	6.6	0.74
i-PrMA	0.85	6.2	0.93	5.8	0.82
t-BuMA	0.30	0.7	0.91	4.5	0.79
t-BuMA	0.50	0.7	0.94	5.6	0.87
t-BuMA	0.70	1.9	0.99	4.9	0.93

[a]$[M_1]_0 + [M_2]_0$, 15 mmole; THF, 15 ml; [n-BuLi], 0.30 mmole; time, 3 to 7 min; f_1 and F_1, mole fractions of M_1 in feed and polymer, respectively.

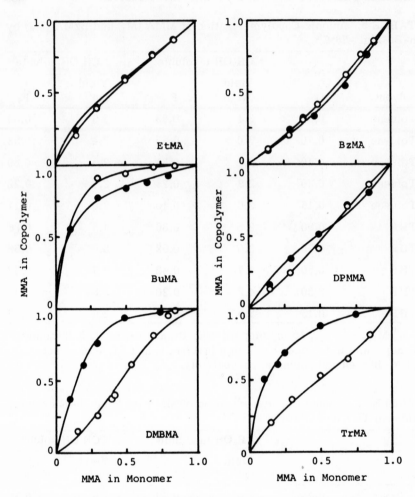

FIG. 1. Copolymer composition curves for anionic copolymeri-zations of MMA (M_1) and methacrylates (M_2) by n-BuLi at -78°C in toluene (●) and in THF (○).

An ideal copolymerization will be expected in the MMA-BzMA system because the product $r_1 \times r_2 = 1$. The methoxy resonances of MMA unit in the NMR spectra of copoly(MMA-BzMA)'s (Fig. 2) were much broader than that of PMMA, and each peak of the spectra shifted gradually to the downfield side with an increase of MMA

TABLE 7. Monomer Reactivity Ratios in the Copolymerizations of MMA (M_1) with Various Methacrylates by n-BuLi at -78°C[a]

M_2	In toluene		In THF	
	r_1	r_2	r_1	r_2
EtMA	1.10 ± 0.10	0.38 ± 0.22	1.13 ± 0.07	0.52 ± 0.18
EtMA			(1.23 ± 0.02)	(0.49 ± 0.05)
i-PrMA	2.75 ± 0.45	0.20 ± 0.37	2.29 ± 0.18	0.42 ± 0.20
i-PrMA	(2.22 ± 0.31)	(0.31 ± 0.26)	(0.97 ± 0.38)	(0.28 ± 0.03)
t-BuMA			36 ± 27	0.43 ± 0.75
t-BuMA	(4.41 ± 2.55)	(0.02 ± 0.05)	(5.07 ± 2.16)	(0.02 ± 0.09)
BzMA	0.59 ± 0.07	1.60 ± 0.42	0.70 ± 0.06	1.46 ± 0.36
BzMA	(0.57 ± 0.01)	(1.49 ± 0.05)	(0.69 ± 0.06)	(1.84 ± 0.36)
DL-MBMA	1.68 ± 0.17	0.78 ± 0.32	2.04 ± 0.16	1.52 ± 0.25
DPMMA	0.57 ± 0.15	0.55 ± 0.61	1.11 ± 0.22	1.57 ± 0.61
DMBMA	19.1 ± 4.0	0.56 ± 0.39	2.59 ± 1.35	2.00 ± 1.10
TrMA	6.28 ± 0.30	0.13 ± 0.07	0.62 ± 0.08	0.62 ± 0.32

[a]Parentheses denote the data from methanol soluble products.

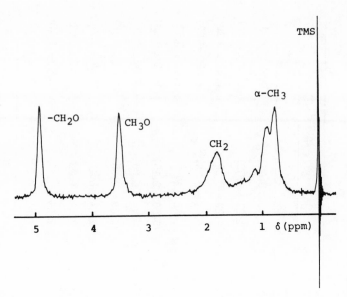

FIG. 2. NMR spectrum of copoly(MMA-BzMA) obtained in THF
(100 MHz, CCl₄, 60°C).

content as shown in Fig. 3. These facts indicate that the arrangement
of the monomeric units in the copolymers was more or less random,
as expected.

The copolymerizations of MMA with MBMA and DMBMA in THF
showed both r_1 and r_2 were greater than unity, indicating block
structures of the copolymers. On the other hand, the copolymerization with DPMMA in toluene and that with TrMA in THF may give
rather alternating structures of the copolymers. These structures
could be confirmed as follows [15]. By the treatment with methanolic
hydrochloric acid the copoly(MMA-DMBMA) and copoly(MMA-TrMA)
yielded quantitatively copolymers of MMA and methacrylic acid (MAA),
where only the hydrolysis of DMB and Tr ester groups proceeded
selectively (Scheme 3). The NMR spectra of α-methyl groups of the
copoly(MMA-MAA) thus obtained are shown in Fig. 4. The complete
triads assignments of the six peaks in the spectrum have been done
by Klesper et al. [16]. Peak 1 contains all of the tactic triads (I, H,
S) of MAA sequences, and peak 6 is assigned to the syndiotactic triad
of MMA sequences. In the spectrum A of copoly(MMA-MAA) derived
from copoly(MMA-DMBMA) the intensities of peaks 1 and 6 were
strong relative to the other peaks. On the contrary, those in the

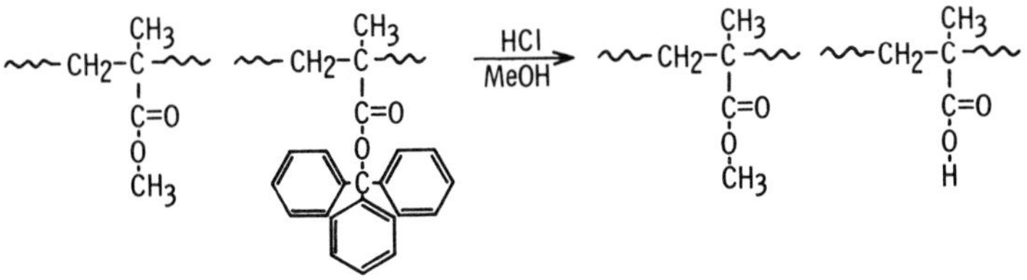

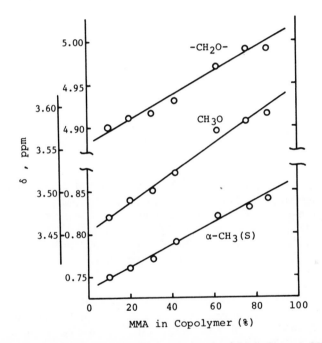

SCHEME 3.

FIG. 3. Chemical shifts of α-CH$_3$(syndio), CH$_3$O, and CH$_2$O of copoly(MMA-BzMA) obtained in THF by n-BuLi at -78°C (100 MHz, CCl$_4$, TMS, 60°C).

spectrum B of the polymer, which was originated from copoly(MMA-TrMA), were weak. These data indicate that copoly(MMA-DMBMA) contained rather long sequences of MMA and of DMBMA, while copoly(MMA-TrMA) had some alternating structure as expected from the r_1 and r_2 values.

The relative reactivity of a monomer toward an anion will

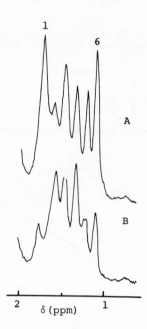

FIG. 4. NMR spectra of copoly(MMA-MAA)s which were derived from copoly(MMA-DMBMA) (A) and copoly(MMA-TrMA) (B) (100 MHz, pyridine-d_5, HMDS, 100°C). Polymerization conditions: $[MMA]_0 = [M_2]_0$, 2.5 mmole; THF, 10 ml; [n-BuLi], 0.25 mmole; -78°C; 24 hr. Yield: 88% (A); 100% (B); [MMA]/[DMBMA] in copolymer, 1.2; [MMA]/[TrMA] in copolymer, 1.0.

depend on the electron density on the β-carbon of the monomer. The logarithm of relative reactivity $(1/r_1)$ of the methacrylates toward MMA anion was plotted against the ^{13}C chemical shift of the β-carbon as shown in Fig. 5. Most data showed a linear relationship, though they are a little scattered. However, the points for DMBMA and TrMA in toluene deviated greatly from the line. The plots of the ^{13}C chemical shift of β-carbon against Taft's σ^* value of the ester group are shown in Fig. 6. A fine linear relationship was observed between the electron density of β-carbon and the polar effect of the ester group. A good linearity was also obtained between the chemical shift of ^1H (cis to the carbonyl group) of the monomer and the σ^* value. Consequently, the reactivity of the methacrylates toward an anion is explained as being governed by the polar effect of the ester group, not only

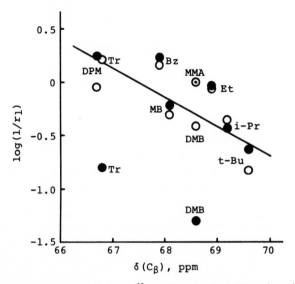

FIG. 5. Plots of $\log(1/r_1)$ vs $^{13}C_\beta$ chemical shifts of methacrylates:
(●) in toluene; (○) in THF.

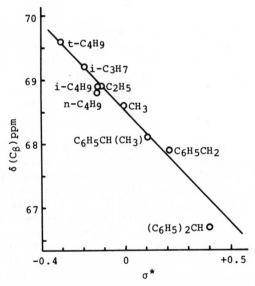

FIG. 6. Plots of Taft σ^* vs $^{13}C_\beta$ chemical shifts of methacrylates.

in THF but also in toluene. The only exception is the extremely low
reactivity of DMBMA and TrMA in toluene, suggesting that the
mechanisms of the copolymerizations of MMA-DMBMA and MMA-
TrMA systems in toluene are distinct from the other mechanisms.

TACTICITY OF COPOLYMER [9, 10]

Few studies have been reported on the stereoregularity of
copolymers. The authors investigated the tacticity of the copolymers
of the above methacrylates with each other in order to learn the
effect of bulkiness of the monomer on the stereoregulation of the
copolymer.

Table 8 shows the stereoregularity of the copolymers formed from
equimolar amounts of feed monomers by n-BuLi in toluene at -78°C.
The isotacticity of the copolymer of MMA and BzMA slightly exceeded
those of poly-MMA and poly-BzMA prepared under similar reaction
conditions. However, the isotacticity of most copolymers lay in
between those of the corresponding homopolymers. In some cases
where the copolymers contained bulky monomers, their isotacticity
was lowered, for example, the copolymers of TrMA-DPMMA,
TrMA-DMBMA, and TrMA-DPEMA.

The data of the copolymers obtained in THF are listed in Table 9.
They are similar to the copolymers obtained in toluene, but in this
case the syndiotacticity of the copolymer was in between those of
the corresponding homopolymers. It must be mentioned that the
isotacticity of the copolymers which contained TrMA as one of the
comonomers increased with increasing bulkiness of another
comonomer, suggesting the steric influence of TrMA on the
stereoregulation of the copolymer in THF.

As stated previously, the methacrylates formed the polymers
with a variety of tacticity even under the same reaction conditions.
It has been postulated that in the course of the stereospecific
polymerization of MMA, the propagation may proceed with a
preferred helical conformation [17]. We can consider a similar
situation in the stereospecific polymerization of other methacrylates.
If the copolymerization of the methacrylates proceeds stereoregularly
also with a helical conformation of the growing chain, the differences
in size and shape of the ester groups may cause the following cases:
1) If once a monomer forms a helical chain in the initial stage of the
copolymerization, it excludes the incorporation of the comonomer,
growing to a stereoregular homopolymer; 2) if both monomers
copolymerize initially, where the helical conformation is disturbed,
both can continue to add to the chain end, forming a less tactic
copolymer; 3) it is more probable that in the copolymerization

both the monomers have tendencies to take individual tacticity in the same polymer chain, producing a copolymer of mean stereo-regularity. In an extreme case, one monomer having a tendency of forming high stereoregularity may force the comonomer to take identical tacticity in the same helical chain, yielding a stereoregular copolymer. These are represented schematically in Scheme 4.

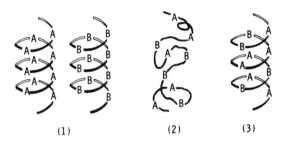

(1)   (2)   (3)

SCHEME 4.

Most of the data shown in Tables 8 and 9 seem to belong to the third case, although higher stereoregularity than those of homo-polymers was found only in copoly(MMA-BzMA) obtained in toluene. The isotactic propagation of TrMA was completely disturbed by the incorporation of a small size comonomer, and to some extent the isotacticity was retained in the copolymeri-zation with a comonomer having similar bulkiness. It was found that the copolymerization of MMA and TrMA in toluene produced a highly isotactic poly-MMA (I:H:S = 83:11:6) and a less stereoregular copolymer simultaneously. The copoly(MMA-TrMA) obtained was converted to copoly(MMA-MAA), which was then fractionated with methanol. The insoluble fraction was proved to be the poly-MMA and the soluble one the copolymer (Fig. 7). A similar result has been observed in the copolymerization of MMA and MPhA by n-BuLi in toluene (Table 10) [18].

The copolymerization of equimolar amounts of these monomers gave only an isotactic homopolymer of MPhA at -78°C, while above 30°C an alternating copolymer, which was completely atactic, was produced. On the other hand, the product at -40°C was found to consist of a highly isotactic poly-MMA and a copolymer. These facts indicate that in the copolymerization with MMA an extremely bulky comonomer can incorporate only in the growing chain with random conformation. The extremely low reactivity of TrMA in

TABLE 8. Stereoregularity of Copolymethacrylates Obtained in Toluene by n–BuLi at –78°C

M$_2$	M$_2$ homo-polymer			MMA				BzMA				DL–MBMA				DPMMA				DMBMA				DPEMA			
	I	H	S	$\frac{m_1}{m_2}$	I	H	S	$\frac{m_1}{m_2}$	I	H	S	$\frac{m_1}{m_2}$	I	H	S	$\frac{m_1}{m_2}$	I	H	S	$\frac{m_1}{m_2}$	I	H	S	$\frac{m_1}{m_2}$	I	H	S
MMA	70	17	13																								
BzMA	81	15	4	1.1	84	10	6																				
DL–MBMA	56	35	9	1.0	66	27	7	1.1	79	15	6																
DPMMA	99	1	0	2.0	80	16	4	0.9	79	16	5	1.1	65	28	7												
DMBMA	68	18	13	5.3	61	25	14	5.9	69	21	10	1.5	65	26	9	1.2	68	23	9								
DPEMA	23	28	49	4.3	60	27	13	-	-	-	-	-	-	-	-	4.3	66	22	12	-	-	-	-				
TrMA	96	2	2	6.4	74	17	9	8.9	81	12	7	4.1	59	28	13	-	72	20	8	0.7	41	37	22	1.0	27	37	36

TABLE 9. Stereoregularity of Copolymethacrylates Obtained in THF by n-BuLi at -78°C

	M2 homo-polymer			MMA				BzMA				DL-MBMA				DPMMA				DMBMA				DPEMA			
M2	I	H	S	$\frac{m_1}{m_2}$	I	H	S	$\frac{m_1}{m_2}$	I	H	S	$\frac{m_1}{m_2}$	I	H	S	$\frac{m_1}{m_2}$	I	H	S	$\frac{m_1}{m_2}$	I	H	S	$\frac{m_1}{m_2}$	I	H	S
MMA	6	40	54																								
BzMA	6	31	63	1.1	8	34	58																				
DL-MBMA	8	32	60	1.0	7	37	56	1.0	8	27	65																
DPMMA	2	11	87	1.2	6	30	64	0.9	6	26	68	0.8	11	23	66												
DMBMA	8	30	62	1.1	7	31	62	0.9	13	40	47	1.0	8	33	59	1.0	7	27	66								
DPEMA	21	46	33	1.0	10	36	54	-	-	-	-	-	-	-	-	1.0	8	28	64	-	-	-	-				
TrMA	94	4	2	1.0	19	40	41	1.0	16	40	44	1.1	22	39	39	-	26	35	39	1.1	38	41	21	1.0	51	34	15

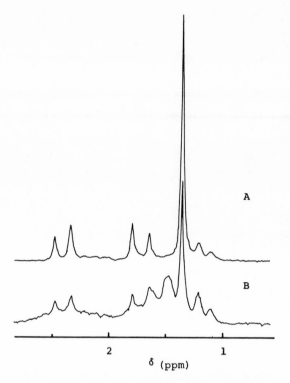

FIG. 7. NMR spectra of fractions of copoly(MMA–MAA) which
was derived from copoly(MMA–TrMA) obtained in toluene by n-BuLi
at -78°C. (A) Methanol-insoluble fraction; (B) methanol-soluble
fraction (100 MHz, pyridine-d_5, 100°C, HMDS).

toluene may be explained by the difficulty of incorporation into
the isotactic polymer chain in the copolymerization.

In this paper the authors have described three topics on the
anionic polymerization of a series of methacrylates: the stereoregulation in the polymerization, the relative reactivity of the
monomers in the copolymerization, and the stereoregularity
of the copolymers. These are correlated with each other and
involve many unsolved problems. Further investigations are
continuing.

TABLE 10. Anionic Copolymerization of MPhA and MMA by n-BuLi in Toluene[a]

Temperature (°C)	Time (hr)	Yield (%)	MPhA unit in copolymer (mole %)	Alternate sequence (mole %)
-78	24	15	97	0
-40	72	68	44	31
0	72	80	47	93
30	96	59	47	100[b]

[a]MPhA, 6.8 mmole; MMA, 6.8 mmole; n-BuLi, 0.68 mmole.
[b]Cotacticity: I, 24.8%; H, 50.4%; S, 24.8%.

ACKNOWLEDGMENT

A part of this work was supported by a Grant-in-aid for Scientific Research from the Ministry of Education.

REFERENCES

[1] H. Yuki, K. Hatada, T. Niinomi, and Y. Kikuchi, Polym. J., 1, 36 (1970).
[2] H. Yuki et al., Unpublished Data.
[3] H. Yuki, K. Ohta, K. Ono, and S. Murahashi, J. Polym. Sci., A-1, 6, 829 (1968).
[4] H. Yuki, K. Ohta, K. Hatada, Y. Okamoto, K. Kamanaru, K. Obayashi, and M. Mochida, To Be Submitted to J. Polym. Sci., Polym. Chem. Ed.
[5] H. Yuki, Y. Okamoto, K. Ohta, Y. Shimada, and K. Hatada, To Be Published.
[6] H. Yuki, K. Hatada, Y. Kikuchi, and T. Niinomi, J. Polym. Sci., B, 6, 753 (1968).
[7] H. Yuki, K. Hatada, T. Niinomi, and K. Miyaji, Polym. J., 1, 130 (1970).

[8] H. Yuki, K. Hatada, T. Niinomi, M. Hashimoto, and J. Ohshima, Ibid., 2, 629 (1971).

[9] H. Yuki, Y. Okamoto, and K. Ohta, Preprints, 23rd Symposium on Macromolecular Science, Japan, Tokyo, 1974.

[10] H. Yuki, Y. Okamoto, K. Ohta, and K. Hatada, Polym. J., 6, 573 (1974); J. Polym. Sci., Polym. Chem. Ed., 13, 1161 (1975).

[11] K. Ito, T. Sugie, and Y. Yamashita, Makromol. Chem., 125, 291 (1969).

[12] J. C. Bevington, D. O. Harris, and F. S. Rankin, Eur. Polym. J., 6, 725 (1970).

[13] P. Vlcek, D. Doskocilova, and J. Trekoval, J. Polym. Sci., Polym. Symp. Ed., 42, 231 (1973).

[14] M. Szwarc, Carbanions, Living Polymers, and Electron-Transfer Processes, Wiley-Interscience, New York, 1968, p. 520.

[15] H. Yuki, Y. Okamoto, K. Ohta, and K. Hatada, Unpublished Results.

[16] E. Klesper and W. Gronski, J. Polym. Sci., B, 7, 661 (1969).

[17] D. L. Glusker, I. Lysloff, and E. Stiles, J. Polym. Sci., 49, 315 (1961).

[18] K. Hatada, J. Ohshima, T. Komatsu, S. Kokan, and H. Yuki, Polymer, 14, 565 (1973).

Initiation Reaction of the Alkali-Metal Catalyzed Polymerization of Dienes Bearing Allylic Protons

HAJIME YASUDA and HISAYA TANI

Department of Polymer Science
Faculty of Science
Osaka University
Toyonaka, Osaka 560, Japan

ABSTRACT

Initiation of the polymerization of dienes bearing allylic protons with metallic alkali metals were investigated. Addition of a tertiary amine into the solution resulted in the complete inhibition of the polymerization to give a mono- or dianionic dienylmetal compound and reduced dimers in quantitative yield. The planar conformations of the isolated dienyl alkali metal compounds, which are considered to be real active species toward the alkali-metal catalyzed polymerizations, were determined by PMR and ^{13}C NMR spectroscopy. The catalytic activity toward polymerization as well as the chemical reactivity of those dienyl alkali metals was investigated.

INTRODUCTION

Anionic polymerizations of dienes, especially those of butadiene, styrene, and isoprene with alkyllithium or arylalkali metals, have been extensively studied by many workers. Rate constants of initiation and propagation of the polymerization, structure of the propagating chain ends, and association behavior of polystyryl-, polyisoprenyl-, or polybutadienyllithium were investigated in detail directly by PMR spectroscopy or by kinetics by Morton [1], Bywater [2], Hsieh [3], Urwin [4], and Makowski [5]. The effect of the alkyl-lithium structure, solvent type, temperature, and initiation level on polymer chemistry were further studied by others [6-10]. The polymerizations of styrene and butadiene in polar media with sodium naphthalene to induce living polymerization were precisely investigated by Szwarc and co-workers [11]. Their technique made it possible to measure the rate of propagation via an ion-pair independent of that via a free ion. In contrast to those spectacular works, the polymerizations of dienes with metallic alkali metals [12, 13], especially initiation of the polymerization of dienes bearing allylic protons, have received very little attention [14]. In the polymerization of dienes catalyzed by alkali metals, dianions derived from diene dimers have been considered to be a real active species. The reaction of butadiene with sodium metal in ammonia to give a butadiene dimer dianion was reported by Ziegler [14]. The corresponding reaction of isoprene with sodium naphthalene to give isoprene dimer dianions was reported by Suga [15]. Dimer dianion [16], as well as tetramer dianion [17] of α-methylstyrene obtained from the monomer and an excess amount of potassium, is also known as a typical carbanion. This paper describes the isolation of a catalytically active species, the preparation and reaction of dienes bearing allylic protons (mainly 1,3-pentadiene) with metallic alkali metals (mainly potassium), and the structure of the dienyl anion of pentadienyl alkali metals. These works were done mainly from the viewpoint of synthetic chemistry.

ISOLATION OF DIENYLALKALIMETAL COMPOUNDS

Polymerization of 1,3-pentadiene with a catalytic amount of an metallic alkali metal can be considered to proceed through a step-wise reaction:

$$C_5H_8 + K \begin{cases} C_5H_8^- M^+ \xrightarrow{\text{coupling}} C_{10}H_{16} \cdot M_2 \rightleftharpoons \text{oligomer} \cdot M_2 \rightleftharpoons \text{polymer} \cdot M_2 \\ C_5H_7M \xrightarrow{C_5H_8} C_{10}H_{15}M \rightleftharpoons \text{oligomer} \cdot M \rightleftharpoons \text{polymer} \cdot M \\ \text{(dimeric)} \end{cases}$$

where M is an alkali metal.

The first one is proceeded by a dianion type initiator formed by coupling of radical monoanions and other is proceeded by a mono-anionic initiator. If this scheme is correct, the initiation product is expected to be caught by increasing the amount of alkali metal and/or by decreasing the reaction temperature. This expectation was realized.

The reaction of 1,3-pentadiene (0.1 mole) with metallic potassium (0.05 mole) dispersed in tetrahydrofuran (20 ml) was carried out at 0°C for 5 hr. By the addition of an excess amount of n-hexane into the reaction mixture, yellowish powdery pentadienylpotassium was obtained as a precipitate. Recrystallization from a mixture of THF and n-hexane gave needle crystals. Unreacted monomer, reduced dimer, and polymer were obtained from the n-hexane soluble fraction using an appropriate method. By increasing the reaction temperature and/or by decreasing the amount of metallic potassium, the yield of polymer increased rapidly. Other dienes also behaved similarly and gave corresponding dienylpotassium compounds (Table 1). 2,3-Dimethyl-1,3-butadiene and isoprene gave only polymers by the reaction with metallic potassium.

The addition of a tertiary amine into the reaction system resulted in the complete inhibition of the polymerization. The use of 1 mole equivalence of triethylamine, triethylenediamine, tetramethylethylene-diamine, or tetramethylpropylenediamine was most effective for the inhibition of polymerization and gave quantitatively a mixture of dienylpotassium and the reduced dimer [18]. Pyridine or dipyridyl was less effective for this purpose. Thus the nature of a tertiary amine becomes a crucial factor for the successful preparation of dienyl alkali metal compounds. In contrast to our method, the reaction of conjugated dienes, 1,3-pentadiene, 2,4-hexadiene, 2-methyl-, and 4-methyl-1,3-pentadiene with n-butyllithium gave only polymers in place of the formation of the 1:1 dienyl metal compounds, even in the presence of a tertiary amine. Preparation of corresponding dienyllithium compounds were available only

TABLE 1. Conversion of Dienes by the Reaction with Metallic Potassium in a 1:1 Mole Ratio in THF for 5 hr[a]

Monomer	Reaction temperature (°C)	K-compound[b] (%)	Dimers[c] (%)	Oligomers[d] (%)	Recovered monomer (%)
C—C≡C—C≡C	-20	20	20	30	30
	0	16	17	58	9
	40	2	2	91	5
C=C—C—C≡C	0	45	45	3	7
C—C=C—C=C—C	0	6	7	69	18
C C C—C=C—C≡C	0	0	0	96	4
	0	45	45	1	9
	0	43	46	1	10
	40	42	49	0	9

[a]Represented by mole % conversion of monomer.
[b]Determined titrimetrically. The compound was isolated by cooling the reaction mixture after addition of an excess of n-pentane.
[c]Detected with glpc.
[d]Calculated by gravimetry.

when unconjugated dienes [19], not conjugated ones [20], were
reacted with n-butyllithium.

Reaction of cis-1,3-pentadiene, trans-1,3-pentadiene, or 1,4-
pentadiene (1.5 mole) with metallic potassium (1.0 mole) in a mixture
of tetrahydrofuran (2.0 mole) and triethylamine (1.0 mole) gave quantitative yields of crystalline pentadienylpotassium tetrahydrofuranate
and a mixture of reduced pentadient dimers following the stoichiometry
represented by

$$
\left.
\begin{array}{l}
\overset{t}{CH_3-CH=CH-CH=CH_2} \\[2mm]
\overset{c}{CH_3-CH=CH-CH=CH_2} \\[2mm]
CH_2=CH-CH_2-CH=CH_2
\end{array}
\right\}
\xrightarrow[\substack{N(C_2H_5)_3}]{K,\ THF}
\frac{1}{2}\ \underset{(K\text{-compound})}{C_5H_7K\cdot THF} + \frac{1}{4}\ \underset{(reduced\ dimer)}{C_{10}H_{18}}
$$

$$(C_5H_8)$$

PMR and CMR spectra showed that the presence or the absence of a
tertiary amine in the reaction system had no effect on the structure
of the pentadienylpotassium compound.

Similarly to the reactions of pentadienes, other dienes such as
2-methyl-1,3-pentadiene, 4-methyl-1,3-pentadiene, four isomers of
n-hexadienes, two isomers of cycloheptadienes, and three isomers of
cyclooctadienes gave the dienyl potassium tetrahydrofuranate and reduced dimers in 90 to 95% yields. In these cases an unconjugated diene
required higher temperature than a conjugated one. For example, in
the case of n-hexadienes, 1,5-hexadiene required a high temperature
(50°C for 24 hr) for completion of the reaction, whereas the reaction
of 1,4- or 2,4-hexadiene was completed at 30 or 0°C, respectively,
in 5 hr:

$$
\left.
\begin{array}{l}
\overset{t}{CH_3-CH=CH}-\overset{t}{CH=C-CH_3} \\[2mm]
\overset{t}{CH_3-CH=CH}-\overset{c}{CH=CH-CH_3} \\[2mm]
CH_2=CH-CH_2-CH=CH-CH_3 \\[2mm]
CH_2=CH-CH_2-CH_2-CH=CH_2
\end{array}
\right\}
\xrightarrow[\substack{NEt_3}]{K,\ THF}
\frac{1}{2}\ C_6H_9K\cdot THF + \frac{1}{4}\ C_{12}H_{22}
$$

$$(C_6H_{10})$$

In contrast to these reactions, a monomer dianion was obtained from the reaction with 2,3-dimethyl-1,3-butadiene. The absence of a tertiary amine in the reaction system gave no 2,3-dimethyl-1,3-butadienylpotassium at all, but the addition of triethylamine produced 2,3-dimethyl-1,3-butadienyldipotassium and a reduced dimer, 2,3,6,7-tetramethyl-2,6-octadiene, in 47 and 50% yields, respectively.

$$\begin{array}{c} CH_3 \quad CH_3 \\ | \qquad | \\ CH_2{=}C{-\!-}C{=}CH_2 \\ (C_6H_{10}) \end{array} \quad \xrightarrow[N(C_2H_5)_3]{K,\ THF} \quad \frac{1}{3}\ C_6H_8K_2\ (THF)_2\ +\ \frac{1}{3}\ C_{12}H_{22}$$

(K-compound) (reduced dimer)

In the cases of 1,3- and 1,4-cyclohexadiene, the dienyl compounds were not obtained by the reaction with potassium or sodium, and those dienes isomerized quantitatively to an equal amount of benzene and cyclohexene by allowing the mixture to stand at room temperature for 24 hr:

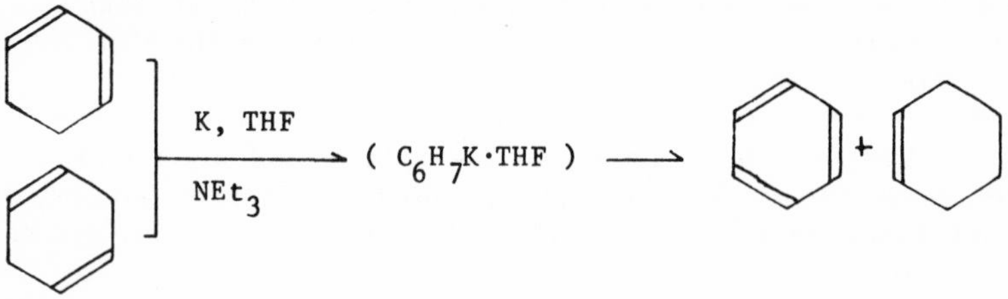

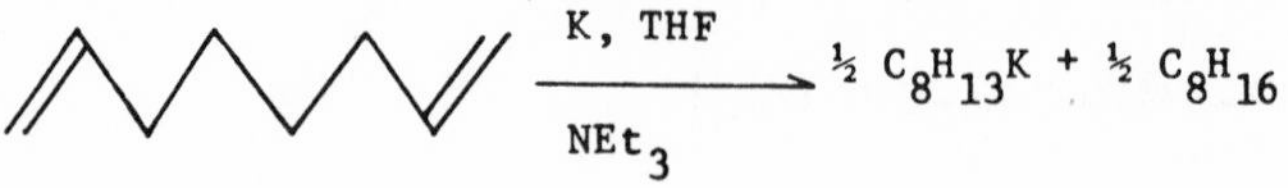

The migration of the double bond occurred only when the reactant has at least two olefinic groups. The reaction of 1-pentene, 2-pentene, 1-octene, or 2-octene with metallic potassium in tetrahydrofuran at 60°C for 4 days did not result in any migration, and the starting materials were recovered. In contrast to this, the double bonds of 1,7-octadiene, as with 1,5-hexadiene, were migrated to form 3,5-octadienyl- and 2,4-hexadienylpotassium, respectively. The active species for the migration of the double bond by metallic alkali metal in tetrahydrofuran-tertiary amine was considered to be the same one as that of the dimsylsodium-dimethylsulfoxide [21] and potassium tert-butoxide—dimethylsulfoxide [22] system. The structures of reduced dimers obtained simultaneously with dienyl alkali metal compounds were determined by IR, mass and NMR spectroscopies. These were composed of the same isomeric mixture of hydrocarbons, irrespective of the positions of the double bonds in the starting hydrocarbons as shown in Table 2. These facts show that the double bond in the unconjugated diene would migrate to a conjugated position prior to reacting it with metallic potassium.

The anionic nature of the dienylpotassium compounds was obtained from the structures of the deuterolyses products (Table 3). All the dienylpotassium compounds were found to be monoanions, except for 2,3-dimethyl-1,3-butadienylpotassium in which the formation of a monomeric dianion was confirmed by deuterolysis. The stoichiometry was represented as referred to above in the equation. It is interesting to note that the potassium compounds of aliphatic 1,3-dienes, which are active monomers for anionic polymerization, gave 1,3-conjugated diene by hydrolyses, while cyclic 1,3-dienes, which are inactive toward anionic polymerization, gave 1,4-unconjugated cyclic dienes. The experimental results referred to above are qualitatively in accordance with our expectation of the polymerization of pentadiene with metallic potassium.

1,3-Pentadiene or 2,3-dimethyl-1,3-butadiene was selected as a typical example of the polymerization of a conjugated diene bearing allylic protons. The crystalline pentadienylpotassium behaved as a catalyst in the polymerization of 1,3-pentadiene to give results quite similar to those obtained with metallic potassium as shown in Table 4. This fact seems to suggest that the polymerization of 1,3-pentadiene catalyzed by metallic potassium in the absence of a tertiary amine proceeds through a monoanion as an initiating species. 2,3-Dimethyl-1,3-butadiene also behaved quite similarly. Thus pentadienylpotassium or 2,3-dimethyl-1,3-butadienylpotassium may be regarded as an active catalyst species.

CHEMICAL REACTIVITY OF DIENYL ALKALI METALS

Alkylations of the dienyl alkali metal compounds by methyliodide and by tertiary-butylbromide were useful for understanding the

TABLE 2. Structure of Reduced Dimers

Diene	Reduced dimer	Mole ratio

TABLE 3. Deuterolysis of Dienylpotassium Compounds

Diene	Deuterolysis

```
C           (cis)
|
C=C—C=C  (trans)

C
||
C—C—C=C
```

```
                    CD
                    |
                    C=C—C=C
                    cis:trans = 38:62
```

```
C        C  (trans-trans)
|        |
C=C—C=C  (trans-cis)

C        C
|        ||
C=C—C—C

C        C
||       ||
C—C—C—C
```

```
                    CD       C
                    |        |
                    C=C—C=C

                    (trans-cis)
```

```
C     C
|     |  (trans)
C=C—C=C

C     C
||    |
C—C=C—C
```

```
                    CD    C
                    |     |      cis
                    C=C—C=C
```

```
  C   C
  |   |
C=C—C=C
```

```
                    CD  CD
                    |   |
                    C=C—C=C
```

TABLE 4. Polymerization of 1,3-Pentadiene and 2,3-Dimethyl-1,3-butadiene[a]

Monomer	Catalyst	Yield of polymer (%)	Microstructure		
			1,2 (%)	1,4 (%)	3,4 (%)
1,3-Pentadiene	Metallic-K	83	20	80	0
	Pentadienyl-K	82	25	75	0
2,3-Dimethyl-1,3-butadiene	Metallic-K	85	35	65	–
	2,3-Dimethylbutadienyl-K_2	85	34	66	–

[a]Catalyst, 1 mole % of monomer; solvent, THF; time, 5 days (30°C).

TABLE 5. Relative Mole Ratio of Alkylated Dienes Reacted at C_1 or C_3 of Dienylpotassium Compounds[a]

Potassium compounds	MeI		tert-BuBr	
	C_1 (%)	C_3 (%)	C_1 (%)	C_3 (%)
Pentadienyl-K	65	35	16	84
Hexadienyl-K	44	56	31	69
Cycloheptadienyl-K	12	88	50	50
Cyclooctadienyl-K	5	95	50	50

[a]Alkyl halides reacted at C_1 and C_3 of the dienylmetal compounds gave 1,3- and 1,4-type alkylated dienes.

chemical properties of the dienyl metal compounds. Although a 1,4- unconjugated diene was rarely detected in the hydrolyzate of the dienyl metal compounds, both alkylated 1,3- and 1,4-dienes were formed by alkylation (Table 5). The yield of 3-alkyl-1,4-pentadiene increased by increasing the bulkiness of an alkyl halide. The formations of both 1,3- and 1,4-dienes are explained by the existence of an equilibrium between two structures A and B. The correlation existing between the bulkiness of an alkyl halide and the yield of 1,4-dienes

(A)   (B)

should be interpreted by an electronic factor rather than by the steric factor of an alkyl halide. The pentadienylpotassium tetrahydrofuranate could be converted easily to other etherates by a donor exchange reaction. Ether-free pentadienylpotassium was obtained by heating it at 50°C for 3 hr in vacuum, and addition of ether resulted in the other etherate. The product was isolated as a crystalline etherate. By taking advantage of this reaction, the

solvent effect on the alkylation of pentadienylpotassium was examined
by using allylbromide as an alkyl halide. Alkylation occurred mainly
at a terminal position to give 1,3,7-octatriene when the reaction
occurred in ether. It proceeded only at a central position in n-hexane
to give 3-allyl-1,4-hexadiene. It is interesting to note that the

Terminal alkyl   Central alkyl

increase in the yield of trans-1,3-pentadiene by hydrolysis corresponds well to the increase in the yield of dienes alkylated at a central
carbon atom (Table 6). These results are reasonably explained by
the existence of an equilibrium between the two extreme structures
A and B.

TABLE 6. Effect of Coordinated Ether on the Alkylation and
Hydrolysis of Pentadienylpotassium

Ethers	Allylation at		Hydrolyzed pentadiene	
	C_1 (%)	C_3 (%)	cis (%)	trans (%)
Diglyme	89	11	78	22
$\langle$ O $\rangle$—CH$_2$OCH$_3$	87	13	82	18
1,3-Dioxane	85	15	89	11
THF	59	41	38	62
(n-Hexane)	1	99	2	98

STRUCTURES OF THE DIENYL
METAL COMPOUNDS

Three different planar conformations can be considered for the delocalized structure of pentadienyl anion (I); i.e., U-shaped (cis, cis), W-shaped (trans, trans), and sickle-shaped (cis, trans) [23-25].

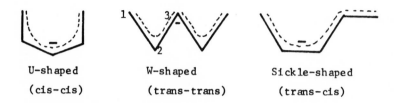

| U-shaped | W-shaped | Sickle-shaped |
| (cis-cis) | (trans-trans) | (trans-cis) |

The structure of the pentadienyl alkali metal compound in solution was, therefore, examined by the use of NMR techniques. CMR spectrum of pentadienylsodium, -potassium, and -rubidium showed the presence of only three peaks (Table 7). The sickle-shaped conformation requires five peaks. This result excludes the sickle-shaped structure. PMR spectrum was next studied (Table 8). The proton-proton coupling constant of the inner bond, J_{23}, is about 11 Hz and corresponds to that of a trans conformation, because the coupling constant J_{23} in a cis form is known to be about 6.5 Hz and that of a trans one is ~12 Hz [19, 26]. The observed coupling constants, J_{12}, of outer bonds, cis (9.3 Hz) and trans (16.2 Hz), correspond well to those of the known coupling constants of cis (7.5 to 9.0 Hz) and trans (~16 Hz) [26]. Thus pentadienylsodium, -potassium -rubidium, or -cesium is concluded to assume a W-shaped conformation. The conformation of 2-methylpenta-dienylpotassium (II) is concluded to also be W-shaped. In contrast to this, the conformation of cycloheptadienyl- (III) or cyclooctadienylpotassium (IV) was determined to be U-shaped

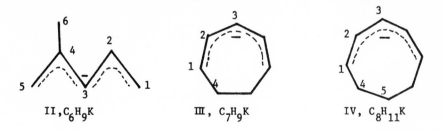

II, C_6H_9K III, C_7H_9K IV, $C_8H_{11}K$

TABLE 7. CMR Chemical Shift of Pentadienyl Alkali Metals[a,b]

	C_1	C_2	C_3	C_4	C_5	C_6
I (K)	78.2	135.6	78.5	-	-	-
I' (Na)	77.2	135.5	77.4	-	-	-
I'' (Rb)	76.4	135.0	76.8	-	-	-
II	79.1	133.5	79.7	142.3	76.0	28.5
III	93.5	131.0	75.0	36.8	-	-
IV	88.7	136.1	69.7	29.0	16.0	-

[a] In ppm downfield from external TMS in d_8-THF (calibrated using the upfield THF peak, assumed to be 25.8 ppm). Data were collected at 25.2 MHz on a Varian XL-100-15 spectrometer with a Digilab Model FTS-NMR-3 Fourier transform accessory.

[b] Peak assignments were made in part from an off-resonance decoupled spectrum, and the data collected at 35°C are shown except for the case of I'(0°).

from the observed value of the coupling constant. This result is in good agreement with the fact that these compounds cannot take a W-shaped conformation due to steric reasons. A general scheme for the initiation reaction of 1,3-pentadiene, 2,4-hexadiene; or other dienes with metallic potassium is reasonably explained by the following equations. As a typical example, the reaction of 1,3-pentadiene with metallic potassium is expressed in the equation. The radical monoanion formed between a diene and metallic potassium couples to give a dianion (Eqs. 1 and 2).

$$CH_3-CH=CH-CH=CH_2 + K \longrightarrow CH_3-\overset{\cdot}{C}H-CH=CH-CH_2^- \cdot K^+ \qquad (1)$$

$$2CH_3-\overset{\cdot}{C}H-CH=CH-CH_2^- \cdot K^+ \longrightarrow$$

$$
\begin{array}{cc}
CH_3 & CH_3 \\
| & |
\end{array} \qquad (2)
$$

$$K^+ \cdot CH_2-CH=CH-CH-CH-CH=CH-CH_2^- \cdot K^+$$

TABLE 8. PMR Chemical Shift of Pentadienyl Alkali Metals[a,b]

	C_1	C_2	C_3	C_4	C_5	C_6	J_{12}	J_{23}
I (K)	3.45	6.30	3.55	–	–	–	16.2 (trans) 9.3 (cis)	10.9
I' (Na)	3.44	6.29	3.54	–	–	–	16.2 (trans) 9.3 (cis)	10.9
II	3.36	6.23	3.49	–	3.59	1.90	16.4 (trans) 9.5 (cis)	10.6
III	3.84	5.86	3.39	2.72	–	–	8.0 (cis)	7.5
IV	3.08	5.87	2.64	3.10	1.11	–	7.8 (cis)	6.8

[a]In ppm downfield from external TMS in d_8–THF (calibrated using the upfield THF peak, assumed to be 1.85 ppm).
[b]Data were collected at 100.0 MHz on a Varian XL–100–15 spectrometer at 35°C.

The dianion rapidly abstracts two hydrogen atoms from two acidic methyl (or methylene) protons of the two unreacted pentadiene molecules to give 1 mole of reduced dimer and 2 moles of pentadienylpotassium (Eq. 3).

$$\overset{\qquad\qquad CH_3 \quad CH_3}{\underset{\qquad\qquad |\qquad\quad |}{K^+ \cdot \bar{C}H_2 - CH = CH - CH - CH - CH = CH - CH_2^- \cdot K^+}}$$

$$\overset{\qquad\qquad\qquad\qquad\qquad CH_3 \quad CH_3}{\underset{\qquad\qquad\qquad\qquad\qquad |\qquad\quad |}{+\ 2CH_3 - CH = CH - CH = CH_2 \longrightarrow CH_3 - CH = CH - \overset{\bullet}{C}H - CH - CH = CH - CH_3}} \qquad (3)$$

$$+\ 2[CH_2 \doteq CH \doteq CH \doteq CH \doteq CH_2]^- K^+$$

The rate of reaction in this step is considered to be very fast, because only monodeuterated 1,3-conjugated diene was obtained by deuterolysis. Any deuterated diene dimers were not detected in every case, irrespective of the presence or the absence of a tertiary amine in the reaction system. The formation of other isomeric diene dimer is explained by cross coupling of the isomeric two kinds of radical monoanions (Eq. 2') and then abstraction of allylic proton from 1,3-pentadiene (Eq. 3')

$$CH_3 - \overset{\bullet}{C}H - CH = CH - CH_2^- \cdot K^+ + \overset{\bullet}{C}H_2 - CH = C\bar{H} - CH - CH_3 \cdot K^+$$

$$\overset{\qquad\qquad\qquad\qquad CH_3}{\underset{\qquad\qquad\qquad\qquad |}{\longrightarrow K^+ \cdot \bar{C}H_2 - CH = CH - CH - CH_2 - CH = CH - C\bar{H} - CH_3 \cdot K^+}} \qquad (2')$$

$$\overset{\qquad\qquad CH_3 \qquad\qquad\qquad\qquad CH_3}{\underset{\qquad\qquad |\qquad\qquad\qquad\qquad\qquad |}{K^+ \cdot CH_2 - CH = CH - CH - CH_2 - CH = CH - CH\ \ K^+}}$$

$$+\ 2CH_3 - CH = CH - CH = CH_2$$

$$\qquad\qquad\qquad\qquad\qquad\qquad\qquad\qquad\qquad\qquad (3')$$

$$\overset{\qquad\qquad CH_3}{\underset{\qquad\qquad |}{\longrightarrow CH_3 - CH = CH - CH - CH_2 - CH = CH - CH_2 - CH_3}}$$

$$+\ 2[CH_2 \doteq CH \doteq CH \doteq CH \doteq CH_2]^- \cdot K^+$$

This finding gave us very important information concerning the catalytically active species; i.e., the dienyl monoanion is a real active species. Diene dimer dianion is not necessarily a catalyst species.

Finally, we should like to turn our attention back to the polymerization of 1,3-pentadiene and 2,3-dimethyl-1,3-butadiene.

The crystalline pentadienylpotassium behaved as a catalyst in the polymerization of 1,3-pentadiene to give results quite similar to those obtained with metallic potassium. This fact seems to suggest that the polymerization of 1,3-pentadiene catalyzed by metallic potassium in the absence of a tertiary amine proceeds via a monoanion as an initiating species. Diene dimer dianions, formed initially as mentioned before, are considered to abstract an allylic proton of the monomer in preference to the initiation of polymerization to give monoanions and reduced diene dimers.

The information about the second step in the polymerization was obtained from the equimolar reaction of 1,3-pentadiene with pentadienylpotassium, which was carried out in tetrahydrofuran at -35°C for 2 hr. Deuterolysis of this reaction mixture with D_2O gave 6-methyl-1,3,7-nonatriene-9-d, in 20% yield, in addition to poly(1,3-pentadiene). The yield of this pentadiene dimer was about 60% in the case of crystalline pentadienylsodium. In contrast to this, the reaction carried out at 30° gave exclusively poly(1,3-pentadiene).

$CH_3-CH=CH-CH=CH_2$

$\downarrow$ K

$CH_3-\overset{.}{C}H-CH=CH-CH_2^-\cdot K^+$

$\downarrow$ coupling

$K^+\ \bar{C}H_2-CH=CH-\overset{\overset{\displaystyle CH_3}{|}}{C}H-\overset{\overset{\displaystyle CH_3}{|}}{C}H-CH=CH-CH_2^-\ K^+$

$\downarrow$

$[CH_2\text{=}\!\text{=}CH\text{=}\!\text{=}CH\text{=}\!\text{=}CH\text{=}\!\text{=}CH_2]\ K^+\ +\ CH_3-CH=CH-\overset{\overset{\displaystyle CH_3}{|}}{C}H-\overset{\overset{\displaystyle CH_3}{|}}{C}H-CH=CH-CH_3$

$\downarrow C_5H_8$

$CH_2=CH-CH=CH-CH_2-\overset{\overset{\displaystyle CH_3}{|}}{C}H-\overset{CH}{\underset{CH_2}{CH}}\ K^+ \xrightarrow{D_2O} CH_2=CH-CH=CH-CH_2-\overset{\overset{\displaystyle CH_3}{|}}{C}H-CH=CH-CH_2D$

$\downarrow\downarrow$

$CH_2=CH-CH=CH-CH_2\!-\!\!\left[\!CH-CH=CH-CH_2\!\right]_n\!\!\overset{\overset{\displaystyle CH_3}{|}}{C}H-\overset{CH}{\underset{CH_2}{CH}}\ K^+$

These experimental results suggest that the second step in the polymerization is mainly the 1,4-addition of 1,3-pentadiene to pentadienyl alkali metal, in good agreement with the observed microstructure of poly(1,3-pentadiene), and that the polymerization proceeds via pentadienyl alkali metal with the monoanionic active species. The addition reaction in the polymerization of 2,3-dimethyl-1,3-butadiene with metallic potassium in tetrahydrofuran is considered to be the formation of its dipotassium compound, and the polymerization is considered to proceed via dienyldipotassium compound on both sides with its dianionic active species [27].

$$
\begin{array}{c}
\underset{|}{CH_3} \quad \underset{|}{CH_3} \\
CH_2{=}C{-}C{=}CH_2
\end{array}
$$

$$
K^+ \quad \overset{CH_2}{\underset{CH_2}{\diagdown}}C{-}C\overset{CH_2}{\underset{CH_2}{\diagup}} \quad K^+
$$

$\downarrow$ 2,3-dimethyl-1,3-butadiene

$$
K^+ \quad \overset{CH_2}{\underset{CH_2}{\diagdown}}C{-}\underset{|}{\overset{CH_3}{C}}{-}CH_2{+}CH_2{-}CH{=}\underset{|}{\overset{CH_3}{C}}{-}\underset{|}{\overset{CH_3}{C}}{-}CH_2\underset{n}{\}}CH_2{-}\underset{|}{\overset{CH_3}{C}}{-}C\overset{CH_2}{\underset{CH_2}{\diagup}} \quad K^+
$$

CONCLUDING REMARKS

The number of monoanionic crystalline dienyl alkali metals obtained during the investigation was not only useful for a better understanding of the initiation mechanism but also useful for a general synthesis of other dienyl metal compounds or organic dienyl derivatives. For example, thermally unstable dipentadinyl-zincs was, at first time, prepared by the reaction with anhydrous dichlorzinc at low temperature. Thermally stable Sn, Al, and Si compounds were also obtained [28]. By the reaction of those dienyl metal compounds with carbonyl compounds or alkyl halides, new dienyl derivatives became available.

ACKNOWLEDGMENTS

The authors express their hearty thanks to Dr. A. Yasuhara and Messrs. Y. Ohnuma, A. Kashiwara, M. Narita, and K. Kanemitsuya for their skillful and enthusiastic experimental support throughout this work.

REFERENCES

[1] M. Morton, E. E. Bostick, and R. G. Clarke, J. Polym. Sci., A1, 475 (1963); M. Morton, E. E. Bostick, and R. A. Livigni, Ibid., 1735 (1963); M. Morton and L. J. Fetters, Ibid., A2, 3311 (1964); M. Morton, L. J. Fetters, R. A. Pett, and J. F. Meier, Macromolecules, 3, 327 (1970).

[2] S. Bywater and D. J. Worsfold, Can. J. Chem., 40, 1564 (1962); D. J. Worsfold and S. Bywater, Ibid., 42, 2884 (1964); S. Bywater, Adv. Polym. Sci., 4, 66 (1965); S. Bywater and D. J. Worsfold, J. Phys. Chem., 70, 162 (1966); A. F. Johnson and D. J. Worsfold, J. Polym. Sci., A3, 449 (1965); D. J. Worsfold, Ibid., A1, 2783 (1967); D. J. Worsfold and S. Bywater, J. Organometal. Chem., 9, 1 (1967); F. Schué and S. Bywater, Macromolecules, 1, 328 (1968); S. Bywater, D. J. Worsfold, and Hollingsworth, Ibid., 5, 389 (1972).

[3] H. L. Hsieh, J. Polym. Sci., A3, 153, 163, 173 (1965); H. L. Hsieh and O. F. Mekinney, J. Polym. Sci., B, 843 (1966).

[4] J. R. Urwin and J. M. Stearne, Eur. Polym. J., 1, 227 (1965); D. N. Cramond, P. S. Lawry and J. R. Urwin, Ibid., 2, 107 (1966).

[5] H. S. Makowski and M. Lynn, J. Macromol. Chem., 1, 443 (1966).

[6] R. S. Stearns and L. E. Forman, J. Polym. Sci., 41, 381 (1959).

[7] H. Morita and A. V. Tobolsky, J. Amer. Chem. Soc., 79, 5853 (1957).

[8] A. W. Langer, Jr., Trans. N. Y. Acad. Sci., 21, 741 (1965).

[9] J. N. Hay, J. F. McCabe, and J. C. Robb, J. Chem. Soc., Faraday I, 68, 1153 (1972).

[10] Y. Minoura, K. Shiira, and M. Harada, J. Polym. Sci., A-1, 6, 559 (1968).

[11] M. Levy and M. Szwarc, J. Amer. Chem. Soc., 82, 521 (1960);
      S. Khanna, M. Levy, and M. Szwarc, Trans. Faraday Soc., 58,
      747 (1962); E. Ureta, J. Smid, and M. Szwarc, J. Polym. Sci.,
      A-1, 4, 2219 (1966); D. N. Bhattacharyya, C. L. Lee, J. Smia,
      and M. Szwarc, J. Phys. Chem., 69, 608 (1965); K. J. Toelle,
      J. Smid, and M. Szwarc, J. Polym. Sci., B, 4, 1037 (1965).
[12] A. V. Tobolsky and C. E. Regers, J. Polym. Sci., 40, 73
      (1959).
[13] F. Schué, Bull. Soc. Chim. Fr., 4, 980 (1965).
[14] K. Ziegler, H. Grimm, and R. Willer, Justus Liebigs Ann.
      Chem., 542, 90 (1940).
[15] K. Suga, S. Watanabe, H. Kikuchi, and T. Watanabe, Can. J.
      Chem., 46, 2619 (1968).
[16] A. Vrancken, J. Smid, and M. Szwarc, Trans. Faraday Soc.,
      58, 2036 (1962); C. L. Lee, J. Smid, and M. Szwarc, J. Phys.
      Chem., 66, 904 (1962).
[17] R. L. Williams and D. H. Richards, Chem. Commun., 1967,
      414.
[18] H. Yasuda, T. Narita, and H. Tani, Tetrahedron Lett., 27,
      2443 (1973).
[19] R. B. Bates, D. W. Gosselink, and J. A. Kaczynski, Ibid., 3,
      199, 205 (1967); R. B. Bates, S. Brenner, W. H. Deines,
      D. A. McCombs, and D. E. Potter, J. Amer. Chem. Soc.,
      92, 6345 (1970); R. B. Bates, L. M. Kroposki, and D. E.
      Potter, J. Org. Chem., 37, 560 (1972).
[20] Allyllithium have been prepared by addition of alkyllithiums
      to 1,3-dienes. (a) W. H. Glaze and P. C. Jones, Chem.
      Commun., 1969, 1434; W. H. Glaze, J. E. Hanicak, M. L.
      Moore, and J. Chaudhuri, J. Organometal. Chem., 44, 39
      (1972); W. H. Glaze, Hanicak, J. Chaudhuri, M. L. Moore,
      and D. P. Duncan, Ibid., 51, 13 (1973). (b) M. Morton,
      R. K. Sanderson, and R. Sakata, J. Polym. Sci., B, 9, 61
      (1971); M. Morton, L. A. Faluo, and L. J. Fetters, J. Polym.
      Sci., Polym. Lett. Ed., 10, 561 (1972). (c) A. Ulrich,
      A. Deluzarche, A. Maillard, F. Schué, and C. Tanieliean,
      Bull. Soc. Chim. Fr., 1972, 2460.
[21] J. Klein, S. Glily, and D. Kost, J. Org. Chem., 35, 1281
      (1970).
[22] D. Devaprabhakara, C. G. Cardenas, and P. D. Gardner,
      J. Amer. Chem. Soc., 85, 1553 (1963).
[23] N. Bauld, Ibid., 84, 4347 (1962).
[24] D. Glass, R. S. Boikes, and S. Winstein, Tetrahedron Lett.,
      10, 999 (1966).

[25] R. Hoffmann and R. A. Olofson, J. Amer. Chem. Soc., 88, 943 (1966).

[26] W. T. Ford and M. Newcomb, Ibid., 96, 309 (1974).

[27] H. Yasuda, A. Yasuhara, and H. Tani, Macromolecules, 7, 145 (1974).

[28] H. Yasuda and H. Tani, 22nd Symposium on Organometallic Chemistry, Kyoto, 1974; H. Yasuda and H. Tani, Tetrahedron Lett., 1, 11 (1975).

Ionic Polymerization of Vinyl Aromatic Monomers

JAMES M. PEARSON

Research Laboratories
Xerox Corporation
Webster, New York 14580

ABSTRACT

The ionic polymerization of vinyl monomers possessing aromatic and heterocyclic functional groups has not been studied
in any systematic fashion. Only in a few isolated cases have
detailed mechanistic and structural studies been reported. The
anionic polymerization of a number of vinylanthracene monomers has recently been investigated and some rationalization
of this system is presented. The cationic and anionic polymerization of the N-, 3-, and 2-vinylcarbazole series of monomers
is discussed in some detail. The important role of vinyl aro-
matic/vinyl heterocyclic monomers, i.e., diphenylethylene and
the vinylcarbazoles, in elucidating the mechanistic aspects of
cationic polymerization, "change transfer" polymerization, and
photoionic polymerization is considered.

INTRODUCTION

Using the refined synthetic and analytical tools available today, the
organic chemist possesses the capability for introducing virtually any

functional group into a vinyl monomer. An even greater challenge, perhaps, is the polymerization of such monomers into high molecular weight polymers and copolymers. Within the class of monomers covered here, i.e., vinyl aromatics/heterocyclics, free radical polymerization methods often give rise to low molecular weight, poorly defined products resulting from facile reactions of the nonselective free radicals with the functional groups present in the system. In certain cases, ionic polymerization procedures provide the only route to the desired polymeric products.

Polymeric materials possessing aromatic (i.e., naphthalene, anthracene, pyrene) and heterocyclic (i.e., indole, carbazole, phenothiazine) groups have not found widespread commercial utility. These are specialty monomers/polymers, and it is as specialty materials that they are finding and will continue to find applications, i.e., organic semiconductors/photoconductors, polymer stabilizers/destabilizers, photoreactive materials, and in pharmacological and medical areas. It is perhaps for this reason that no systematic studies of the ionic polymerization of these monomers have been reported. The effect of aromatic ring structure and of heterogroups on the polymerization reactions have been elucidated only in a few isolated cases.

It is only possible here to select in a subjective manner and review some of the more novel and interesting systems and to speculate on the future trends of this area of ionic polymerization. I would like to emphasize the need to consider the chemical, material, and mechanistic aspects of polymerization processes, but unfortunately few systems have been subjected to such detailed analyses.

The pioneering and elegant studies of the Syracuse group have established anionic polymerization on a sound base. The kinetic and mechanistic aspects of the polymerization processes are well understood, except in a few isolated cases. The role of solvent, counterion, etc. can be predicted, and it is possible to exercise a high degree of control over most polymerization reactions. The situation is considerably more complicated in the field of cationic polymerization and no unified mechanistic picture has emerged. Here the individual reaction steps are complex and the polymerization is not readily amenable to control. Although living-type polymers have been demonstrated for ring-opening polymerization, i.e., THF, no examples of a living carbenium ion system have yet been found.

ANIONIC POLYMERIZATION

Perhaps one of the most intriguing unsolved problems in this area is the polymerization of 9-vinylanthracene (1). This monomer has yielded only low molecular weight products, $\overline{M}_n \sim 1000$ to 5000, by all known polymerization techniques [1]. The anionic polymerization of the other members of the series, i.e., vinyl naphthalenes, vinyl phenanthrenes, and vinyl pyrene, have all been reported to proceed in a conventional manner.

The anomalous results for 9-vinylanthracene have not been explained. Several plausible mechanisms have been proposed: 1) an extremely efficient chain (electron) transfer process, 2) a through-the-ring propagation, and 3) a "wrong" monomer addition. These are shown in Fig. 1.

The synthesis and the anionic polymerization of 9-vinyl-10-methylanthracene (1a), 2-vinylanthracene (2) and 2-propenyl-2-anthracene (2a) have recently been investigated in our laboratory in an attempt to resolve these issues [2] (Fig. 2).

With the 2-vinylanthracenes it has been possible to achieve high molecular weight, $\overline{M}_n \sim 10^5$ to 10^6, soluble polymers. IR analysis is indicative of a conventional 1,2 vinyl polymer structure with no involvement of the anthracene rings.

The spectra of the propagating anions of monomers 2 and 2a are shown in Figs. 3 and 4, respectively. Identical spectra were obtained

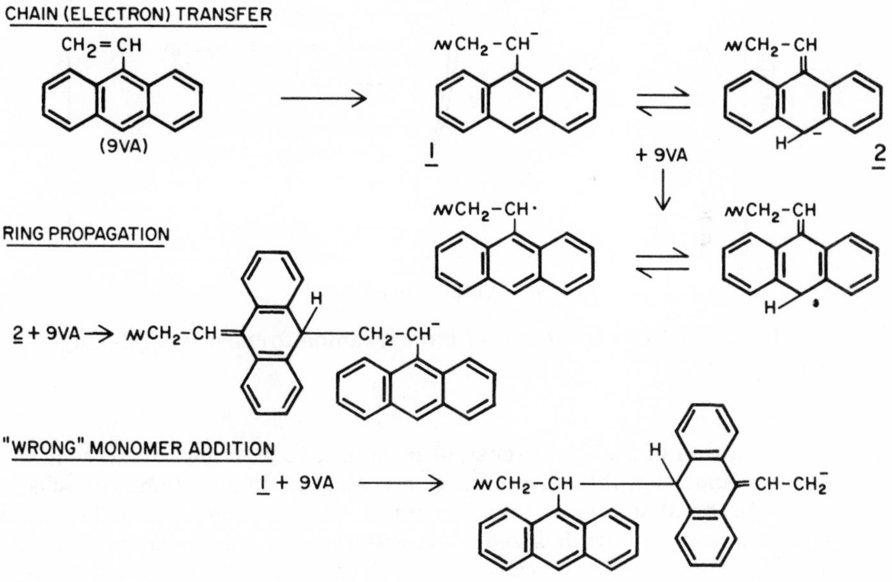

FIG. 1. Anionic polymerization of 9-vinylanthracene.

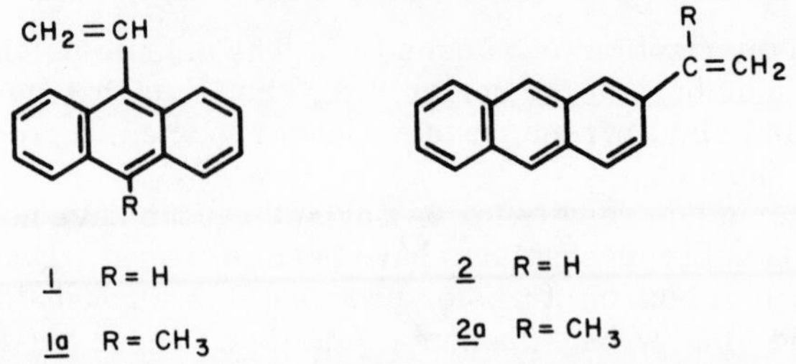

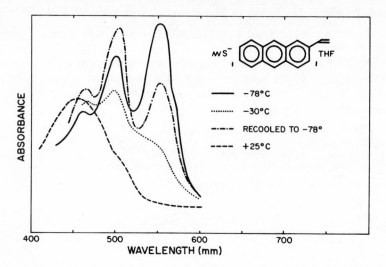

FIG. 2. 9-Vinylanthracene (1), 9-vinyl-10-methylanthracene (1a), 2-vinylanthracene (2), and 2-propenyl-2-anthracene (2a).

FIG. 3. Optical spectrum of the carbanion formed from 2-vinyl-anthracene.

from addition of a slight excess of monomer to a living polystyrene solution and from the polymerization reaction. In an attempt to sub-stantiate the structure of this carbanion, the model compound 1-propenyl-2-anthracene was synthesized. The spectrum of the carbanion from this olefin was essentially identical to that from 2-vinylanthracene shown in Fig. 3. All these anions exhibit instability at elevated temperatures, decomposing to give some species with an absorption maxima around 460 nm which is not effective in initiating polymerization.

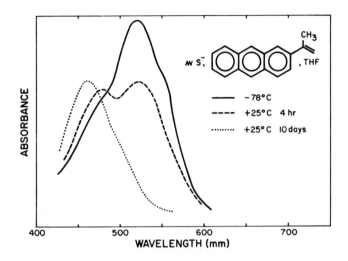

FIG. 4. Optical spectrum of the carbanion formed from 2-propenyl-2-anthracene.

All attempts to duplicate these results for the 9-vinylanthracene monomers were unsuccessful. The spectrum of the species generated by addition of 9-vinylanthracene to living polystyrene at -78°C is shown in Fig. 5 (9-vinyl-10-methylanthracene produces a very similar spectrum), and it exhibits some of the features anticipated for the conventional $\sim CH_2-CH^-$ carbanion. The spectrum of the 9-anthrylmethide
A
anion has been reported [3], and Fig. 6 shows the anion formed from the 1-propenyl-9-anthracene model compound. The species produced from the olefins at low temperatures are highly unstable and possibly rearrange to produce the ring-type carbanion. The spectra of the species existing at ambient temperatures are not inconsistent with such a structure, c.f., the adducts formed between the styrene anion and anthracene ($\sim$450 nm) pyrene ($\sim$500 nm) and benz[e]pyrene ($\sim$600 nm) [4].

Although the results from these investigations are not totally definitive, it has been clearly demonstrated that anionic polymerization of 2-vinylanthracene monomers to high molecular weight polymers can be achieved under proper reaction conditions. The 9-vinylanthracene system remains a paradox but the present spectroscopic analysis casts very serious doubts on the reliability of the published work on this monomer.

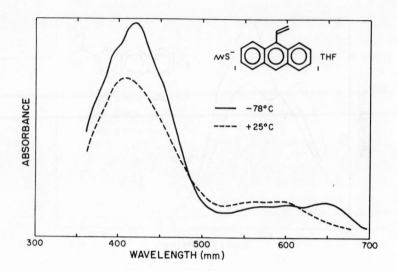

FIG. 5. Optical spectrum of the species formed in the reaction of 9-vinylanthracene with "living" polystyrene in THF.

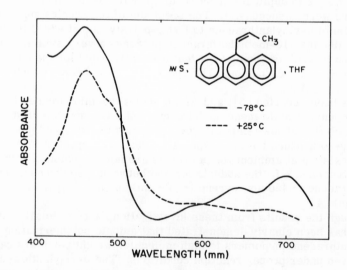

FIG. 6. Optical spectrum of the species formed in the reaction of 1-propenyl-9-anthracene with "living" polystyrene in THF.

These vinylanthracene studies serve to illustrate the complexities resulting from participation of the aromatic moiety in the polymerization reaction. Carbanion addition and electron transfer to the pyrene nucleus was also demonstrated in the copolymerization of 1-vinyl pyrene [5]. The anionic homopolymerization of this monomer exhibited all the characteristics of a "living" polymer system. In the formation of block copolymers, however, anomalous reactions and products were shown to result from the addition and electron transfer processes illustrated by Fig. 7.

Such complications can be used to advantage [6]. The dianionic adducts formed between "living" polymers and aromatic/heterocyclic molecules can be coupled according to the scheme of Fig. 8 using a variety of suitable difunctional reagents. In this case the products are copolymers with the aromatic/heterocyclic moieties incorporated into the polymer backbone.

In the case of vinyl heterocyclic monomers only a few systems have been investigated in enough detail to warrant consideration. A very interesting intramolecular solvation/coordination phenomenon has been found in the polymerization of 2-vinylpyridine and 2-vinylquinoline [7].

The anionic polymerization of N-vinylcarbazole (3) has also been studied [8] and it was concluded that the high electron density at the $>$N—CH=CH$_2$ site precluded formation of a stable anion. However, we have recently shown [9] that the corresponding 3-vinyl and 2-vinyl-N-alkylcarbazole monomers (4 and 5) could be polymerized anionically (Fig. 9).

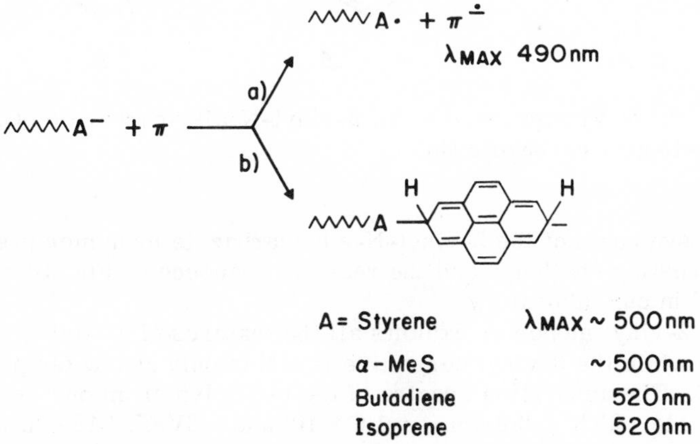

FIG. 7. Reactions of living polymer anions with pyrene.

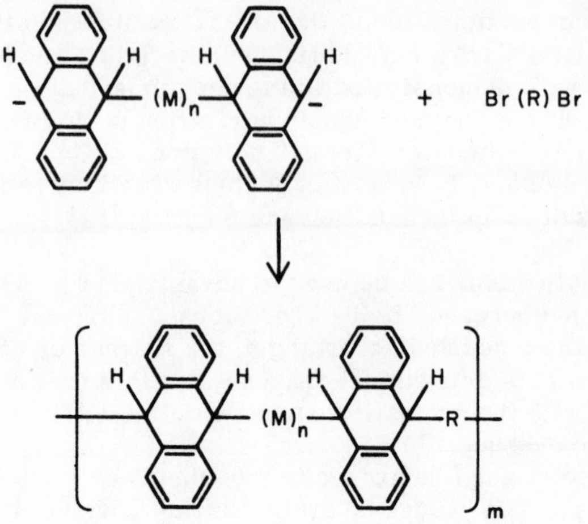

FIG. 8. Copolymer formation via carbanion coupling reactions.

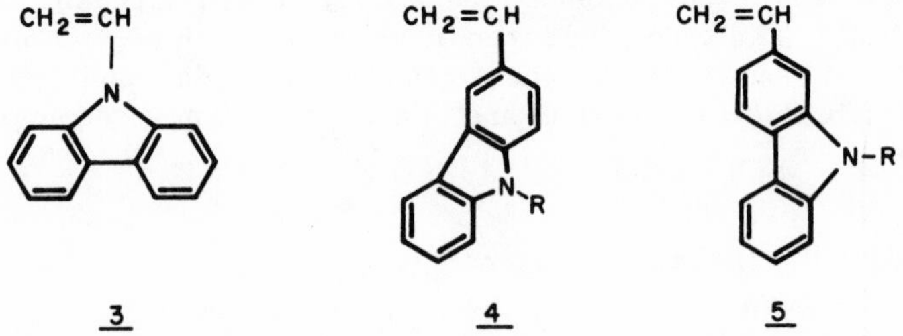

FIG. 9. N-Vinylcarbazole (3), 3-vinyl-N-alkylcarbazole (4), and 2-vinyl-N-alkylcarbazole (5).

The synthesis of the 2-vinyl-N-ethylcarbazole monomer presented an interesting challenge and the reaction sequence of Fig. 10 was developed in our laboratory.

The 2-vinyl monomer exhibits all the features of a "living" polymer system while the 3-vinyl derivative is stable only at low temperatures, $<-50°C$. The absorption spectra of the two polymer anions are shown in Fig. 11; ~2VK⁻, 498 nm, $\epsilon = 1.6 \times 10^4$ and ~3VK⁻, 545 nm, $\epsilon = 1.5 \times 10^4$. Unlike the N-vinyl- and 3-vinylcarbazoles, the vinyl function in 5 is not in direct conjugation with the electron-rich nitrogen moiety and the carbanion is stable enough to effect polymerization. In its

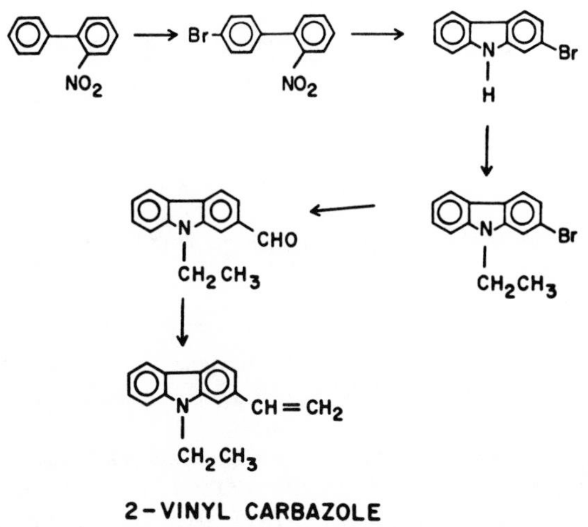

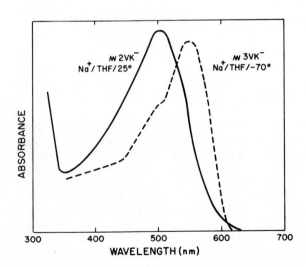

FIG. 10. Synthesis of the 2-vinyl-N-ethylcarbazole monomer.

FIG. 11. Optical spectra of the carbanions formed from 2-vinyl-
N-ethylcarbazole.

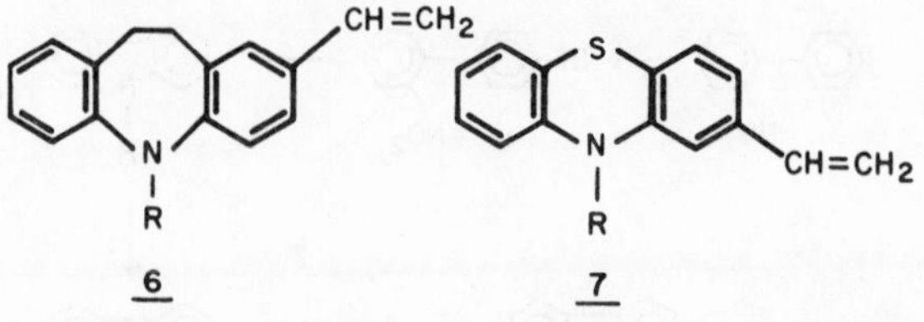

FIG. 12. N-Methyl-2-vinyliminobibenzyl (6) and N-ethyl-2-vinylphenothiazine (7).

anionic polymerization reactions, 5 behaves like a substituted styrene. The 3-vinyl monomer, 4, exhibits a behavior intermediate between the 2- and N-vinyl systems.

Two other interesting vinyl heterocyclic monomers have recently been shown to polymerize by anionic methods [10], N-methyl-2-vinyliminobibenzyl (6) and N-ethyl-2-vinylphenothiazine (7) (Fig. 12).

Here again, as with the 2-vinylcarbazole, the molecular structure inhibits direct conjugation of the vinyl group with the electron-rich center and permits facile, controlled anionic polymerization. Although many other monomers are known [11], there have been no serious studies of the oxygen- or sulfur-containing vinyl heterocyclic monomers.

CATIONIC POLYMERIZATION

There have been very few detailed studies of the cationic polymerization of vinyl aromatic monomers other than styrene and its derivatives which I will exclude here. Virtually all vinyl monomers possessing polynuclear aromatic hydrocarbon groups can be polymerized cationically, but the absence of information on polymer properties and structure in the literature is alarming. The complex structure claimed for the polymer (oligomer) of 9-vinylanthracene [12] shows the complications which can result from unstable and reactive carbenium ions (Fig. 13).

We have recently re-investigated [2] the cationic polymerization of the 9- and 2-vinylanthracene monomers. Polymers with molecular weights around 2×10^4, which are readily soluble, were obtained, and spectroscopic analysis favors a conventional 1,2-vinyl polymer structure. These polymerizations were critically dependent on monomer purity, monomer and polymer solubility, and the mechanism is still not completely established.

In the vinyl heterocyclic monomer case the opposite picture is found.

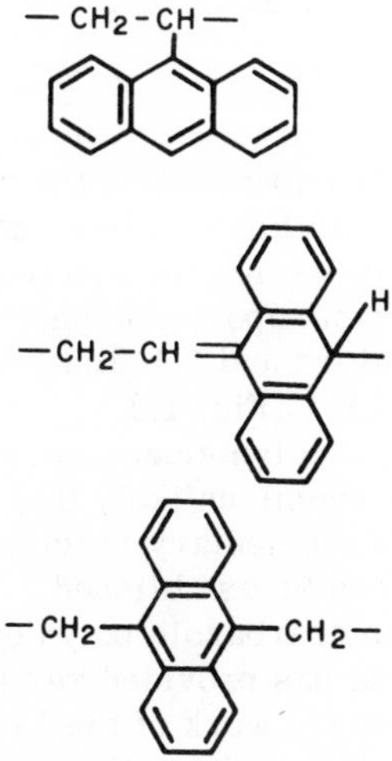

FIG. 13. Structure of poly-9-vinylanthracene.

The monomer N-vinylcarbazole is probably one of the most widely used in cationic polymerization studies. However, here again no systematic studies of variations in the heterocyclic ring structures have been made.

One of the major problems in any cationic polymerization is to establish the nature and the concentration of the active carbenium ion species. Studies with the model vinyl monomer, 1,1-diphenylethylene, where the complications arising from propagation, etc. are precluded, illustrate the complex nature of the initiation reaction.

In methylene chloride as solvent and $SbCl_5$ as initiator, the following reactions can occur [13] (D is 1,1-diphenylethylene, $D^{+\cdot}$ the olefin radical cation, D^+ the diphenylethyl carbenium ion):

$$D + SbCl_5 \rightleftharpoons [D\ SbCl_5\ complex] \rightleftharpoons D^{+\cdot},\ SbCl_5^{-\cdot} \tag{1}$$

$$2D^{+\cdot},\ SbCl_5^{-\cdot} \rightarrow SbCl_5^{-\cdot};\ ^+DD^+\ SbCl_5^{-\cdot}$$
$$(electron\ transfer)$$

$$D + SbCl_5 \rightarrow ClD^+,\ SbCl_4^- \tag{2}$$
$$(Cl^+\ transfer)$$

$$D + SbCl_5 + (H_2O) \rightarrow HD^+, SbCl_5 OH^- \qquad\qquad (3)$$

(cocatalyst)

The reaction is strongly dependent on the monomer/initiator ratio and on the nature of the initiator; no dimerization was observed using $TiCl_4$ [14]. Further evidence for the electron transfer mode of initiation, i.e., Eq. (1), has been reported using the stable triarylaminium salts, $Ar_3N^{+\cdot}X^-$, as initiators and 1,1-diphenylethylene and N-vinylcarbazole as monomers [15] (Fig. 14).

Clearly, each individual polymerization system has to be considered on its own merits, and it seems unlikely that any general reaction scheme comprising a sequence of elementary reactions (as for free radical and anionic polymerization) can be established.

Investigations of N-vinylcarbazole polymerization initiated with stable carbenium ion salts has provided key insights into the propagation reaction. The pioneering work of the Liverpool group on this system has permitted kp^+ values to be estimated for the first time [16]; k_p^+ (NVK) $\sim 10^5$ to 10^6 mole^{-1} l sec^{-1} in the temperature range -25 to 25°C. The mechanism has been well established, e.g., with tropylium ion [16] (Fig. 15).

The role of solvent, counterion, etc., even in this relatively "clean" polymerization system, are still far from being resolved. Parallel studies using the 2- and 3-vinylcarbazole monomers may help clarify some of these outstanding issues.

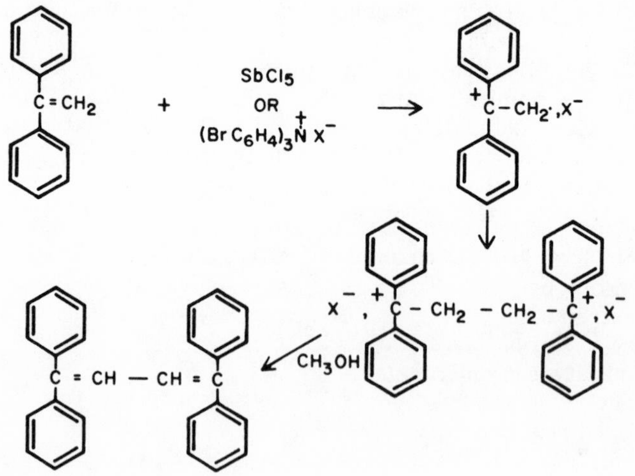

FIG. 14. Electron transfer mode of initiation.

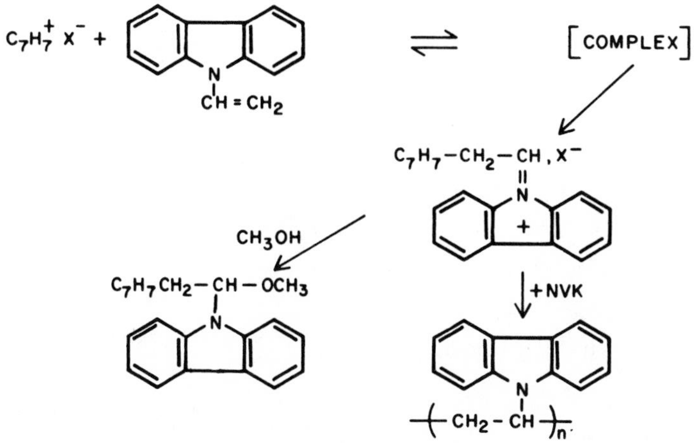

FIG. 15. Addition mode of initiation of cationic polymerization of N-vinylcarbazole.

One of the most extensively studied areas in this field is that of the so-called "charge-transfer" polymerization [17]. In principle, monomers exhibiting donor or acceptor functions should be polymerizable via charge-transfer interaction with a suitable partner. Olefins possessing electron-withdrawing groups are not widely known and few such systems have been characterized. On the other hand, monomers having electron-donor characteristics have been widely investigated and N-vinylcarbazole (NVK) has been a favorite donor monomer with a wide variety of electron acceptors including polymerizable and copolymerizable monomers. In spite of the intensive work with NVK, there is still considerable conflict between various groups. The possible reactions between NVK and typical electron acceptor molecules are summarized in Fig. 16.

The course of the reaction in any particular system will be determined by the solvent, impurities (for example, O_2 and acid), temperature, and activation by light. Too many investigators have failed to appreciate the extreme sensitivity of these systems with the result that the literature is full of conflicting data.

The initiation of cationic polymerization via charge-transfer complexes can be illustrated by the NVK/tetracyanoethylene (TCNE) system. An idealized reaction scheme is presented in Fig. 17.

However, the mechanism of this polymerization has been shown to be considerably more complex [18, 19]. In many instances polymerization results from the acidic impurity tricyanoethenol [20]. Excluding

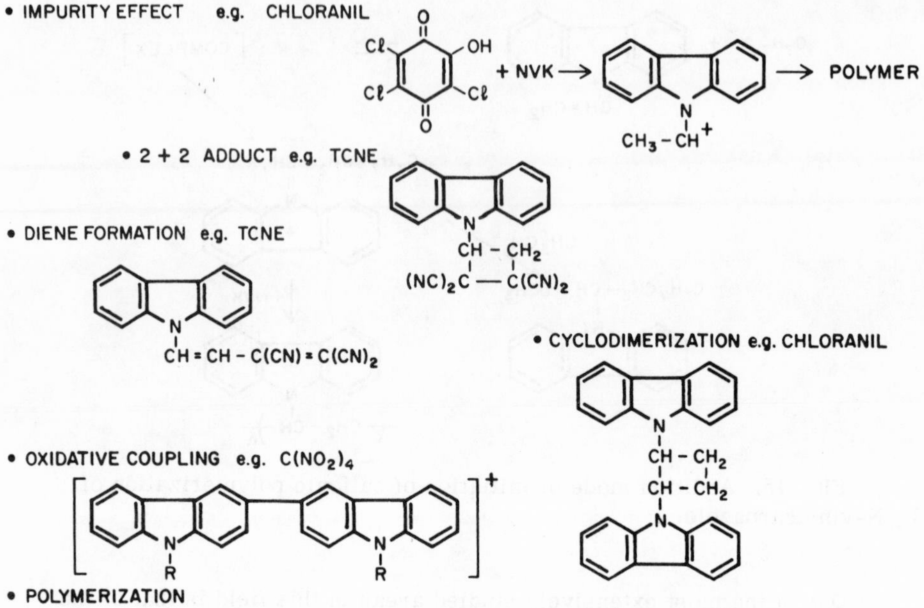

FIG. 16. Reactions between NVK and electron acceptors.

this effect, the reaction scheme of Fig. 18 has been proposed to account for the reaction products found in the NVK/TCNE system.

This is one of the few cases where there is solid evidence to substantiate the formation of olefin cation radicals from thermal activation of charge-transfer complexes. In all too many instances the appearance of charge-transfer absorption bands is taken as evidence for the participation of the complexes and the ion radical products as reaction intermediates.

Photochemical activation of charge-transfer complexes can be utilized to initiate polymerization [21]. It is extremely important to recognize the nature and formation of the various possible excited states. These are described in the scheme illustrated by Fig. 19.

Here, both direct excitation of the ground state complex and interactions between locally excited donor or acceptor with ground state partner are considered. Furthermore, singlet or triplet excited states may be involved in the reactions. There have been numerous studies of the photoinduced reactions of NVK and the chloranil system is perhaps the most clearly understood [22]. Most of the controversial

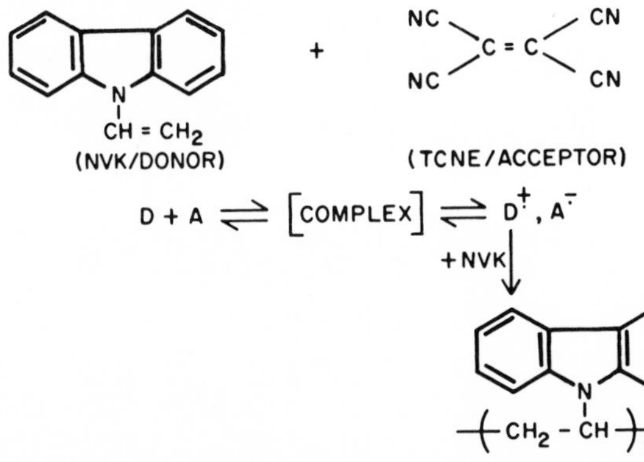

FIG. 17. Initiation of polymerization by charge-transfer complexes.

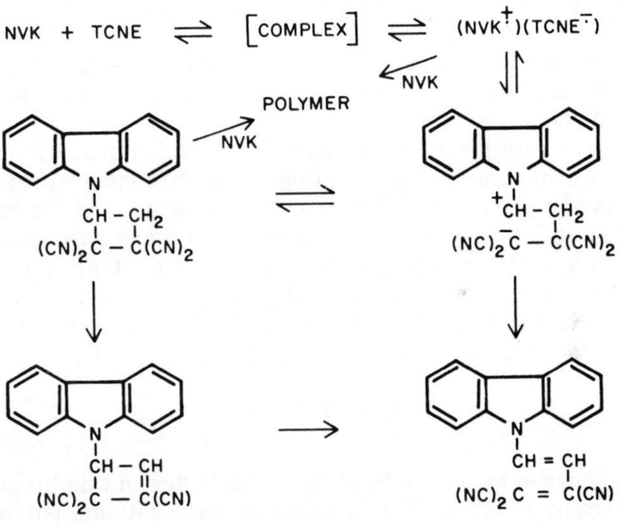

FIG. 18. Source of the reaction products found in the NVK/TCNE system.

issues arise from experimental methodology, i.e., system purity, role of solvent, effect of oxygen, etc., and this serves to re-emphasize the critical need to develop "clean" reaction systems. The extreme sensitivity of NVK to oxidation and the possible duality of polymerization

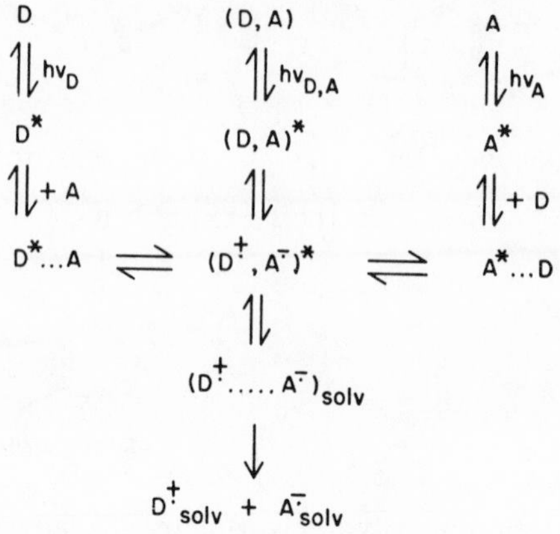

$$D \qquad\qquad (D, A) \qquad\qquad A$$

$$\Big\Updownarrow hv_D \qquad \Big\Updownarrow hv_{D,A} \qquad \Big\Updownarrow hv_A$$

$$D^* \qquad\qquad (D, A)^* \qquad\qquad A^*$$

$$\Big\Updownarrow {+A} \qquad \Big\Updownarrow \qquad \Big\Updownarrow {+D}$$

$$D^*...A \;\rightleftarrows\; (D^{\dot{+}}, A^{\bar{\ }})^* \;\rightleftarrows\; A^*...D$$

$$\Big\Updownarrow$$

$$(D^{\dot{+}}.....A^{\bar{\ }})_{solv}$$

$$\downarrow$$

$$D^{\dot{+}}_{solv} + A^{\bar{\ }}_{solv}$$

FIG. 19. Photoinduced ionic polymerization.

mechanism (radical and cationic) are particular disadvantages associated with this monomer, and a search for a superior model monomer would seem desirable.

Some very exciting work is appearing on the primary processes involved in photoionic polymerization. Ultrafast laser spectroscopic techniques are being used [23] to probe the nature and the reactions of the excited states involved in the initiation process. Early studies using α-methylstyrene will probably be extended to other monomer systems and will contribute significantly to our understanding of the fundamental processes involved in initiation of cationic polymerization.

CONCLUSIONS

I have touched on only a fraction of the topics in this broad field. A considerable effort is still required to place cationic polymerization on the same level of understanding as anionic. The goal of "living" carbenium ion polymers may yet be achieved. In a real sense it is the polymeric products of these vinyl aromatic polymerizations that are important. Can they leap the gap between laboratory curiosities and practical materials?

The range of monomers that can be polymerized ionically, the combinations that can be achieved, and the manner in which they can be combined (i.e., alternating, block, graft, stereocontrolled, etc.)

gives the technique almost unlimited potential. It is to be hoped that specialty applications will continue to develop [24]. In my opinion, the unique morphologies of many of these materials, and the materials properties that can be combined, have yet to be exploited. Their potential in the electronic, photochemical, medical, and other fields is great. It remains to be seen whether it will be realized.

ACKNOWLEDGMENT

I would like to thank my many colleagues at Xerox for their invaluable contributions to several of the research activities presented in this review and for their stimulating discussions on the ever-challenging topic of ionic polymerization.

REFERENCES

[1] A. Rembaum and A. Eisenberg, Macromol. Rev., 1, 57 (1966).
[2] M. Stolka, J. F. Yanus and J. M. Pearson, Presented at the North East Regional Meeting, American Chemical Society, Rochester, 1973; Submitted for Publication.
[3] J. M. Pearson, D. J. Williams, and M. Levy, J. Amer. Chem. Soc., 93, 5478 (1971).
[4] J. Jagur-Grodzinski and M. Szwarc, Ibid., 91, 7594 (1969).
[5] J. J. O'Malley, J. F. Yanus, and J. M. Pearson, Macromolecules, 5, 158 (1972).
[6] F. J. Burgess, A. V. Cunliffe, D. H. Richards, and P. Shadbolt, Eur. Polym. J., 10, 193 (1974).
[7] A. Rigo, M. Szwarc, and G. Sackmann, Macromolecules, 4, 622 (1971).
[8] A. Rembaum, A. M. Hermann, and R. Haack, J. Polym. Sci., B, 5, 407, (1967).
[9] W. W. Limburg, J. F. Yanus, A. Goedde, D. J. Williams, and J. M. Pearson, J. Polym. Sci., Polym. Chem. Ed., 13, 1133 (1975).
[10] P. Hyde, L. J. Kricka, A. Ledwith, and K. C. Smith, Polymer, 15, 387 (1974).
[11] K. Takemoto, J. Macromol. Sci.—Revs. Macromol. Chem., C5, 29 (1970).
[12] R. H. Michel, J. Polym. Sci., A2, 2533 (1964).
[13] B. E. Fleischfresser, W. J. Cheng, J. M. Pearson, and M. Szwarc, J. Amer. Chem. Soc., 90, 2172 (1968)

[14] G. Sauvet, J. P. Vairon, and P. Sigwalt, Bull. Soc. Chim. Fr., 1970, 4031.

[15] C. E. H. Bawn, F. A. Bell, and A. Ledwith, Chem. Commun., 1968, 599; D. C. Sherrington and A. Ledwith, Unpublished Results.

[16] P. M. Bowyer, A. Ledwith, and D. C. Sherrington, Polymer, 12, 509 (1971).

[17] L. P. Ellinger, Polymer, 6, 549 (1965).

[18] C. E. H. Bawn, A. Ledwith, and M. Sambhi, Ibid., 12, 209 (1971).

[19] T. Nakamura, M. Soma, T. Onishi, and K. Tamura, Makromol. Chem., 135, 341 (1970).

[20] S. Akoi, R. F. Tarvin, and J. K. Stille, Macromolecules, 3, 472 (1970); 5, 663 (1972).

[21] A. Ledwith, Acc. Chem. Res., 5, 133 (1972).

[22] K. Tada, Y. Shirota, and H. Mikawa, Macromolecules, 6, 9 (1973).

[23] M. Irie, H. Masuhara, K. Hayaski, and N. Mataga, J. Phys. Chem., 78, 341 (1974).

[24] J. M. Pearson, Polym. Preprints, 12, 68 (1971).

Anionic Polymerization of Styrene in Binary Solvents

NORIO ISE

Department of Polymer Chemistry
Kyoto University
Kyoto, Japan

ABSTRACT

Experimental findings on the triple ion formation were examined by adopting more elaborate purification method for solvents. For polystyryllithium in dimethoxyethane (DME) and benzene mixtures, the overall rate constant of propagation (k_p) was confirmed to <u>increase</u> with increasing living end concentration under some conditions, in contrast with cases for polystyryllithium in tetrahydrofuran-benzene mixtures and for polystyryl sodium, potassium and cesium in DME-benzene mixtures. The propagation by the inter-molecular triple ions proposed earlier in order to elucidate the kinetic "anomaly" mentioned above was briefly discussed.

It had been firmly established by the pioneering work of Szwarc and Schulz on the kinetics of living anionic polymerizations of styrene in ethereal solvents that the propagation proceeds through

two types of the growing chain ends, namely free ions and ion pairs
[1]:

$$k_p' \downarrow \text{ + monomer} \qquad\qquad k_p'' \downarrow \text{ + monomer}$$

ion pair   free ion

If one denotes the dissociation constant by K and the propagation
rate constants of the ion pairs and of the free ions by k_p' and k_p'',
respectively, the mass action law gives the following equation for
the overall rate constant of propagation k_p:

$$k_p = k_p' + k_p'' K^{1/2} [LE]^{-1/2} \tag{1}$$

where $[LE]$ is the concentration of the growing chain ends. It has been
demonstrated that Eq. (1) holds for a variety of living anionic systems.
The binary solvent systems have not been exceptions. Figure 1 shows
one of the examples obtained for polystyryllithium in tetrahydrofuran
(THF)-benzene mixtures [2]. Clearly the intercept of the k_p-$[LE]^{-1/2}$
plot is k_p' and the slope is $k_p'' K^{1/2}$. Because these constants have
definite physical significances, the intercept and slope under consideration should be positive. This has usually been the case.

In continuation of our research project, we studied binary
solvent mixtures of dimethoxyethane (DME) and benzene for
polystyryllithium [3] in order to compare their behavior with
that in THF-benzene mixtures studied earlier [2]. The DME-
benzene mixtures showed unexpectedly anomalous behavior in
the kinetic pattern, as is shown in Fig. 2. At low concentrations
of DME and at high concentrations of the growing chain ends, the
k_p value <u>increased</u> with decreasing $[LE]^{-1/2}$. If a linear approximation is allowed, the slope is negative (instead of positive),
which is physically impossible. This anomaly was attributed in
the previous study [3] to formation of the intermolecular triple
ions. At the same time, however, it was also inferred that another

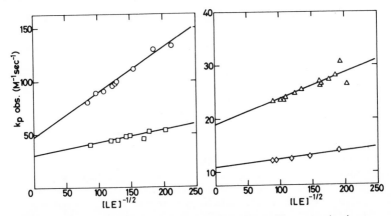

FIG. 1. Dependence of the apparent propagation constant on polystyryllithium concentration in THF-benzene mixtures at 25°C. THF contents in percent: (○) 60; (□) 50; (△) 40; and (◇) 30.

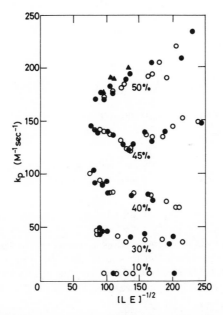

FIG. 2. Dependence of the apparent propagation constant on polystyryllithium concentration in DME-benzene mixtures at 25°C. The DME contents are given as percentages in the figure. The filled and open symbols indicate data obtained by independent runs at 5(▲), 3(●), and 0(○) kV/cm.

factor or other factors were responsible for the observed deviation from the two-state mechanism. The most serious case is the possible contribution of impurities which could not be removed from the polymerization systems by our previous purification methods. Thus it was intended to examine the purification procedures and to find a better method for purification. After almost 4 years of trial and error processes, we finally managed to arrive at a "new" purification method [4] which is compared with the old one in Table 1. The important points in the new purification method are the argon gas and the repeated drying-deaeration cycles. According to the manufacturer, the argon contained a smaller amount of oxygen than the nitrogen. We suspected that the oxygen might have a vitiating effect to the kinetic data. Furthermore, the deaeration-drying cycle in the previous method was broken when the solvent (DME or THF) started showing a dark blue color. In the revised procedure, this criterion was abandoned and the cycle was further continued until bubbling, which could be noticed when the stoppers of the vacuum line were opened instantaneously, became weak and constant.

TABLE 1. Purification Methods of Solvents

Step	In 1970	In 1974
1	Solvents were dried over Na for:	
	1 week	1 month
2	Dried solvents were refluxed over CaH_2 and distilled off into a vessel into which Na-K alloy was introduced under an atmosphere of:	
	Nitrogen (O_2: 0.005%)	Argon (O_2: 0.0005%)
3	The glass vessels were connected with a vacuum line and cooled down to -78°C, and the stopper was opened instantaneously to deaerate the solvent and closed again. The cycle was repeated:	
	Until dark blue color developed	For about 3 weeks until degassing became weak and constant

By using such a more elaborate purification procedure, the observed kinetic data became slightly larger. For example, at 25°C and a DME content of 40% for polystyryllithium, k_p'' is 5.1×10^4 $M^{-1}sec^{-1}$ in the 1974 work whereas it was 2.7×10^4 $M^{-1}sec^{-1}$ in the 1971 work. The k_p' became larger from 32 $M^{-1}sec^{-1}$ to 68 $M^{-1}sec^{-1}$. At a DME content of 50%, k_p' is 87 $M^{-1}sec^{-1}$ whereas it was 78 $M^{-1}sec^{-1}$, and k_p'' is 5.4×10^4 $M^{-1}sec^{-1}$ where it was 3.3×10^4 $M^{-1}sec^{-1}$ [5]. In spite of such a numerical discrepancy, the basic feature of the concentration dependence of k_p was not essentially affected; that is, k_p increased with an increasing con-

centration of living ends at low DME contents, or the Szwarc-Schulz plots again showed a negative slope as was found in the 1971 work.

Thus it seems that the triple ion contribution is highly possible. According to the electrochemistry, not only free ions and ion-pairs, but also higher ionic aggregates can exist in weak electrolyte solutions, especially in low dielectric media. If the dielectric constant is lowered, triple ions are formed in addition to free ions and ion-pairs. If the dielectric constant is further lowered, quadruples may be formed. The consideration of the potential energy of a system containing three ions makes understanding easier:

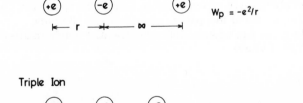

Suppose that we have two oppositely charged ions and one positively charged ion, all having equal size. If the first two are separated by a distance r and the third by infinity (or if an ion-pair is formed), the potential energy is $W_p = -e^2/r$. If the three ions are separated from each other by r from the nearest neighbor (or if a triple ion is formed), the potential energy of the system is $W_t = -3e^2/2r$. The triple ions are more stable than the ion-pairs.

Next we want to discuss the conductance equation by Fuoss and Kraus [6] for solutions containing free ions, ion-pairs, and triple ions. In this case we have three dissociation equilibria,

$$AB \rightleftharpoons A^+ + B^- \qquad\qquad K \tag{2}$$

$$(A_2 B)^+ \rightleftharpoons A^+ + AB \qquad k_1 \tag{3}$$

$$(AB_2)^- \rightleftharpoons AB + B^- \qquad k_2 \tag{4}$$

the respective dissociation constants being K, k_1, and k_2. If we assume $k_1 = k_2 = k$, or, in other words, if we assume bilateral triple ion formation with the same stability, the mass action law gives

$$1 - \gamma - 3\gamma_3 = C\gamma^2 K^{-1} \qquad \text{for ion-pairs} \tag{5}$$

and

$$C\gamma(1 - \gamma - 3\gamma_3) = \gamma_3 k \qquad \text{for triple ions} \tag{6}$$

where C is the electrolyte concentration, and γ and γ_3 are the fractions of free ions and ion-pairs, respectively. For low dielectric media, we can assume that γ and γ_3 are much smaller than unity. Then we have

$$\gamma = (C/K)^{-1/2} \tag{7}$$

and

$$\gamma_3 = (CK)^{1/2} k^{-1} \tag{8}$$

Since the total conductance Λ is the sum of the conductances of the free ions and triple ions, we have

$$\Lambda (= \gamma\Lambda_0 + \gamma_3\lambda_0) = \Lambda_0 K^{1/2} C^{-1/2} + \lambda_0 K^{1/2} C^{1/2} k^{-1} \tag{9}$$

where Λ_0 is the sum of the limiting conductances of free cations and anions, and λ_0 the sum of the conductances of free ions and triple ions. At lower concentration, Eq. (9) reduces to

$$\Lambda = \Lambda_0 K^{1/2} C^{-1/2} \tag{10}$$

whereas at higher concentrations it simplifies to

$$\Lambda = \lambda_0 K^{1/2} k^{-1} C^{1/2} \tag{11}$$

In other words, Λ decreases at first with increasing concentration, and then increases through a minimum at a concentration C_{min}. From Eq. (9) we have

$$C_{min} = k \Lambda_0 / \lambda_0$$

and the conductance at C_{min}, Λ_{min}, is given by

$$\Lambda_{min} = 2(\gamma \Lambda_0)_{min} = 2(\gamma_3 \lambda_0)_{min}$$

In other words, the conductance of the triple ions is equal to that of the free ions at this concentration.

The situation is well understood by Fuoss and Kraus data for tetraisoamylammonium nitrate in H_2O-dioxane mixtures [6], which are shown in Table 2. With increasing concentration, the fraction of free ions (γ) decreases whereas that of triple ions (γ_3) increases. The observed conductance, which is in good agreement with the value calculated by Eq. (9), decreases and increases after passing a minimum. At $C = 8.0 \times 10^{-5}$, the fractions of the free ions and of the triple ions are equal. At $C = 24 \times 10^{-5}$, the conductance shows a minimum. An important point is that the triple ion contribution cannot be overlooked even in the concentration range below C_{min}, where the conductance decreases with increasing concentration.

In this respect it is interesting to examine the concentration dependence of the electric conductance of the living polymer systems. As shown in Fig. 3, log Λ decreases linearly with log[LE] [4]. The slopes of the linear relationship, summarized in Table 3, are smaller than $+0.5$. These values are, however,

TABLE 2. Comparison of γ and γ_3

$C \times 10^5$ (M)	$\gamma \times 10^5$	$\gamma_3 \times 10^5$	$\Lambda \times 10^4$	
			Calc	Obs
1.5	2.4	0.5	7.7	7.5
3.0	1.7	0.7	5.8	5.8
8.0	1.1	1.1	4.4	-
10	0.95	1.2	4.05	4.03
24	0.61	1.9	3.68	-
30	0.55	2.1	3.75	3.68
100	0.30	3.8	4.70	5.25

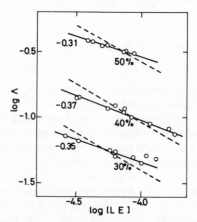

FIG. 3. Dependence of the equivalent conductance Λ on the living end concentration in DME-benzene mixtures at $0°C$.

definitely larger than -0.5 which is indicated in Fig. 3 by dashed lines. The deviation from -0.5 implies that the polymerization systems studied are in an intermediate state between those which can be described by Eq. (10) and by Eq. (11), respectively. In other

TABLE 3. Slope Values of the log Λ–log[LE] Plots for Polystyryl-lithium in DME–Benzene Mixtures [4]

DME content (%)	Temperature (°C)	Slope
30	0	-0.3_5
40	0	-0.3_7
	25	-0.4_0
	35	-0.4_4
50	0	-0.3_1
	25	-0.4_6

words, there exist not only free ions and ion-pairs, but also triple ions in the polymerization systems.

It should be noted that the previous work [2] also gave slope values for log Λ–log[LE] plots larger than -0.5.

For justification of the presence of the triple ions, the following consideration would be useful. First, the Fuoss-Kraus theory [7] showed that the critical concentration (C_0), at which the free ion—ion-pair equilibrium starts to fail completely, can be given by

$$C_0 = 3.2 \times 10^{-7}D^3 \tag{12}$$

for 1-1 electrolytes and 25°C, where D is the dielectric constant. The C_0 values for the DME–benzene mixtures were found to be 2.6×10^{-5}, 2.2×10^{-5}, and 1.8×10^{-5} M for DME contents of 50, 45, and 40%, respectively. It is to be noted that almost all of our kinetic measurements were carried out at higher concentrations (2×10^{-5} to 2×10^{-4}) than these C_0 values. Second, it is interesting to compare C_{min} and the concentrations employed in the kinetic study. Since the mobility of the gegenions (Li^+ in the present case) is expected to be larger than those of bulky polystyryl free ions and triple ions, we can assume $\Lambda_0 \approx \lambda_0$. Then $C_{min} \approx k$. The dissociation constants k determined from the conductance data (and hence the

C_{min} values) are 6.7×10^{-5}, 1.8×10^{-4}, and 4.8×10^{-4} M at 0°C for DME contents of 30, 40, and 50%, respectively. It is to be remembered that the kinetic and conductance measurements were mostly carried out in a concentration range below these C_{min} values.

Thus we may expect Λ to decrease with increasing concentration, even when we have triple ions in addition to free ions and ion-pairs.

The next question is the structure of the triple ions in the living anionic systems. Naturally, we have two possibilities; $\oplus \ominus \oplus$ and $\ominus \oplus \ominus$. Because the negative charge is located at the end of the growing chains, it would not be easy to accept that the triple ion $\oplus \ominus \oplus$ is stable. Furthermore, if it exists, its reactivity would be expected to be smaller than that of ion-pairs (both solvent-separated or contact). The observed fact, however, is the argumentation of the overall rate constant (k_p) with increasing living end concentration. Thus we may exclude the contribution of the triple ions $\oplus \ominus \oplus$.

If the above argument is acceptable, we have only $\ominus \oplus \ominus$ triple ions; in other words, the triple ion formation is unilateral [8]. Then, if precise data analysis is our aim, Eq. (9) cannot be applied to the present living anionic systems because this equation was derived for bilateral triple ion systems. Thus we have to use Wooster's equation for unilateral systems, which reads

$$\Lambda \left(1 + \frac{C}{k}\right)^{1/2} = \Lambda_0 K^{1/2} C^{-1/2} + \lambda_0 K^{1/2} k^{-1} C^{1/2} \tag{13}$$

This equation can be rearranged into

$$C\Lambda^2 = \Lambda_0^2 K + (2\Lambda_0 \lambda_0 - \Lambda_0^2)(K/k)C \tag{14}$$

which shows that $C\Lambda^2$ is a linear function of C. The conductance data [4] of polystyryllithium in DME-benzene mixtures are shown in Fig. 4 according to Eq. (14). As expected, the plot gives linear relationships. From the slope and intercept, the K and k values are estimated using the conductance values, which are determined by the Walden product, and by the assumption that $\Lambda_0 = \lambda_0$.

When the propagation process proceeds through the three-state mechanism, namely by free ions, ion-pairs, and triple ions, the

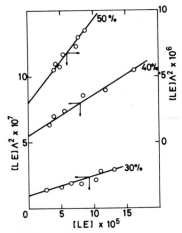

FIG. 4. Wooster plot for triple ion formation of polystyryllithium in DME-benzene mixtures at 0° [4].

overall rate constant (k_p) is a function of k_p', k_p'', and k_p''' (the rate constant of the triple ions):

$$k_p = (1 - \gamma - 2\gamma_3)k_p' + \gamma k_p'' + \gamma_3 k_p''' \tag{15}$$

In nonpolar solvents, such as those used in our study, we may assume $\gamma \ll 1$ and $\gamma_3 \ll 1$. Thus we have

$$k_p = k_p' + k_p''K^{1/2}\left(1 + \frac{[LE]}{k}\right)^{-1/2}[LE]^{-1/2}$$

$$+ k_p'''K^{1/2}\left(1 + \frac{[LE]}{k}\right)^{-1/2}k^{-1}[LE]^{1/2} \tag{16}$$

Obviously, k_p is a fairly complicated function of the living end concentration. In other words, k_p is not generally a linear function of $[LE]^{1/2}$ or of $[LE]^{-1/2}$. Thus the graphical method employed by Szwarc and Schulz [1] for determination of the rate constants cannot be applied. Furthermore, to be exact, it is not allowed to draw a straight line in the k_p-$[LE]^{-1/2}$ plot such as is shown in Fig. 2.

Thus we have to employ a numerical calculation method in order to estimate k_p', k_p'', and k_p''' as follows.

First, Eq. (16) was expanded in series of $[LE]/k$:

$$k_p = k_p' + K^{1/2}k_p''[LE]^{-1/2} + K^{1/2}\left(k_p''' - \frac{k_p''}{2}\right)k^{-1}[LE]^{1/2} + \cdots \quad (17)$$

A first approximation of k_p''' was substituted into Eq. (17), which was roughly estimated by a "linear" approximation for the k_p-$[LE]^{1/2}$ plot [11]. Then $k_p - K^{1/2}k^{-1}(k_p''' - \frac{1}{2}k_p'')[LE]^{1/2}$ was plotted against $[LE]^{-1/2}$ to obtain a second approximation for k_p''. In this plot, $\frac{1}{2}k_p''$ was estimated by using the first approximation for k_p'', which was obtained from the "positive" slope of Fig. 2, when possible. The $[k_p - K^{1/2}k_p''[LE]^{-1/2}]$ was plotted against $[LE]^{1/2}$ by using the second approximation of k_p'' to evaluate a second approximation for k_p'''. This k_p''' value was used for a third approximation for k_p'', which was further recycled for higher approximations for k_p'' and k_p'''. The calculation cycle was continued until k_p'' and k_p''' became constant. The final results thus obtained are compiled in Table 4.

It should be reminded that, for polystyryllithium in THF-benzene mixtures, the k_p-$[LE]^{-1/2}$ plot gave quite a normal behavior as shown in Fig. 1. Likewise, polystyryl sodium, potassium, and cesium demonstrated the usual kinetic pattern [12]. The anomalous behavior was originally found [3] and subsequently confirmed [4] exclusively for polystyryllithium in DME-benzene mixtures. Consequently, we would like to confirm our previous claim that the triple ions in the present system are formed as a consequence of the molecular geometry of DME and of its solvating power which is specific to lithium ions, probably because of the small Pauling radius. Furthermore, the triple ions under consideration are of the solvent-separated intermolecular type, and, for this reason, their reactivity (k_p''') is larger than that of the free ions (see Table 4 [13], as was pointed out earlier [3].

TABLE 4. Dissociation and Reactivity of Lithium Salts of Living Polystyrene in Dimethoxyethane and Benzene Mixtures [4]

DME content (%)	Temperature (°C)	D	Λ_0 (cm^2 ohm^{-1} equiv^{-1})	$K \times 10^{10}$ (M)	$k \times 10^4$ (M)	k'_p (M^{-1} sec^{-1})	$k''_p \times 10^{-4}$ (M^{-1} sec^{-1})	$k'''_p \times 10^{-4}$ (M^{-1} sec^{-1})
30	0	3.60	36.2	0.76	0.67	15	2.1	4.5
40	0	4.07	37.7	3.9	1.8	40	2.1	6.5
	25	3.81	57.7	0.43	2.0	68	5.1	10
	35	3.70	68.0	0.15	2.3	80	7.6	13
50	0	4.67	39.3	19	4.8	120	2.2	7.4
	25	4.33	59.5	2.5	5.1	87	5.4	11

ACKNOWLEDGMENTS

The present work was accomplished by painstaking efforts of
H. Hirohara, T. Makino, K. Takaya, M. Nakayama, H. Tatsuta, and
H. Yamauchi.

REFERENCES

[1] For the historical development of the kinetic investigation
     of living anionic systems, see M. Szwarc, Carbanions, Living
     Polymers and Electron Transfer Processes, Wiley-
     Interscience, New York, 1968.
[2] N. Ise, H. Hirohara, T. Makino, and I. Sakurada, J. Phys.
     Chem., 72, 4543 (1968).
[3] N. Ise, H. Hirohara, T. Makino, K. Takaya, and M. Nakayama,
     Ibid., 74, 606 (1970).
[4] K. Takaya and N. Ise, J. Chem. Soc., Faraday I, 74, 1338 (1974).
[5] The disagreement is disturbing. However, it is to be noted that
     k_p' and k_p'' for other systems also became much larger. We
     know the deep grief of Macbeth who said long ago, "All our
     yesterdays have lighted fools the way to dusty death."
[6] R. M. Fuoss and C. A. Kraus, J. Amer. Chem. Soc., 55,
     2387 (1933).
[7] R. M. Fuoss and F. Accascina, Electrolytic Conductance,
     Interscience, New York, 1959, Chap. 18.
[8] The term "unilateral" was first coined by Wooster thirty-
     seven years ago [9]. This author discussed monosodium
     benzophenone in liquid ammonia, which formed negatively
     charged triple ions. Though this unilateral triple ion is
     due to a chemical bond, he correctly pointed out [10] that
     "coulombic forces alone might give rise to a situation which
     closely simulates unilateral triple ion formation ..., if one
     of the simple ions is much larger than the other ...." He
     also analyzed the triple ion formation in aqueous hydrofluoric
     acid solutions.
[9] C. B. Wooster, J. Amer. Chem. Soc., 59, 377 (1937).
[10] C. B. Wooster, Ibid., 60, 1609 (1938).

[11] Under some conditions, Eq. (15) reduces to a much simpler form. First, if $[LE]/k$ is smaller than unity, we have

$$k_p = k_p' + k_p''K^{1/2}[LE]^{-1/2} + k_p'''K^{1/2}k^{-1}[LE]^{1/2} \tag{A}$$

If k_p'' is large or k is large, the third term of the left-hand side of Eq. (A) can be ignored and we have

$$k_p = k_p' + k_p''K^{1/2}[LE]^{-1/2} \tag{B}$$

which is nothing else than Eq. (1). Obviously, the case considered by Szwarc and Schulz [1] is a limiting one of Eq. (16). Furthermore, if k_p''' is large or if k is small, we may have

$$k_p = k_p' + k_p'''K^{1/2}k^{-1}[LE]^{1/2} \tag{C}$$

In other words, if the triple ion contribution is overwhelming, k_p' is a linear function of $[LE]^{1/2}$. The kinetic data for the polystyryllithium in DME-benzene mixtures under consideration showed that k_p was very approximately a linear function of $[LE]^{1/2}$. From this linear approximation, the first approximation of k_p''' was conveniently evaluated and introduced into the numerical calculation mentioned in the text.

Concerning Eq. (1), it should be emphasized, the two-state mechanism leads us to Eq. (1) but the linearity between k_p and $[LE]^{-1/2}$ does not necessarily imply that there exist only free ions and ion pairs. Even if we have concurrently the free ions, ion pairs, and triple ions, the observed rate equation apparently becomes identical to Eq. (1) when the triple ions happen to have a low reactivity.

[12] K. Takaya, S. Yamauchi, and N. Ise, J. Chem. Soc., Faraday I, 74, 1330 (1974).

[13] As pointed out in the text, the triple ions are more reactive
      than the free ions. We are not claiming, however, that this
      is always the case. The triple ions of a contact type, if they
      exist, should be less reactive than the free ions. The reac-
      tivity of various kinds of growing ends is expected to depend
      sensitively on their solvation. Naturally, nonsolvated free
      ions should be most reactive.

The Polymerization of Heteroaromatic Monomers

YOSHIO IWAKURA

Department of Industrial Chemistry
Faculty of Engineering
Seikei University
Tokyo, Japan

and

FUJIO TODA

Department of Synthetic Chemistry
Faculty of Engineering
University of Tokyo
Tokyo, Japan

ABSTRACT

Vinyl monomers containing a heteroaromatic substituent
such as thiophene, thiazole, oxazole, and pyridine have
been synthesized and radical copolymerization with styrene
of these monomers was studied. Monomer reactivity ratios
(r_1, r_2) and Alfrey-Price Q-e values were determined.
The reactivities of these monomers ($1/r_1$) are generally
related to localization energy of the β-carbon of monomers
(L_β). In the case of anionic polymerization of isopropenyl
monomers, a considerable amount of monomer remained
in the living polymerization system. The equilibrium
monomer concentrations were determined at different
temperatures, and the heats and entropies of polymerization

were obtained. We also obtained propagation rates for these monomers by the capillary flow method. The effect of hetero-aromatic compounds on the polymerization of styrene in THF initiated by the sodium salt of α-methylstyrene was studied.

INTRODUCTION

Szwarc and his co-workers carried out an extensive kinetic study of living polymerization. Few papers have appeared dealing with the living anionic polymerization of vinyl or isopropenyl monomers containing a heteroaromatic substituent except for that regarding 2-vinyl pyridine. Vinyl or isopropenyl monomers having an aromatic heterocycle such as oxazole, thiazole, and pyridine are thought to be very similar to styrene both with respect to the structure and electronic properties, since they possess a cyclic structure with a stable 6-π electron conjugation system like benzene. Therefore, it was expected that these monomers would give living polymers under appropriate conditions. From these viewpoints, a systematic study of the polymerization of such monomers has become of special interest to us. First of all, radical copolymerization of hetero-aromatic monomers with styrene was carried out partly for the purpose of comparing polymerization behaviors of radical and anionic polymerization.

RADICAL COPOLYMERIZATION WITH STYRENE

Radical copolymerization of heteroaromatic monomers (M_2) with styrene (M_1) was carried out by the conventional method. The conversion was always controlled within 10%, and the copolymer compositions were determined by elemental analysis. Monomer reactivity ratios r_1 and r_2 were determined both by the Finemann-Ross and the Mayo-Lewis methods, and the Alfrey-Price's Q and e values were calculated from them. The values are summarized in Table 1 together with those of the monomers reported in the literature.

It is convenient to compare the monomer reactivity by $1/r_1$ since it stands for the relative rate of the addition of the monomer to benzyl radical. From Table 1, it is evident that all the hetero-aromatic monomers tried are more reactive than styrene since $1/r_1$ is always larger than unity. Moreover, monomers with two heteroatoms are always more reactive than those with one hetero-atom except for 3-isopropenyl-5-methylisoxazole, e.g., all

TABLE 1. Monomer Reactivity Ratios and Alfrey-Price's Q and e Values

No.	Monomer (M_2)	r_1	r_2	Q	e	$1/r_1$
	2-Vinyl furan [6]	0.24	1.9	1.9	-0.8	4.2
	2-Vinyl thiophene [7]	0.35	3.1	3.0	-0.8	2.9
	2-Vinyl pyridine [8]	0.55	1.1	1.1	-0.1	1.8
	2-Isopropenylpyridine	0.42	0.95	1.1	+0.16	2.4
I-2	2-Vinyl-4, 5-dimethyloxazole	0.18	1.3	2.2	+0.40	5.6
I-3	2-Isopropenyl-4, 5-dimethyloxazole	0.15	2.2	2.8	+0.24	6.7
I-4	2-Vinyl-4-isobutyl-5-methyloxazole	0.10	3.0	4.2	+0.30	10
I-5	2-Isopropenyl-4-isobutyl-5-methyloxazole	0.10	3.4	4.1	+0.24	10
II-1	2-Vinyl thiazole [3]	0.14	3.3	3.5	+0.08	7.1
II-2	2-Isopropenylthiazole	0.09	3.8	4.8	+0.24	11
II-3	2-Vinyl-4-methylthiazole	0.15	2.8	3.2	+0.13	6.7
III-2	3-Isopropenyl-5-methylisoxazole	0.67	0.98	0.88	-0.15	1.5
III-1	2-Isopropenyl-5-methyl-1, 3, 4-oxadiazole	0.16	0.83	2.0	+0.62	6.2

oxazole monomers are more reactive than furan or pyridine monomers, and thiazole monomers are also more reactive than thiophene
or pyridine monomers. Koton studied the polymerization rates of
vinyl compounds containing furan, thiophene, and pyridine, and
concluded that the polymerizabilities of these monomers increased
by the introduction of a heteroatom into the ring or by the addition
of condensed ring [1]. It was further reported that in the series
styrene, 2-vinyl pyridine, 2-vinyl quinoline, and 4-vinyl pyridine,
the reactivities increased from 1.00 to 1.82 to 2.04 to 5.88 [2].
From this fact it was concluded that the introduction of a second
heteroatom further enhanced the reactivity of the monomer.

These empirical generalizations are quite consistent with the
tendency found here, and they can be extended to the oxazole and
the thiazole monomers.

The reactivity of 3-isopropenyl-5-methylisoxazole turned out to
have a different trend from the others. This may be due to the
poorer conjugation of the isoxazole ring compared to the other
heteroaromatics, since, in the former compound, heteroatoms
locate adjacent to each other and prevent the radical from
delocalizing throughout the ring. The difference is clearly shown
by the number of the contributing structures.

$$X = O, S \; ; \; Y = C, N$$

A similar explanation was given for the different reactivities of
2-vinyl- and 4-vinylthiazoles by Schilling et al. [3]. The different
trend of the isoxazole monomer is also reflected in their Q and e
values. The e values of all the other monomers listed in Table 1
are larger than that of styrene, i.e., -0.8, showing the electron-
withdrawing nature of such rings. The introduction of a second
heteroatom to the ring apparently increased the positive character
of double bond.

The polymerizabilities of vinyl monomers have been investigated
from the theoretical viewpoint and were explained in terms of the
localization energy [4] or the resonance stabilization energy (ΔE_{rs})
between the attacking radical and the monomer [5]. The author

calculated the energy levels of the molecular orbitals of each monomer and those of the radical derived from it by the simple Hückel method. The parameters of heteroatoms used in the calculations are shown in Table 2. Values of L (localization energy of β-carbon) are shown in Table 3. In Fig. 1, $\log(1/r_1)$ is plotted

TABLE 2. Parameters Used in the Molecular Orbital Calculations [9]

Heteroatom (X)	a	b	1
—O—	2	0.2	0.6
—N—	0.6	0.1	1
—S—	0	0	0.5
—CH$_3$	3	-0.1	1

[a]$\alpha_X = \alpha + a\beta$, α_X: Coulomb integral of heteroatom X. $\alpha_{adj} = \alpha + b\beta$, α_{adj}: Coulomb integral of carbon adjacent to X. $\beta_{c-x} = 1\beta$, β_{c-x}: Exchange integral between carbon and X.

TABLE 3. Relative Monomer Reactivities and Localization Energies

No.	Monomer (M$_2$)	$1/r_1$	$L(-\beta)$
1.	Styrene	1.00	1.70
2.	2-Vinyl pyridine	1.8	1.69
3.	2-Vinyl thiophene	2.9	1.61
4.	2-Vinyl furan	4.2	1.57
5.	2-Vinyl-4, 5-dimethyloxazole	5.6	1.53
6.	2-Vinyl-4-methylthiazole	6.7	1.55
7.	2-Vinyl thiazole	7.1	1.58
8.	2-Vinyl-4-isobutyl-5-methyloxazole	10	1.55

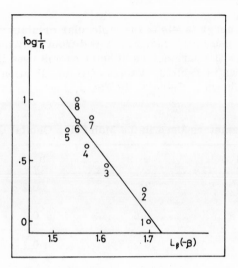

FIG. 1. Log $(1/r_1)$ vs $L_\beta(-\beta)$.

against L_β. Figure 1 shows that the reactivities of these monomers are generally related to L_β and that the monomer reactivity increases in the decreasing order of L_β. Although the relation is rather rough, it is thought to be satisfactory for the qualitative discussion taking into account that many assumptions have been made in the orbital calculations, and that $1/r_1$ has been taken as a parameter for the monomer reactivity. Thus empirical generalizations, 1) the introduction of a heteroatom into the ring enhances the monomer reactivity, and 2) a second heteroatom further enhances the reactivity, are supported from the theoretical standpoint.

KINETICS OF ANIONIC POLYMERIZATION OF HETEROAROMATIC MONOMERS BY THE CAPILLARY FLOW METHOD

The anionic polymerization in THF of various heteroaromatic vinyl and isopropenyl monomers was accomplished by a use of sodium naphthalene as an initiator. All monomers but furan monomers gave living polymers.

Anionic polymerization is generally too rapid for the measurement of its polymerization rate by the method employing a sampling technique. We therefore applied the capillary flow method which was used for the investigation of anionic polymerization by Szwarc [10]. The apparatus used in this study is shown in Fig. 2.

Two cylindrical reservoirs contained the solution of sodium naphthalene and of the monomer respectively. The capillary was immersed in a beaker containing methanol. The solution from each reservoir was pressured into the capillary by dry and purified argon. The conversion to polymer was determined by gas chromatography to estimate the concentration of the residual monomer using o-xylene as an internal standard.

A typical time-conversion curve and the first-order kinetic treatment are shown in Fig. 3.

In some cases of isopropenyl monomers, conversion curves were saturated and never reached 100%. In these cases the propagation constant K_p was determined by the slope of the tangent line at the origin in the conversion curve. The propagation rate constants of heteroaromatic monomers are summarized in Table 4.

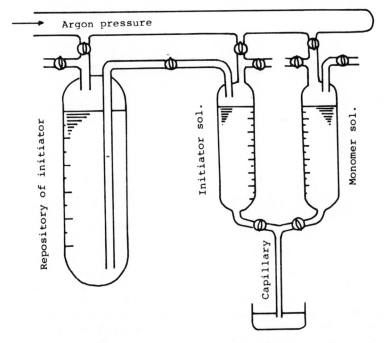

FIG. 2. Apparatus of the capillary flow method for measuring rates of fast reactions.

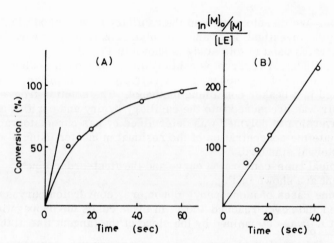

FIG. 3. Polymerization of 2-vinylthiophene at 25°C. (A) Conversion curve. (B) Rate of polymerization. $[LE] = 0.90 \times 10^{-2}$ mole/liter; $[M]_0 = 7.3 \times 10^{-2}$ mole/liter.

EQUILIBRIUM ANIONIC POLYMERIZATION OF ISOPROPENYL MONOMERS CONTAINING AROMATIC HETEROCYCLES

In this article, the anionic polymerization of 2-isopropenyl-4,5-dimethyloxazole (I), 2-isopropenylthiazole (II), and 2-isopropenylpyridine (III) in THF are described.

These monomers produced red colored living polymers after the addition of sodium naphthalene as an initiator. It was observed, however, that a considerable amount of monomer remained in the polymerization system at room temperature in each case. The conversion of the monomer into polymer increased at lower temperatures. Such phenomena are well known in the case of α-methylstyrene and are interpreted as equilibrium polymerization.

TABLE 4. Polymerization Rate Constants Measured by the Capillary Flow Technique

Monomer	$[LE] \times 10^2$ (mole/liter)	k_p [liter/(mole)(sec)]
Styrene	0.53	510
2-Vinyl thiophene	0.90	6
2-Vinyl pyridine	0.63	3560
2-Isopropenylpyridine	1.29	60
2-Vinyl-4, 5-dimethyloxazole	0.76	4.8×10^5
2-Isopropenyl-4, 5-dimethyloxazole	0.86	700
2-Vinyl-4-methylthiazole	0.42	5600
2-Vinyl-4, 5-dimethylthiazole	0.90	9850
2-Isopropenylthiazole	0.81	240
2-Isopropenyl-4-methylthiazole	1.00	770
2-Isopropenyl-4, 5-dimethylthiazole	0.52	1560
2-Isopropenyl-5-methyl-1, 3, 4-oxadiazole	0.90	340

The equilibrium polymerization of α-methylstyrene has been extensively studied by McCormick [11], Bywater [12], and Szwarc [13]. The relation between the equilibrium monomer concentration, M_e, and the equilibrium constant, K_∞, was given by Tobolsky [14] as shown in Eq. (1), assuming that the equilibrium constant K_n in Eq. (2) is independent of n:

$$(M_0 - M_e)/P^*_{total} = K_\infty M_e/(1 - K_\infty M_e) \tag{1}$$

$$P^*_n + M_e \xrightleftharpoons{K_n} P^*_{n+1} \tag{2}$$

where P^*_{total}, M_0, and P^*_n denote the total living end concentration, initial monomer concentration, and living n-mer concentration, respectively. Szwarc modified Eq. (1) and obtained the more general expression (3) [15] which is applicable to the system in which the equalities $K_n = K_{n+1}$ are assumed to be valid only for n exceeding some value s + 1 (s > 1), whereas for n ≦ s the respective K_n's may differ from K_∞ and from each other:

$$(M_0 - R_s - M_e)/P^*_{total} - Q^*_s = K_\infty M_e(1 - K_\infty M_e) \tag{3}$$

where Q^*_s and R_s denote the concentration of living polymers which have a degree of polymerization less than s and the amount per unit volume of the monomer incorporated in these polymers, respectively.

For a high number-average degree of polymerization, the approximation $K_\infty = 1/M_e$ is obtained both from Eqs. (1) and (3). Thus the determination of the temperature dependence of the equilibrium constant is reduced to the analytical problem of measuring the equilibrium monomer concentrations at different temperatures. McCormick [11], Bywater [12], and Szwarc [13] independently measured the residual monomer concentration of α-methylstyrene in equilibrium with its living polymer. Bywater also studied the styrene-living polystyrene system [16]. In the latter case, the polymerization was carried out in benzene or cyclohexane rather than THF, and n-butyllithium was used as initiator to avoid side reactions since the residual concentration of styrene proved to be much lower compared to α-methylstyrene. A higher temperature (100 to 150°C) was required to measure the equilibrium monomer concentration even by the use of UV spectroscopy for analyses.

With respect to the monomers studied here, the equilibrium monomer concentrations were comparatively high and could be determined by gas chromatography using an internal or external standard.

2-Isopropenyl-4, 5-dimethyloxazole (I) was polymerized using sodium naphthalene or living α-methylstyrene tetramer as initiator. The experiments were carried out under three different initial conditions to check the reproducibility of the data. The results are shown in Table 5.

Approximating $K_\infty = 1/M_e$, the thermodynamic equation for polymerization is expressed as

$$\ln(1/M_e) = -\Delta H/RT + \Delta S°/R \tag{4}$$

where ΔH, $\Delta S°$, and R denote the heat and the entropy of polymerization and gas constant, respectively. In Fig. 4, $\ln(1/M_e)$ is plotted

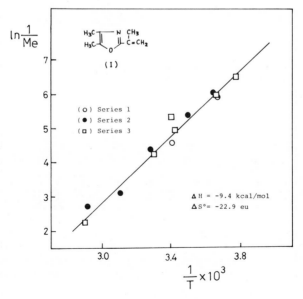

FIG. 4. Plot of $\ln(1/Me)$ vs $1/T$ for Monomer I.

TABLE 5. Equilibrium Monomer Concentration of (I) at Different Temperatures

Series[a]	Temperature (°K)	Time (hr)	[LE] (mole/liter)	[M₀] (mole/liter)	[Mₑ] (mole/liter)
1	273.9	65	0.0046	0.20	0.0026
	286.1	65	0.0046	0.20	0.0049
	304.7	65	0.0046	0.20	0.0133
	323.1	65	0.0046	0.20	0.0471
	343.1	65	0.0046	0.20	0.0630
2	273.5	65	0.0058	0.15	0.0029
	293.2	65	0.0058	0.15	0.0104
3	265.2	65	0.0098	0.73	0.0016
	273.5	65	0.0098	0.73	0.0028
	292.2	65	0.0098	0.73	0.0067
	294.4	65	0.0098	0.73	0.0046
	303.2	65	0.0098	0.73	0.0146
	343.2	65	0.0098	0.73	0.1030

[a]Series 1 and 2 were initiated by living α-methylstyrene tetramer, while series 3 was initiated by sodium naphthalene.

against the reciprocal of temperatures (°K). The data obtained from three series of experiments gave a consistent straight line, suggesting that Eq. (4) is valid for the temperature range of the experiments, i.e., equilibrium polymerization occurred. The slope of the line and the intercept on the y-axis were calculated by the least squares method, and from their values ΔH and $\Delta S°$ were obtained as -9.4 kcal/mole and -22.9 eu, respectively.

Alternatively, by substituting -9.4 kcal/mole for ΔH in Eq. (4), $\Delta S°$ can be calculated for each plot. The results are shown in Table 6. Table 6 shows that the constancy of $\Delta S°$ is excellent throughout the three experimental series. The values calculated in this way are in good agreement with those calculated by the intercept.

The same procedures were carried out with respect to the monomers II and III, and the results are shown in Figs. 5 and 6. In all cases, a linear relation was obtained between $\ln(1/M_e)$ and $1/T$.

TABLE 6. S° Calculated from the Slope and K_∞. Monomer (I)

Series	Temperature (°K)	K_∞ (liter/mole)	$\Delta S°$ (eu)
1	273.9	385	-22.5
	286.1	204	-22.3
	304.7	75.2	-22.2
	323.1	21.2	-23.1
	343.1	15.9	-21.9
2	273.5	345	-22.8
	293.2	96.2	-23.0
3	265.2	625	-22.6
	273.5	357	-22.7
	292.2	149	-22.2
	294.4	217	-21.3
	303.2	68.5	-22.6
	343.2	9.71	-22.8

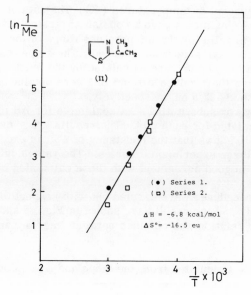

FIG. 5. Plot of ln(1/Me) vs 1/T for Monomer II.

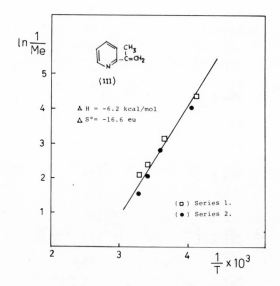

FIG. 6. Plot of ln(1/Me) vs 1/T for Monomer III.

From Figs. 5 and 6, ΔH and $\Delta S°$ were obtained in the same manner as in the case of monomer I. Those values are summarized in Table 7 along with those of other monomers reported in the literature.

Table 7 shows that the values obtained for monomers I, II, and III are close to those for α-methylstyrene. The marked difference in ΔH between α-methylstyrene and styrene was explained by the steric hindrance exerted by the α-methyl substitution [17]. Steric hindrance is common to monomers I, II, and III, and their values for the heats of polymerization appear reasonable. Their entropies of polymerization are a little smaller than that of α-methylstyrene.

At room temperature, α-methylstyrene gives only low molecular weight oligomers, the concentrated solution of which gives only a precipitate when it is poured into a large volume of methanol. On the other hand, monomers I, II, and III gave powdery polymers even at room temperature, showing that their anionic polymerizabilities are better than that of α-methylstyrene. This conclusion was supported by their ceiling temperatures as shown in Table 7.

TABLE 7. Heats and Entropies of Polymerization and Ceiling Temperatures.

Monomer	ΔH_{ss} (kcal/mole)	$\Delta S°_{ss}$ (eu)	$T_c{}^a$ (°C)	Ref.
I	-9.4	-22.9	137	
II	-6.8	-16.5	139	
III	-6.2	-16.6	100	
α-Methylstyrene	-6.96	-24.8	8	11
	-8.02	-28.75	6	12
	-7.47	-26.5	9	13
Styrene	-17.2	-28.7	326	16
Methyl methacrylate	-12.9	-29.5	164	25
Isobutyraldehyde	-3.7	-17.6	-63.4	26
Chloral	-3.5	-12.4	11	26

$^a T_c$ is defined by $\Delta H/\Delta S°$.

At temperatures higher than 60°C, some side reaction other than polymerization-depolymerization seems to take place slowly in the polymerization of monomers II and III, since the plots obtained from the sample, which was maintained at 60°C for many hours and then allowed to stand at room temperature for enough time to restore the equilibrium before killing the living ends, deviated considerably from the plot for a sample which was polymerized at room temperature from the first. Polymerization occurred much faster than the side reaction, however, and the system attained its equilibrium before the side reaction became appreciable, at least at temperatures lower than 60°C.

ANIONIC POLYMERIZATION OF STYRENE IN THE PRESENCE OF PYRIDINE

Some papers have been published regarding the effect of additives on the rates of anionic polymerization of styrene [18, 19]. Welch reported [20] that the rate of polymerization of styrene initiated by n-butyllithium in benzene was greatly accelerated by small quantities of Lewis bases, such as ethers or amines, and retarded by Lewis acids, such as zinc or aluminum alkyls. In general, the marked increase in polymerization rate by Lewis bases is explained to be due to the complex formation of the type $R^-[Li\ 2B]$, resulting from the activation of the carbanion. On the other hand, Lewis acids form complexes of the type $[RA]^-Li^+$ as a result of the deactivation of the carbanion.

Szwarc reported [21] the polymerization of styrene in THF in the presence of anthracene, which formed a complex with living polystyrene and greatly retarded the polymerization rate. In this case, polymerization proceeded to completion, and the living polymer complexed with anthracene was called a "dormant polymers" because it had a potential ability to grow. The effect of glymes were also reported by Szwarc and his co-worker [22].

In this paper we report the marked effect of pyridine for decreasing the polymerization rate of styrene in THF initiated by the sodium salt of α-methylstyrene tetramer. When a THF solution of styrene with a small quantity of pyridine as an additive was mixed with the initiator, the red color of the initiator immediately disappeared and a yellow color developed. Polymerization proceeded gradually, and finally all the monomer was consumed. Without any additive, the anionic polymerization of styrene in THF is a very fast process, and a skillful technique is required to determine its rate. In the presence of pyridine, however, the polymerization rate decreased so much that it could be determined by the simple method described in the experimental section. Polymerization obeyed a simple first-order rate equation:

$$\ln[M]_0/[M] = k_p[LE]t \qquad (5)$$

Plots of $\log[M]_0/[M]$ against time, shown in Fig. 7, resulted in a straight line passing through the origin.

From the slope of the line a, k_p (apparent) was calculated to be 0.19 mole^{-1}liter sec^{-1}, which was very much smaller than that in the absence of pyridine. In the case of line a, pyridine was first mixed with the initiator and then monomer solution was added to this mixture. On the other hand, in the case of the line b, pyridine

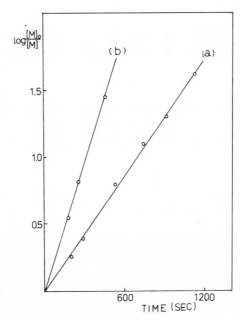

FIG. 7. Plots of $\log([M]_0/[M]$ vs time. Polymerization conditions: (a) $[LE] = 0.019$ mole/liter, $[Py]/[LE] = 3$, $[M]_0 = 0.65$ mole/liter; (b) $[LE] = 0.018$ mole/liter, $[Py]/[LE] = 2$, $[M]_0 = 0.65$ mole/liter. Initiator: $Na^+, (\alpha\text{-MeSt})_4^{2-}, Na^+$ in both cases. Methods of mixing: (a) (pyridine + initiator) + monomer; (b) (pyridine + monomer) + initiator.

TABLE 8. Molecular Weights of Polystyrenes Obtained in the Presence of Pyridine

Run	$\dfrac{[LE]}{(mole/liter)}$	$\dfrac{[M]_0}{[LE]}$	$\dfrac{[Py]^a}{[LE]}$	$[\eta]^b$	$MW,^c$ obs	$MW,^d$ calc	Method^e
1	0.11	10	14	0.047	–	–	Argon
2	0.096	15	21	0.072	–	–	Argon
3	0.018	42	28	0.097	9800	8900	Vacuum
4	0.018	43	55	0.098	9800	9100	Vacuum
5	0.048	60	83	0.124	15000	13000	Argon
6	0.029	97	14	0.140	18000	20000	Argon
7	0.015	194	28	0.230	35000	40000	Argon

^{a}Py: Pyridine.

^{b}Measured in benzene at 30°C.

^{c}Calculated from the relation, $[\eta] = 1.0 \times 10^{-4} [M]^{0.74}$.

^{d}Calculated by MW $= 2[M]_0/[LE] \times 104 + 470$. Initiator: Na$^+$, $(\alpha\text{-MeSt})_4^{2-}$, Na$^+$.

was first mixed with monomer and the mixture was added to the
initiator. In the latter case the line also passed through the origin,
indicating that the complex formation of carbanion with pyridine
was much faster than polymerization.

The intrinsic viscosities of the polystyrene prepared in the
presence of pyridine are shown in Table 8.

Table 8 shows that intrinsic viscosities increased in parallel
with the ratio of initial monomer concentration to that of living
ends, i.e., $[M]_0/[LE]$. The molecular weights of the polymers
calculated from the intrinsic viscosities are compared to those
calculated on the assumption of typical living polymerization
(Table 8, MW columns). Both values are approximately the same,
showing that no significant chain transfer or termination had
occurred during the polymerization process. Thus it was concluded that pyridine had a marked effect on the polymerization
rate but not on the molecular weight of the polymer. This system
is thought to be an example of dormant polymers.

Changes in the electronic spectrum are shown in Fig. 8. On
addition of pyridine to living polystyrene in THF, the absorbance at
340 nm diminished at once, indicating that a drastic change
had occurred at the living ends.

It was reported that n-butyllithium formed a σ-complex with
pyridine [23]:

The structure of the complex was confirmed by its NMR spectrum.
Considering this fact and the drastic decrease of the absorbance at
340 nm, the complex between living polystyrene and pyridine may
be close to a σ-complex, which does not have absorption at wavelengths longer than 300 nm. The remaining weak absorption may
be due to the polymerization of the active species. However, since
the absorption maximum shifted from 340 to 330 nm, pyridine must
have some weak interaction with this species. Accordingly, we
consider that pyridine has two types of interaction with living
polystyrene. The nature of the complex is not yet clear, and
further study is now in progress.

We have recently shown that dipyridyls are formed in the
reaction of the sodium salt of α-methylstyrene tetramer with
pyridine [24]. In the case of the reaction of living polystyrene

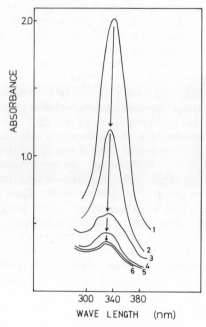

ABSORBANCE

300   340   380
WAVE LENGTH   (nm)

FIG. 8. Changes in the electronic spectrum. (1) Spectrum of
living polystyrene sodium. (2)-(6) Spectra when pyridine solution
(2 ml, 0.25 mole/liter) was added portionwise to the solutions of
living polystyrene sodium (20 ml, 0.017 mole/liter). Measurement
of the spectra was carried out in vacuum using a 1-mm cell with a
0.9-mm spacer.

sodium with pyridine, however, the complex seemed to be stable and
neither the formation of dipyridyls nor the isomerization of the living
polystyrene sodium was observed.

The polymerization of 2-vinyl pyridine in the presence of pyridine
was also tried. However, all the monomer was consumed in several
seconds and the presence of pyridine did not show any effect on the
polymerization rate. Moreover, no change was observed in the
electronic spectrum of the solution of living poly(2-vinyl pyridine)
sodium by the addition of pyridine. These facts are illustrated by

Na^+, $(\alpha\text{-MeSt})_4^{2-}$, Na^+ + pyridine ———→ complex

$\qquad\qquad\qquad\qquad\qquad\qquad\qquad\qquad\qquad\qquad$ ↘

$\qquad\qquad\qquad\qquad\qquad\qquad\qquad\qquad\qquad\qquad$ dipyridyls

Poly St$^-$, Na^+ + pyridine ———→ complex

Poly(2-vinylpyridine)$^-$, Na^+ + pyridine ———→ no reaction which indicates that the reactivities of the carbanions are greatly affected by the substituents.

REFERENCES

[1] M. M. Koton, J. Polym. Sci., 30, 331 (1958).
[2] G. E. Ham, Copolymerization, Wiley-Interscience, New York, 1964, p. 509.
[3] C. L. Schilling, Jr. and J. E. Mulvaney, Macromolecules, 1, 445 (1968).
[4] Takayuki Fueno, Teiji Tsuruta and Junji Furukawa, J. Polym. Sci., 40, 487 (1959).
[5] K. Hayashi, T. Yonezawa, C. Nagata, S. Okamura, and k. Fukui Ibid., 26, 311 (1957).
[6] C. Aso, 16th Annual Meeting of Japan Chemical Society, 1963, Preprint p. 333.
[7] C. Walling, E. R. Brigs, and K. B. Wolfstern, J. Amer. Chem. Soc., 70, 1543 (1948).
[8] C. G. Overberger and F. W. Michebott, Ibid., 80, 988 (1958).
[9] T. Yonezawa et al., Ryoshikagaku Nyumon, Kagakudojin, Kyoto, 1963, p. 55.
[10] C. Geacinton, J. Smid, and M. Szwarc, J. Amer. Chem. Soc., 84, 2508 (1962).
[11] H. W. McCormick, J. Polym. Sci., 25, 488 (1957).
[12] D. J. Warsfold and S. Bywater, Ibid., 26, 299 (1957).
[13] A. Vrancken, J. Smid, and M. Szwarc, Trans. Faraday Soc., 58, 2036 (1962).
[14] A. V. Tobolsky, J. Polym. Sci., 25, 220 (1957).
[15] M. Szwarc, Proc. Roy. Soc., Ser. A, 279, 260 (1964).
[16] S. Bywater and D. J. Warsfold, J. Polym. Sci., 58, 571 (1962).
[17] M. Szwarc, Adv. Polym. Sci., 4, 457 (1967).
[18] S. Bywater and D. J. Warsfold, Can. J. Chem., 40, 1564 (1962).
[19] F. J. Welch, J. Amer. Chem. Soc., 81, 1345 (1959).
[20] F. J. Welch, Ibid., 82, 6000 (1960).
[21] S. N. Khanna, M. Levy, and M. Szwarc, Trans. Faraday Soc., 58, 747 (1962).
[22] M. Shinohara, J. Smid, and M. Szwarc, J. Amer. Chem. Soc., 90, 2175 (1968).
[23] G. Frankel and J. C. Cooper, Tetrahedron Lett., 1968, 1825.
[24] K. Yagi, F. Toda, and Y. Iwakura, J. Polym. Sci., Polym. Lett. Ed., 10, 113 (1972).
[25] S. Bywater, Trans. Faraday Soc., 51, 1267 (1955).
[26] I. Mita, I. Imai, and H. Kambe, Makromol. Chem., 137, 169 (1970).